T0301784

A Course in the Large Sample Theory of Statistical Inference

This book provides an accessible but rigorous introduction to asymptotic theory in parametric statistical models. Asymptotic results for estimation and testing are derived using the "moving alternative" formulation due to R. A. Fisher and L. Le Cam. Later chapters include discussions of linear rank statistics and of chi-squared tests for contingency table analysis, including situations where parameters are estimated from the complete ungrouped data. This book is based on lecture notes prepared by the first author, subsequently edited, expanded and updated by the second author.

Key features:

- Succinct account of the concept of "asymptotic linearity" and its uses
- Simplified derivations of the major results, under an assumption of joint asymptotic normality
- Inclusion of numerical illustrations, practical examples and advice
- Highlighting some unexpected consequences of the theory
- Large number of exercises, many with hints to solutions

Some facility with linear algebra and with real analysis including 'epsilon-delta' arguments is required. Concepts and results from measure theory are explained when used. Familiarity with undergraduate probability and statistics including basic concepts of estimation and hypothesis testing is necessary, and experience with applying these concepts to data analysis would be very helpful.

W. J. ("Jack") Hall was Professor at the University of Rochester from 1969 to his death in 2012. He was instrumental in founding the graduate program in Statistics. His research interests included decision theory, survival analysis, semiparametric inference and sequential analysis. He worked with medical colleagues to develop innovative statistical designs for clinical trials in cardiology.

David Oakes is Professor and a former department chair at the University of Rochester. His areas of research interests include survival analysis and stochastic processes.

Chapman & Hall/CRC Texts in Statistical Science

Recently Published Titles

Design and Analysis of Experiments and Observational Studies using R
Nathan Taback

Statistical Theory
A Concise Introduction, Second Edition
Felix Abramovich and Ya'acov Ritov

Applied Linear Regression for Longitudinal Data
With an Emphasis on Missing Observations
Frans E.S. Tan and Shahab Jolani

Fundamentals of Mathematical Statistics
Steffen Lauritzen

Modelling Survival Data in Medical Research, Fourth Edition
David Collett

Applied Categorical and Count Data Analysis, Second Edition
Wan Tang, Hua He, and Xin M. Tu

Geographic Data Science with Python
Sergio Rey, Dani Arribas-Bel, and Levi John Wolf

Models for Multi-State Survival Data
Rates, Risks, and Pseudo-Values
Per Kragh Andersen and Henrik Ravn

Spatio–Temporal Methods in Environmental Epidemiology Using R, Second Edition
Gavin Shaddick, James V. Zidek, and Alex Schmidt

A Course in the Large Sample Theory of Statistical Inference
W. Jackson Hall and David Oakes

For more information about this series, please visit:
https://www.routledge.com/Chapman–HallCRC-Texts-in-Statistical-Science/book-series/CHTEXSTASCI

A Course in the Large Sample Theory of Statistical Inference

William Jackson Hall and David Oakes

CRC Press is an imprint of the
Taylor & Francis Group, an **informa** business
A CHAPMAN & HALL BOOK

First edition published 2024
by CRC Press
6000 Broken Sound Parkway NW, Suite 300, Boca Raton, FL 33487-2742

and by CRC Press
4 Park Square, Milton Park, Abingdon, Oxon, OX14 4RN

CRC Press is an imprint of Taylor & Francis Group, LLC

Library of Congress Cataloging-in-Publication Data
Names: Hall, W. J. (William Jackson), author. | Oakes, David (Statistician), author.
Title: A course in the large sample theory of statistical inference /
William Jackson Hall and David Oakes.
Description: First edition. | Boca Raton, FL : CRC Press, 2024. |
Series: Chapman & Hall/CRC texts in statistical science |
Includes bibliographical references and index.
Identifiers: LCCN 2023028294 (print) | LCCN 2023028295 (ebook) |
ISBN 9781498726061 (hardback) | ISBN 9781032595238 (paperback) |
ISBN 9780429160080 (ebook)
Subjects: LCSH: Statistical hypothesis testing–Asymptotic theory.
Classification: LCC QA277 .H353 2024 (print) | LCC QA277 (ebook) |
DDC 519.5/4–dc23/eng/20231011
LC record available at https://lccn.loc.gov/2023028294
LC ebook record available at https://lccn.loc.gov/2023028295

ISBN: 9781498726061 (hbk)
ISBN: 9781032595238 (pbk)
ISBN: 9780429160080 (ebk)

DOI: 10.1201/9780429160080

Typeset in Palatino
by codeMantra

Contents

Preface

This book is based on lecture notes prepared by W.J. ("Jack") Hall for his graduate-level statistics course in large sample theory, which he taught at the University of Rochester for many decades until shortly before his death in 2012. The course provided an accessible, but rigorous introduction to asymptotic theory in parametric models, using the moving alternative formulation due to R.A. Fisher and L. LeCam. Jack was a superb teacher, and his notes reflect his thoughtful approach to exposition. He was the first recipient of the Lifetime Achievement in Graduate Teaching Award from the University of Rochester, bestowed upon him in 2004. He had a deep knowledge of the relevant literature and an unusual ability to explain difficult mathematical concepts in simple terms. Although most of the results are well known, Jack's approach is often novel and not available in other comparable textbooks. Jack included many examples and exercises to illustrate the general theory as well as much sage advice as to how to use it in applications.

In transforming the course notes that he developed into a book, I have aimed to keep Jack's expository style and have largely kept to his choice of topics. I have added explanation and introduced further examples. Some short numerical calculations and illustrations are given, now programmed in R. Chapter 1 presents introductory material on probability, univariate and multivariate distribution functions, moments, moment-generating functions, quantile functions, and stochastic ordering. An Appendix derives the properties of the gamma functions and its derivatives. Chapter 2 reviews the fundamental notions of weak convergence (convergence in distribution) the O_p/o_p notation, the continuous mapping theorem and the delta method. Several applications of Skorohod embedding are included. Tools from analysis including Taylor's theorem and the (nonstochastic) O/o notation are summarized in an Appendix. Chapter 3 introduces the notion of asymptotic linearity of statistics, including a careful exposition of the Bahadur representation of a sample quantile. Methods of deriving asymptotic representations are described, and properties of influence functions and U-statistics are discussed.

Chapter 4 introduces the key notions of moving alternatives and local asymptotic normality, leading to discussion of LeCam's third lemma, which connects asymptotic distributions under fixed and moving parameters.

After a preliminary discussion of unbiasedness of estimators in finite samples and the associated Cramér-Rao inequality, Chapter 5 develops the notion of "regularity" of an estimator as a large-sample analog and explores its implications. Chapter 6 addresses the role of nuisance parameters and the associated concepts of effective scores and effective information. Estimation of functions of a parameter vector and under constraints on the parameter vector are also discussed. Block matrix formulas, needed for handling problems with multi-dimensional parameters are discussed in an Appendix. A second Appendix discusses the properties of the "profile 1ikelihood" function.

Chapter 7 describes parallel results in the hypothesis testing problem, beginning, as in Chapter 5, with the finite-sample case. The duality between hypothesis testing and estimation is stressed, and examples are given of the use of hypothesis tests to construct confidence regions. Practical guidance is given regarding the use of p-values and sample size calculations. This chapter concludes with some more technical materials on conditions for asymptotic efficiency of test statistics. Chapter 8 discusses rank statistics and estimation, extending material on U-statistics and Hájek projection in Chapter 3. The final chapter gives an introduction to multinomial chi-squared tests, including cases where parameters must be estimated, as in testing the fit of a normal distribution to grouped data. Chapters 3–6 form the heart of the book and could form the basis for a single semester course, with preparatory material from Chapters 1 and 2 and supplementary selections from Chapters 7–9 at the instructor's discretion.

Exercises are included in each chapter, many with copious hints. Prerequisites include some familiarity with linear algebra and with real analysis including the use of "epsilon-delta" arguments and the notion of uniform convergence. Concepts and results from measure theory are explained when they are needed. The reader should have studied probability and statistics at least at undergraduate level and ideally should have some practical experience in data analysis.

My thanks go first of all to Jack for his friendship and continuing inspiration, and to his wife Nancy Hall, without whose support this book would not have been possible. Long before I knew Jack, Stephen Portnoy introduced me to the beauty of U-statistics. Anthony Almudevar contributed ideas and an interesting example. Christine Brower assisted with file transfer. Anonymous reviewers gave useful comments on early drafts. I thank Jack Angus for valuable suggestions on the final manuscript. Acquiring Editors, John Kimmel and Lara Spieker at Taylor & Francis provided essential encouragement and advice.

By contract, all royalties received for the book will be contributed to the Willam Jackson Hall Graduate Fellowship, a fund established by Jack's family, friends and colleagues in support of graduate education in statistics and biostatistics at the University of Rochester.

David Oakes
10/20/2023

1

Random Variables and Vectors

This chapter provides a brief review of some basic facts and methodology about random variables and random vectors and their distributions. It also provides a fuller introduction to three topics: *quantile functions, stochastic ordering* and *multinormal distribution* and discusses a useful *change-of-measure* identity. These topics will all be needed in later chapters, but some are not often discussed in introductory classes in statistical theory. We end the chapter with a brief discussion of software, especially the program R, to be used in later chapters. Other topics reviewed are well covered in intermediate-level texts in mathematical statistics such as BICKEL & DOKSUM (2015) and CASELLA & BERGER (2002). Characteristic functions (needed only rarely here) are discussed in probability texts including SHORACK (2000) and the classic CRAMÉR (1946).

1.1 Random Variables, Their Distributions and Their Attributes

We start by considering a single numerical random variable X. Mathematically, a random variable is defined as a (measurable) function on a probability space defined by the possible outcomes of a random experiment, and we will occasionally need to refer to this structure in proving theorems. More intuitively, X denotes some quantitative data with possible values x and whose actual value x^* will become available eventually, but is currently uncertain. We model this uncertainty by specifying a probability distribution over the values of x. This specification may be based on theoretical and/or empirical considerations. In this chapter, we focus only on the mathematical aspects once such a distribution is assumed. Later chapters address questions of inference—of what can be said about this distribution based on the data that are observed. To recapitulate, we use X to represent the random variable, or potential value of a data item, x for a typical value and x^* for the actual realized value.

A probability distribution for X can be defined in several ways. A general definition is through a *probability measure*: $P(A) \equiv \mathrm{pr}(X \in A)$, defined for every (Borel) subset A of the real line R, and satisfying the probability axioms:

DOI: 10.1201/9780429160080-1

$P(A) \geq 0$, $P(\mathcal{R}) = 1$, $P(\bigcup A_j) = \sum P(A_j)$ for any countable collection of disjoint Borel sets A_1, A_2, ... of $\mathcal{R}$.

Fortunately, a probability measure is determined uniquely by specification of $P(I)$ for every interval I in a manner consistent with these axioms. Here we must sometimes distinguish between open intervals $(a, b) = \{x : a < x < b\}$, closed intervals $[a, b] = \{x : a \leq x \leq b\}$ and the half-open variants $(a, b]$ and $[a, b)$ defined similarly.

It is ordinarily more convenient to specify a distribution through a *distribution function* F: a monotone nondecreasing and right-continuous function from $\mathcal{R}$ to $[0, 1]$ satisfying $\lim_{x\to-\infty} F(x) = 0$ and $\lim_{x\to+\infty} F(x) = 1$. Here right-continuity is defined as $\lim_{\epsilon\downarrow 0+} F(x + \epsilon) = F(x)$ for every x. The relationship between the probability measure and its interpretive meaning is given by

$$F(x) = P((-\infty, x]) = \text{pr}(X \leq x),$$

the probability that the realized value of X is at most x. The increase in F over an interval is the probability of that interval. The interior of intervals over which F is flat has zero probability of occurring. Discontinuities in F occur at values of x having positive probability, and these probabilities correspond to the magnitudes of the associated jumps in F.

The complementary function $\bar{F}(x) \equiv 1 - F(x) = \text{pr}(X > x)$ is sometimes more convenient; it is often called the *survivor function*, terminology borrowed from the context when X represents an age at death.

Still more convenient in most settings, especially applied ones, and almost as general, is a density function. We consider three special cases. A distribution is *absolutely continuous* if its distribution function is absolutely continuous, and hence, by the fundamental theorem of calculus, equal to the integral of its derivative. Its derivative is then called the *density function*. Sometimes we say 'density with respect to Lebesgue measure', with the implication that integrals are in the sense of Lebesgue rather than 'ordinary' Riemann integrals; but the latter are sufficient for our purposes.

To define an absolutely continuous distribution, it is sufficient to specify a density function f: a nonnegative function on $\mathcal{R}$ which integrates to unity. Then $F(x) = \int_{-\infty}^{x} f(t)dt$ and $f(x) = (d/dx)F(x)$, and $P(A) = \int_A f(t)dt$. Indeed, we write

$$\text{pr}\{X \in [x, x + dx)\} = f(x)dx = dF(x), \tag{1.1}$$

in that, dividing by dx and letting $dx \to 0$, we have an exact formula. Densities provide the common way of specifying the absolutely continuous distributions common in statistical inference, such as the normal, exponential, gamma, beta and Pareto; see Section 1.5 below. We sometimes refer to the set $\mathcal{X}$ on which f is positive, that is, where F is strictly increasing, as the *support* of X.

A second special case is that of *discrete distributions*; for these, the distribution function is flat except at a countable set of points $\mathcal{X}$, often called *mass points*, on which it jumps. This set of mass points, also called the *support* of F, is usually some subset of the integers $\mathcal{N}$ or the nonnegative integers $\mathcal{N}^+$. The *density function* specifies the magnitudes of these jumps and is defined to be zero elsewhere. To define a discrete distribution by a density, it is sufficient to specify the mass points $\mathcal{X}$ and non-negative values for f at the mass points, values which sum to unity. Examples of discrete distributions include the binomial, Poisson, negative binomial and discrete uniform distributions.

The relationship of such a density with the distribution function and probability measure, and its interpretive definition, is

$$f(x) \;=\; F(x) - F(x-) \;=\; \text{pr}(X = x) \tag{1.2}$$

where $F(x-) = \lim_{\epsilon \downarrow 0+} F(x - \epsilon)$. Contrast (1.2) with (1.1). To obtain the distribution function or probability measure from f, summation is required:

$$F(x) = \sum_{t \in \mathcal{X} \cap (-\infty, x]} f(t) \quad \text{and} \quad P(A) = \sum_{t \in \mathcal{X} \cap A} f(t).$$

In measure-theoretic terminology, such a density is called a density with respect to *counting measure.*

The third case is a mixture of the first two types. The *mixed-type* density has the form

$$f(x) = p \cdot f_{AC}(x) + q \cdot f_D(x)$$

where $p \in (0, 1)$, $q = 1 - p$, f_{AC} is the density of an absolutely continuous distribution, describing the smoothly increasing part of F, and f_D is the density of a discrete distribution (on $\mathcal{X}$, say), describing the jumps of F. It is often more convenient to omit the p and q factors and let f_{AC} and f_D be *subdensities*, that is, densities of *subdistributions*, the first integrating to p and the second summing over $\mathcal{X}$ to q (rather than to unity). The distribution function and probability measure are each given by a similar mixture formula. In this second form, the probability of any set A is obtained by integrating f_{AC} over A, summing f_D over the mass points in A, and adding.

It is proved in probability texts that there can exist a *singular distribution*, with a distribution function that is everywhere continuous but nowhere differentiable! A common construction uses the so-called *Cantor set*, the subset of the interval $[0, 1]$ consisting of points whose ternary expansion includes the digits 0 and 2 but not the digit 1. However, such sets and distributions do not arise in practice. It can also be proved that the most general distribution function is a mixture of these three types: absolutely continuous, discrete and singular. If we ignore the possibility of singular distributions, then the most general distribution is the mixed-type, defined above.

To unify these special cases, we find it convenient to use *Stieltjes notation*, writing $F(x) = \int_{-\infty}^{x} dF(t)$, with the integral representing summation for discrete cases or components.

Another way to specify a distribution is through the *quantile function*, the inverse of the distribution function; this is described in the next section.

Before proceeding further, we need to recall the concept of *expectation*[1]. For a given function g on $\mathcal{R}$, the expected value of $g(X)$ is defined as the Riemann-Stieltjes (or, more generally, Lebesgue-Stieltjes) integral

$$E\{g(X)\} \equiv \int g(x)dF(x) \;=\; \int gdF,$$

when this exists. Hence, when F is of mixed type with sub-densities f_{AC} and f_D, we have

$$E\{g(X)\} \;=\; \int g(x)f_{AC}(x)dx + \sum_{x\in\mathcal{X}} g(x)f_D(x)$$

if the integral and sum are absolutely convergent; only the integral or the sum is needed in the case of an absolutely continuous or discrete distribution, respectively. By the definition of convergence of an integral, existence of $E\{g(X)\}$ requires the existence of $E|g(X)|$.

Of special interest are the *moments* $\alpha_r = E(X^r)$ and the *central moments* $\mu_r = E\{(X-\mu)^r\}$ where $\mu = E(X) = \alpha_1$, the *mean* (when it exists); here, r is a positive integer. The *variance* is defined as $\text{var}(X) = \sigma^2 = \mu_2$, and its positive square root σ is called the *standard deviation.*

Finally, yet another way to specify a distribution is through a mathematical transform. The most powerful is the *characteristic function*, defined, for all real t, as

$$\psi(t) \equiv \; E(e^{itX}) = E\{\cos(tX)\} + iE\{\sin(tX)\}, \;\; t \in \mathcal{R}$$

where $i = \sqrt{(-1)}$ is the imaginary unit. It turns out that knowledge of $\psi(t)$ for all t near 0, i.e. for $|t| < \epsilon$ for any given $\epsilon > 0$, characterizes the distribution. Probability texts give inversion formulas whereby the distribution function or density can be recovered from the characteristic function.

A mathematically simpler transform is the *moment generating function*:

$$M(t) \;\equiv\; E(e^{tX}).$$

The set on which $M(t)$ is finite is an interval; if the interval includes zero as an interior point, then the moment generating function uniquely determines the distribution. Although there is no explicit inversion formula, the uniqueness

[1]See also the discussion of the "Expectation Identity" in Section 1.4.

property is still very useful as will be illustrated below. The logarithm of the moment generating function, called the *cumulant generating function*:

$$C(t) \;\equiv\; \log M(t).$$

is often more convenient for calculations. Coefficients in Taylor expansions of M and of C around $t = 0$ are the *moments* and *cumulants*, respectively, of the distribution. The first four cumulants are the mean, the variance, μ_3 and $\mu_4 - 3\sigma^4$, respectively. The third and fourth cumulants, when standardized to dimensionless forms by dividing by a power of σ, are referred to as the *coefficient of skewness* $SK \equiv \mu_3/\sigma^3$ and the *coefficient of excess kurtosis* $KUR \equiv (\mu_4/\sigma^4) - 3$. We will see below that these are both zero for a normal distribution, so they serve as useful descriptive measures of departure from normality.

No finite set of moments determines a distribution; the complete set of moments ($r = 1, 2, \ldots$) usually do determine the distribution, but there are exceptions: one such is the lognormal distribution described in *Exercise 10.*

To illustrate the foregoing, the reader might consider the following discrete distributions: Bernoulli, binomial, Poisson and discrete uniform (on integers $1, 2, \ldots, m$), and derive the corresponding cumulant generating functions; similarly, consider absolutely continuous distributions: normal, uniform, gamma, exponential, beta and do likewise, although some of the formulas may not be too convenient. Among these, only the Bernoulli, uniform and exponential distributions have simple forms for their distribution functions. The Cauchy distribution provides an example of a distribution for which moments, even the first, and the moment generating function do not exist, but it has simple forms for the characteristic function and distribution function. Some of these distributions are described in Section 1.5; more details can be found in various textbooks or the books of JOHNSON, KOTZ & BALAKRISHNAN (1994, 1997) and JOHNSON, KEMP & KOTZ (2008). Other distributions we sometimes use for illustrations include the geometric, negative binomial, censored exponential, Laplace, Weibull and Pareto.

We now turn to the inverse of the distribution function.

1.2 The Quantile Function

A distribution function F is nondecreasing and right continuous, with $F(-\infty) = 0$ and $F(\infty) = 1$. The quantile function Q is essentially the inverse of the distribution function, obtained by transposing the horizontal and vertical axes. If F is strictly increasing (over a finite or infinite interval), a unique inverse function is well-defined. Even where F jumps, a unique inverse function is easy to define. But where F is flat, there is an interval of possible values of the inverse, and some convention is needed in order to make the definition of an inverse function unique. Here we choose the mathematically most convenient choice, which can be thought of as the smallest such inverse. However, in applied settings, some central inverse, for example the middle value, may be more suitable. In any case, it cannot much matter since such intervals, at least their interiors, have zero probability.

To cover all possible cases, the *inverse distribution function*, or *quantile function*, is defined by the formula

$$Q(u) \equiv F^{-1}(u) = \inf\{x | F(x) \geq u\}, \ u \in (0,1). \tag{1.3}$$

Here, $\inf A$, the "infimum" of a set A of real numbers, denotes its greatest lower bound, for example, the infimum of the open interval $1 < x < 2$ and of the closed interval $1 \leq x \leq 2$ both equal 1. In the latter case, the infimum is contained in the set in the former case not. The "supremum" ($\sup A$) is defined analogously. It follows that Q is nondecreasing and left-continuous, with jumps where F is flat, and flat where F jumps. See Figure 1.1 for a simple example.

The choice of Q as the smallest possible inverse of F implies that F is the largest possible inverse of Q. The following properties of the function $Q(u)$ are proved in *Exercise 2.* They follow quickly from the monotonicity and right-continuity of the function $F(\cdot)$:

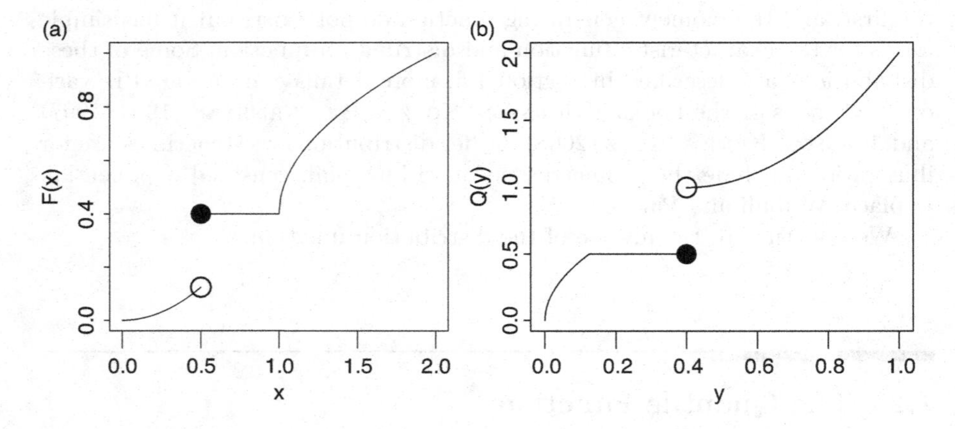

FIGURE 1.1
A distribution function (a) and the associated quantile function (b).

i. $Q\{F(x)\} \leq x$;

ii. $F\{Q(u)\} \geq u$;

iii. $Q(\cdot)$ is left-continuous;

The following lemma is frequently useful.

Lemma 1.1 $Q(u) \leq x \Leftrightarrow u \leq F(x)$.

Proof First, note that

$$Q(u) \leq x \Rightarrow \inf\{x' | F(x') \geq u\} \leq x \Rightarrow x \in \{x' | F(x') \geq u\} \Rightarrow F(x) \geq u,$$

where the middle step holds because of the right-continuity of F. Conversely,

$$F(x) \geq u \Rightarrow x \in \{x' | F(x') \geq u\} \Rightarrow x \geq \inf\{x' | F(x') \geq u\} \Rightarrow x \geq Q(u). \square$$

An immediate consequence (*Exercise 3*) of Lemma 1.1 is the formula $F(x) = \sup\{u : Q(u) \leq x\}$. Indeed, we may reverse the argument: any nondecreasing left-continuous function Q on (0,1) uniquely determines a distribution function. One application of Lemma 1.1 is as follows. Consider a distribution function F with associated quantile function Q. Let U represent a random variable uniformly distributed on $(0, 1)$, that is an absolutely continuous random variable with density $\mathbf{1}_{(0,1)}(x)$ and distribution function $\text{pr}(U \leq u) = u$ for $u \in [0, 1]$. Here $\mathbf{1}_A(x) = \mathbf{1}[x \in A]$ is the *indicator function*, equal to one for $x \in A$ and zero elsewhere. The *quantile transform* of U is defined as $V \equiv Q(U)$. Lemma 1.1 implies that $\text{pr}\{Q(U) \leq x\} = \text{pr}\{U \leq F(x)\}$ which equals $F(x)$. So V has distribution function F. This provides a useful mathematical tool: We have *constructed* a random variable V, from a simple uniformly distributed random variable U, to have a specified distribution function F. For applications of other such *Skorokhod constructions* see Section 2.7 and Chapter 2. This transform has practical implications for Monte Carlo studies: If Q is a computationally convenient function, a random variable with distribution function F can be generated by generating a uniform (pseudo-)random variable, for which computers have algorithms, and transforming from U to V.

Any distribution with a convenient form for its distribution function is a candidate for one with a convenient quantile function. This includes the exponential, Weibull, Cauchy and Pareto distributions but not the normal and gamma. However, these quantile functions are available in R or other software packages, see Section 1.9. Quantiles of special interest include the *median* (50%) and lower and upper *quartiles* (25% and 75%, respectively). Occasionally, we refer to tertiles (defining the upper, middle and lower thirds of the distribution) and deciles (defining tenths).

1.3 Random Vectors, Their Distributions and Properties

We now consider several random variables simultaneously, modeling situations in which several measurements are to be taken or a batch of data collected. We consider a *random vector* $\mathbf{X}$, a column vector of k components $X_1, \ldots, X_k$; equivalently, we consider k random variables, the components of $\mathbf{X}$, simultaneously. The distribution of $\mathbf{X}$ equivalently, the *joint distribution* of the components of $\mathbf{X}$, may be specified in one of several ways. A *probability measure* $P(A) = \text{pr}(\mathbf{X} \in A)$ specifies the probability in each (Borel) subset A of the k-dimensional product space $\mathcal{R}_k$ and satisfies the probability axioms (as before, with $\mathcal{R}$ replaced by $\mathcal{R}_k$).

Most of the relevant concepts can be developed with the case $k = 2$, that is, a random point on the plane or a pair of random variables. In particular, illustrative sketches are easy to construct, and the reader is encouraged to do so.

A *(joint) distribution function (distribution function)* $F(\mathbf{x})$ (i) has a non-negative kth-order difference everywhere (if A is a k-dimensional rectangle with the 2^k corners at $\mathbf{x}+\boldsymbol{\delta}$, where the jth coordinate of $\boldsymbol{\delta}$ is either 0 or some $\delta_j > 0$, then $P(A)$ is a kth-order difference of F at $\mathbf{x}$), (ii) is right-continuous in each coordinate and (iii) is zero when any coordinate tends to $-\infty$ and unity when all coordinates tend to $+\infty$. The relationship with the probability measure is

$$F(\mathbf{x}) = P((-\infty, x_1]\times(-\infty, x_2]\times\cdots\times(-\infty, x_k]) = \text{pr}(X_1 \leq x_1, \ldots, X_k \leq x_k)$$

where $\mathbf{x} = (x_1, \ldots, x_k)^T$ ('T' for transpose). *Marginal distributions* of sub-vectors of $\mathbf{X}$ omitting one or more coordinates, are obtained by substituting $+\infty$ in F for each omitted coordinate.

A *(joint) density function* f for an absolutely continuous distribution is the kth-order partial derivative of the absolutely continuous distribution function F, and F is recovered from f by integration; more generally,

$$P(A) = \int_A f(\mathbf{x})d\mathbf{x},$$

or simply

$$\text{pr}(X_1 \in [x_1, x_1 + dx_1), \ldots, X_k \in [x_k, x_k + dx_k)) = f(x_1, \ldots, x_k)dx_1 \ldots dx_k. \tag{1.4}$$

The density of a discrete distribution assigns non-negative probabilities, summing to unity, to a countable set $\mathcal{X}$ of mass points in $\mathcal{R}_k$. Besides absolutely continuous and discrete distributions, there are *singular distributions.* In dimensions $k > 1$ these become useful in practice. A singular distribution is, conceptually, a continuous distribution of probability over a subset of dimension less than k. For example, in two dimensions, a continuous distribution over a line, or a curve, is a singular distribution. From a practical perspective, we deal with such distributions by transforming to a lower dimension, where the distribution may be absolutely continuous. As in the univariate (one-dimensional) case, the most general distribution is a mixture of the three special types: absolutely continuous, discrete and singular. Densities of marginal distributions are obtained by integrating out, or summing over, the extraneous arguments.

The *expectation* of a scalar function $h(\mathbf{X})$ of $\mathbf{X}$ is defined by integration. In Stieltjes notation, we write

$$E\{h(\mathbf{X})\} \equiv \int h(\mathbf{x})dF(\mathbf{x})$$

where an integral dF is the multiple integral of $f(\mathbf{x})d\mathbf{x}$ in the absolute case and the sum of $f(\mathbf{x})$ over $\mathbf{x} \in \mathcal{X}$ in the discrete case (and a mix of integration and summation in the mixed case). The expectation exists finitely if and only if the expectation of $|h(\mathbf{X})|$ is finite. Such integrals and sums can be done iteratively in any order. A special expectation is the variance:

$$\text{var}\{h(\mathbf{X})\} \equiv E[h(\mathbf{X}) - E\{h(\mathbf{X})\}]^2.$$

There is no direct analog of the quantile function in higher dimensions. However, there is another way of specifying a distribution, namely through the use of *conditional distributions.* We discuss briefly only the absolutely continuous case for bivariate distributions; the discrete case, and higher dimensional cases, are quite analogous. (A general definition is difficult without some measure-theoretic considerations.)

The notion of a conditional distribution depends on that of *conditional probability*: the conditional probability of event A, given event B, is defined as $P(A|B) \equiv P(A \cap B)/P(B)$ (when $P(B) > 0$ and arbitrarily otherwise). Let $f(x, y)$, $f(x)$ and $g(y)$ denote the joint density of (X, Y) and the marginal densities of X and Y, respectively. Then, if (X, Y) is discrete, the conditional probability of $X = x$ given $Y = y$ is

$$f(x|y) \equiv \frac{f(x,y)}{g(y)} \tag{1.5}$$

whenever the denominator is positive. Using Equations (1.1) and (1.4), we take this formula (1.5) to define the conditional density of X given Y generally, not just in the discrete case. Heuristically, in the absolutely continuous case, $f(x|y)$ in Equation (1.5) is $\text{pr}(X \text{ near } x|Y \text{ near } y)/dx$. The conditional distribution function at x is the integral (or sum) of the density over $(-\infty, x]$. The conditional expectation of a function $h(X, Y)$ of (X, Y) given $Y = y$ is the integral (or sum) over x of the function (with Y set equal to y) times the conditional density; the resulting expression is a function of y. Thus,

$$E\{h(X,Y)|Y = y\} = \int h(x,y)f(x|y)dx = H(y),$$

say, a function of y. Conditional moments, means, variances, moment generating functions, etc., are thereby defined in the same way as are the marginal quantitites, but using the conditional distribution. The conditional distribution of Y given $X = x$ is defined analogously.

There are useful formulas relating unconditional means and variances of a function of X and Y to conditional moments. These are referred to as *iterated expectation* and *iterated variance formulas.* The iterated mean (expectation) formula is

$$E\{h(X,Y)\} = E\big[E\{h(X,Y)|Y\}\big] \;=\; E\{H(Y)\}$$

where $H(y) \equiv E\{h(X,Y)|Y = y\}$. The proof (in the continuous case, the discrete case is similar) is immediate:

$$\begin{aligned} E\{h(X,Y)\} &= \int\int h(x,y)f(x,y)dxdy = \int\int h(x,y)f(x|y)g(y)dxdy \\ &= \int H(y)g(y)dy = E\{H(Y)\}. \end{aligned}$$

The iterated variance formula (*Exercise 4*) has two terms, namely the mean of the conditional variance and the variance of the conditional mean:

$$\mathrm{var}\{h(X,Y)\} = E\big[\mathrm{var}\{h(X,Y)\big|Y\}\big] \;+\; \mathrm{var}\big[E\{h(X,Y)\big|Y\}\big].$$

By symmetry of the roles of X and Y, conditioning can be done on X instead of on Y.

We say X and Y are *stochastically independent*, or simply *independent*, if the conditional distribution of one, given the other, is identical to the corresponding marginal distribution. It is necessary and sufficient that the joint distribution function $F(x,y)$ factor into a function of x only times a function of y only; these functions may be taken to be the two marginal distribution functions. Equivalently, the joint density function factors. Also, when X and Y are independent, the expectation of a product of a function of X times a function of Y is the product of the two expectations.

All these conditional concepts extend in a straightforward manner to vector-valued $\mathbf{X}$ and/or vector-valued $\mathbf{Y}$. We define *(mutual) independence* of the coordinates of $\mathbf{X}$ by factorization of the distribution function or density function into products of the marginal quantities. A case of primary interest is that of a *random sample* $\mathbf{X}$ of *size* n, that is of independent and identically distributed $X_1, \ldots, X_n$. In this case

$$F_{\mathbf{X}}(x_1,\ldots,x_n) \;=\; \prod_{i=1}^{n} F(x_i),$$

where F is the common marginal distribution function of each X_i, and similarly with F replaced by f. We say that $\mathbf{X}$ is a *random sample* from a *population* with distribution function F (or density f).

The joint *moment generating function* and *cumulant generating function* of a random vector $\mathbf{X}$ are defined, with argument being a vector $\mathbf{t}$, respectively, as as

$$M(\mathbf{t}) \equiv E\{\exp(\mathbf{t}^T\mathbf{X})\} \quad \text{and} \quad C(\mathbf{t}) \equiv \log M(\mathbf{t}),$$

(with the superscript T denoting *transpose*) and are said to exist whenever the moment generating function is finite in a neighborhood of $\mathbf{t} = \mathbf{0}$. The *joint characteristic function* is defined as $\psi(\mathbf{t}) = E\{\exp(i\mathbf{t}^T\mathbf{X})\}$. Each of these, when it exists, uniquely determines the joint distribution. For a characteristic function, its definition in a neighborhood of the origin ($\| \mathbf{t} \| < \epsilon$, with $\| \cdot \|$ denoting the square root of the sum of squares of the coordinates) is sufficient. Marginal transforms are obtained by setting other coordinates of $\mathbf{t}$ to 0. Independence is characterized by factorization of the moment generating function (and characteristic function) into the product of corresponding marginal quantities, or representing the cumulant generating function as the sum of corresponding marginal quantities.

The mean vector $\boldsymbol{\mu}$ is the vector of marginal means, and written $\boldsymbol{\mu} = E(\mathbf{X})$, with expectation taken elementwise. The *variance matrix* Σ is the matrix of variances and covariances, that is, of elements $E\{(X_i - \mu_i)(X_j - \mu_j)\}$; we write $\Sigma = E\{(\mathbf{X} - \boldsymbol{\mu})(\mathbf{X} - \boldsymbol{\mu})^T\}$, again with expectation taken elementwise. Means, variances and covariances can be obtained by equating the coefficients in a Taylor expansion of the joint cumulant generating function; but the means and variances are more easily derived directly from the marginal distributions. The more descriptive *correlation coefficients* are defined as

$$\rho(X_i, X_j) \equiv \text{cov}(X_i, X_j)/\sqrt{\{\text{var}(X_i)\text{var}(X_j)\}},$$

a dimensionless quantity providing a measure of the linear dependence between X_i and X_j.

Finally, the covariance (matrix) between two random vectors $\mathbf{X}$ (of dimension d_X) and $\mathbf{Y}$ (of dimension d_Y), written $\text{cov}(\mathbf{X}, \mathbf{Y})$, is the $d_X \times d_Y$ matrix of pairwise covariances $\text{cov}(X_i, Y_j)$.

A convenient way of characterizing some joint distributions is through the so-called *Cramér-Wold device.* Consider all linear combinations of the coordinates of $\mathbf{X}$, namely

$$Y \equiv Y_{\mathbf{a}} \equiv \mathbf{a}^T\mathbf{X} = a_1X_1 + a_2X_2 + \cdots + a_kX_k$$

for each $\mathbf{a} \in \mathcal{R}^k$, or just for unit vectors $\mathbf{a}$ (that is, for which $\| \mathbf{a} \| = \sqrt{(\sum a_j^2)} = 1$). Then the characteristic function of Y is

$$\psi_Y(s) \equiv E\{\exp(isY)\} = E\{\exp(is\mathbf{a}^T\mathbf{X})\} = E\{\exp(i\mathbf{t}^T\mathbf{X})\} = \psi_{\mathbf{X}}(\mathbf{t})$$

with $\mathbf{t} = s\mathbf{a}$, and 'for all $|s| < \epsilon$ and all $\| \mathbf{a} \| = 1$' is equivalent to 'for all $\| \mathbf{t} \| < \epsilon$'. Here, we have used the 'expectation identity' described in the next section. Hence, specifying the distribution of $Y = Y_{\mathbf{a}}$ for every unit vector $\mathbf{a}$ is equivalent to specifying the distribution of $\mathbf{X}$. The k-dimensional distribution is replaced by infinitely many one-dimensional distributions.

1.4 Transformations; Functions of Random Variables

Consider one or more functions of a random variable or random vector, say $Y = h(X)$ where either or both of X and Y may be vector-valued. The context may be that of transforming from one coordinate system to another, perhaps nonlinearly, or that interest lies in some statistic(s) Y with X representing a random sample. It is sometimes convenient to continue to express distributional characteristics in terms of the distribution of X, the distribution function of Y may not be needed. For example, if we are interested in the mean of Y, we can either derive the distribution function of Y and then calculate its mean or we can integrate $h(x)$ with respect to the distribution of X.

This latter equivalence[2] is so fundamental that we digress to restate it, and give it a name:

$$\text{Expectation identity}: \text{ if } Y = h(X), \text{ then } E(Y) = E\{h(X)\},$$

that is $\int y dF_Y(y) = \int h(x) dF_X(x)$ And this is so no matter what the dimension of X. Or more generally, $E\{g(Y)\} = E[g\{h(X)\}]$, providing alternative computational opportunities. We used this identity implicitly when defining the variance of a function of X in Section 1.1 above. The proof in the discrete case is immediate, for one-to-one transformations of absolutely continuous distributions the result follows quickly from Equation (1.6). The general case requires concepts from measure theory.

There are several ways of finding the distribution of Y from that of X. We illustrated the *distribution function method* when considering the quantile transform in Section 1.2. Later we consider sums of independent random variables by the *transform method.* We first review the so-called *density method* and allow vector-valued variables.

Suppose that X has an absolutely continuous distribution with density f, and Y, defined by $y = h(x)$, is one to one in X. Then the density of Y is

$$g(y) = |J^{-}(y)| \cdot f(x)\Big|_{x=h^{-1}(y)}, \tag{1.6}$$

where $J(x)$ is the Jacobian of the transformation, the determinant of the matrix of partial derivatives of h in its coordinates and $J^{-}(y)$ is the Jacobian of the inverse transformation, the determinant of the matrix of partial derivatives of the inverse function. If there are multiple inverses, $f(x)$ is replaced by a sum over these inverses, with the Jacobian term being common.

If Y is of lower dimension than X, we proceed through the following three steps: (i) fill out the transformation by introducing additional Y-coordinates so that the resulting transformation is one to one or perhaps two to one, etc.; (ii) apply (1.6) to obtain the joint density of the Y's; and (iii) integrate out the additional coordinates. We give a simple example at the end of the section.

---

[2]Sometimes called "The Theorem of the Unconscious Statistician".

A discrete version is similar, except there is no Jacobian factor in Equation (1.6).

A linear transformation is a simple but important case. The resulting mean and variance formulas are often useful. With $Y = \mathbf{a}^T\mathbf{X} = \mathbf{X}^T\mathbf{a}$,

$$E(Y) = \mathbf{a}^T E(\mathbf{X}) = \mathbf{a}^T \boldsymbol{\mu}$$

and

$$\text{var}(Y) = E\{\mathbf{a}^T(\mathbf{X} - \boldsymbol{\mu})(\mathbf{X} - \boldsymbol{\mu})^T\mathbf{a}\} = \mathbf{a}^T \text{var}(\mathbf{X})\mathbf{a} = \mathbf{a}^T \Sigma \mathbf{a}.$$

Since a variance is non-negative, the variance matrix Σ of $\mathbf{X}$ is positive semi-definite and symmetric. If we simultaneously consider l linear combinations, writing $\mathbf{W} = A^T\mathbf{X}$ for some $k \times l$ matrix A, then

$$E(\mathbf{W}) = A^T E(\mathbf{X}) \quad \text{and} \quad \text{var}(\mathbf{W}) = A^T \Sigma A.$$

For nonlinear functions, the iterated mean and variance formulas from Section 1.3 are often useful.

Now let us consider the case of Y being the sum of k independent random variables, $Y = X_1 + \cdots + X_k$ say. We will work with the moment generating function although we could equally well use the cumulant generating function or characteristic function. By substituting $\sum X_i$ for Y and using the independence of the X_i, the moment generating function $E\{\exp(tY)\}$ of Y is seen to equal the product of the moment generating functions of the component X_i. Equivalently, the cumulant generating function of Y is the sum of the component cumulant generating functions. If the resulting transform is recognized as that of a known distribution, then the uniqueness of transforms identifies this as the distribution of Y. A corollary is that each cumulant of Y is the sum of the corresponding cumulants of the X_i.

The distribution of such a sum is said to be the *convolution* of the marginal distributions. For the sum of just two independent random variables, $Y = X_1 + X_2$, the density method leads to formula (1.7) below (or to the same formula with the subscripts 1 and 2 interchanged):

$$f_Y(y) = \int f_{X_1}(y - x)dF_{X_2}(x). \tag{1.7}$$

Here is the derivation: (i) Write $Y_1 = Y$ and introduce $Y_2 = X_2$, so that $X_1 = Y_1 - Y_2$ and $X_2 = Y_2$; (ii) the Jacobian is unity, so the joint density of (Y_1, Y_2) is $g(y_1, y_2) = f_1(y_1 - y_2) \cdot f_2(y_2)$ (with f_i the density of X_i); (iii) integrating out y_2 yields (1.7) (with a change of notation). Integrating (1.7) yields a convolution formula for the distribution function of Y; simply replace each f by the corresponding F.

1.5 Some Common Univariate Distributions

A number of distributions appear in later chapters, and in exercises, as examples for constructing random sampling models. Others, especially the normal and chi-squared distributions, appear as approximating the distributions of statistics of interest in large-sample methods. We list some of these distributions here, with some pertinent facts summarized or derived. We start with some discrete distributions.

The *binomial distribution* $B(n,p)$, where $0 \le p \le 1$ and n is a positive integer has density $C(n,x)p^x q^{n-x}$ $x \in 0,1,\ldots,n$. Here $q = 1-p$ and $C(n,x) = n!/\{x!(n-x)!\}$. For $n = 1$, it is the *Bernoulli distribution.* The moment generating function is $(q+e^t p)^n$, easily calculated by recognizing a binomial expansion of $(a+b)^n$, and the cumulant generating function is $n\log(q+e^t p)$. The multiplicativity property of transforms shows that the sum of n independent identically distributed Bernoulli random variables is a binomial random variable.

The *Poisson* density is $e^{-\lambda}\lambda^x/x!$ for $x \in 0,1,2,\ldots$. The cumulant generating function, after recognizing an exponential expansion, is found to be $C(t) = \lambda(e^t - 1)$. It follows that adding independent Poisson random variables with parameters $\lambda_1, \lambda_2, \ldots, \lambda_n$ gives a Poisson random variable with parameter $\lambda \equiv \sum \lambda_i$ (why?).

Now we consider some common absolutely continuous distributions:

A *standard normal*, or $N(0,1)$, random variable, often denoted by Z, has density

$$\varphi(z) = \frac{1}{\sqrt{(2\pi)}}\exp\big(-\tfrac{1}{2}z^2\big),$$

and distribution function denoted by

$$\Phi(z) = \int_{-\infty}^{z} \varphi(u)du.$$

This cannot be expressed explicitly in terms of any standard functions; nor can its inverse, the standard normal quantile function. These are readily available in R and other software packages. Evaluations use numerical approximations; see Hastings (1955) or Abramowitz & Stegun (1964). See *Exercise 8* for proof that the density integrates to unity. Using this result the corresponding cumulant generating function is easily found to be $\frac{1}{2}t^2$ (*Exercise 9*). It is clear by symmetry that $E(Z) = 0$ and also $E(Z^{2n+1}) = 0$ for all positive integers n. We also have $E(Z^2) = 1$ and $E(Z^4) = 3$, using the fact that $x\varphi(x) = -d/dx\{\varphi(x)\}$ and integration by parts. Repeated (or inductive) use of partial integration allows the indefinite integral of $x^n\varphi(x)$ to be expressed in terms of elementary functions when $n > 0$ is odd and in terms of $\Phi(x)$ when $n > 0$ is even.

A random variable X with a general *normal distribution*, $N(\mu,\sigma^2)$, having mean μ and variance σ^2, in compact notation $X \sim N(\mu,\sigma^2)$, has density

$(1/\sigma)\varphi\{(x-\mu)/\sigma\}$, distribution function $\Phi\{(x-\mu)/\sigma\}$ and cumulant generating function $\mu t + \frac{1}{2}\sigma^2 t^2$; setting $\mu = 0$ and $\sigma = 1$ yield the standard normal case. Note that all cumulants of order exceeding two are zero. Such an X may conveniently be defined by the transformation $X = \mu + \sigma Z$. That is, for $\sigma > 0$, $X \sim N(\mu, \sigma^2)$ if and only if $Z = (X-\mu)/\sigma \sim N(0,1)$.

From the cumulant generating function, *additivity* is seen to hold. Summing independent normal random variables leads to a normal random variable, with mean and variance obtained by summing the individual means and variances.

The *gamma distribution*: The gamma distribution, with *scale parameter* $\alpha > 0$ and *shape parameter* $\beta > 0$, is an absolutely continuous distribution on $(0, \infty)$ with density

$$f(x) \;=\; \frac{\alpha^\beta}{\Gamma(\beta)} e^{-\alpha x} x^{\beta-1} \mathbf{1}_{(0,\infty)}(x).$$

Here $\Gamma(\beta) = \int t^{\beta-1}\exp(-t)dt$ is the quantity needed to to make $\int f(x)dx = 1$. It is easily shown that $\Gamma(1) = 1$ and the $\Gamma(\beta) = (\beta-1)\Gamma(\beta-1)$ so that, inductively, for integer n, $\Gamma(n) = (n-1)!$. The transformation $y = x^2/2$ in the normal integral shows that $\Gamma(\frac{1}{2}) = \sqrt{\pi}$. See *Exercise 8.* The cumulant generating function is $C(t) = -\beta\log(1-t/\alpha)$, defined for $t < \alpha$. When $\beta = 1$, the distribution reduces to the *exponential distribution*; when $\alpha = 1$ also, it is the *standard exponential.*

The gamma distribution function is

$$F(x) = \frac{\alpha^\beta}{\Gamma(\beta)} \int_0^x e^{-\alpha y} y^{\beta-1} dy = \frac{1}{\Gamma(\beta)} \int_0^{\alpha x} e^{-z} z^{\beta-1} dz \;\equiv \Gamma_{\alpha x}(\beta), \tag{1.8}$$

the *(normalized) incomplete gamma integral*, reducing to $1 - \exp(-\alpha x)$ for the exponential case. However, Equation (1.8) may be converted to an infinite series by successive integration by parts; specifically,

$$\Gamma_z(\beta) \;=\; \sum_{i=0}^{\infty} e^{-z} \frac{z^{\beta+i}}{\Gamma(\beta+i+1)}, \tag{1.9}$$

and numerical evaluation is possible by truncating the series; but even then a computational algorithm for the gamma function, appearing in the denominator, is needed. Again, there is no explicit formula for the quantile function (except for the exponential case). Fortunately R and many other software packages include commands for calculating both the gamma function and the incomplete gamma integral.

All the distributions mentioned above belong to the class of *exponential families.* A distribution of this class has density function of the form

$$f(x, \theta) = \exp\{t(x)c(\theta) - d(\theta) - v(x)\}\mathbf{1}_A(x), \tag{1.10}$$

where $t(x)$, $c(\theta)$, $d(\theta)$ and $v(x)$ are specified functions, the first two possibly vector-valued. The support A of the density must not depend on θ, and the function $d(\theta)$ is determined by the requirement that the density integrates or sums to unity. The parameter θ may be scalar or vector-valued and often $c(\theta)$ has the same dimension as θ. When $c(\theta) \equiv \theta$ the family is said to be in natural form, this can usually be achieved by reparameterization. The natural parameters are $\theta = \log\{p/(1-p)\}$ for the binomial (with the index n assumed known), $\theta = \log \lambda$ for the Poisson, $\theta = (\mu/\sigma^2, 1/\sigma^2)$ for the normal distribution and $\theta = (\alpha, \beta)$ for the gamma distribution. See *Exercise 7*. The usual parameterization of the normal distribution with $\theta = (\mu, \sigma)$ corresponds to the nonnatural form with $c(\theta_1, \theta_2) = (\theta_1/\theta_2^2, 1/\theta_2^2)$. The *Weibull* distribution, with distribution function $F(x, \rho, \theta) = 1 - \exp\{-(\rho x)^\theta\}$ and density $f(x, \rho, \theta) = \theta\rho(\rho x)^{\theta-1}\exp\{-(\rho x)^\theta\}$ is not of exponential family form, since the power θ inside the argument of the exponential precludes any representation of this density in the form (1.10).

The *(central) chi-squared distribution*: The chi-squared distribution with ν (a positive integer) *degrees of freedom* is a gamma distribution with $\alpha = \frac{1}{2}$ and $\beta = \frac{\nu}{2}$. Hence, its density is

$$f(x, \nu) = \frac{1}{2^{\frac{\nu}{2}}\Gamma(\frac{\nu}{2})} e^{-\frac{1}{2}x} x^{\frac{\nu}{2}-1} \quad \text{for } x > 0, \tag{1.11}$$

its distribution function is $\Gamma_{x/2}(\frac{1}{2}\nu)$, and its cumulant generating function is $-\frac{\nu}{2}\log(1-2t)$ $(t < \frac{1}{2})$. This distribution arises as the distribution of the sum of ν independent squared standard normal random variables. Also, with $\nu = n - 1$, it is the distribution of the sum of squared deviations from the sample mean in a sample of size n from a $N(\mu, 1)$ population.

As an introduction to the distribution considered next, let us derive the first of those two facts. The necessary steps are: (i) to find the cumulant generating function of Z^2 with Z standard normal; (ii) to multiply by ν to convert it to the cumulant generating function of the convolution of ν such variables; (iii) to recognize the latter cumulant generating function. For (i), we calculate

$$M_{Z^2}(t) \equiv E\left(e^{Z^2 t}\right) = \int e^{z^2 t}\frac{1}{\sqrt{(2\pi)}}e^{-\frac{1}{2}z^2}dz = \sigma \cdot \int \frac{1}{\sqrt{(2\pi)}\sigma}e^{-\frac{1}{2\sigma^2}z^2}dz = \sigma$$

say, where σ is a function of t given by $\sigma^2 = 1/(1-2t)$ $(t < \frac{1}{2})$. The last equality follows because the integrand is the density of a normal distribution with mean zero and standard deviation σ. Taking logs, we have $C_{Z^2}(t) = -\frac{1}{2}\log(1-2t)$, existing for $t < \frac{1}{2}$. For (ii), multiplying by ν yields the cumulant generating function of the sum $Z_1^2 + \cdots + Z_\nu^2$, and finally (iii) this is recognized as the cumulant generating function of a chi-squared distribution with ν degrees of freedom.

The *noncentral chi-squared distribution*: The noncentral chi-squared distribution with *noncentrality parameter* λ^2 (> 0) and ν *degrees of freedom* is defined as the distribution of the sum, X, of ν independent squared normal random variables, each with unit variance, and with means $\mu_1, \ldots, \mu_\nu$, not all zero. We sometimes write $\boldsymbol{\mu}$ for the vector $(\mu_1, \ldots, \mu_\nu)^T$ and $\lambda^2 \equiv \|\boldsymbol{\mu}\|^2 = \sum_{j=1}^{\nu} \mu_j^2$ for the square of its Euclidean norm. It turns out (see below) that the distribution of X depends on the μ_j's only through λ^2.

We first extend the derivation just given for the central chi-squared distribution. But when we get to step (iii), we find we do not have a recognizable cumulant generating function. However, we will have determined that the distribution depends on the various μ_j's only through the sum of their squares λ^2. We are also able to show that a certain mixture of central chi-squared distributions has the same moment generating function as the non-central chi-squared distribution. This leads to simple formulas for moments and to series expansions for the density and distribution functions of the latter.

We first derive the cumulant generating function of $X = \sum_{j=1}^{\nu} Y_j^2$ with the Y_j's independent $N(\mu_j, 1)$ random variables. As for central chi-squared, the first step is to determine the moment generating function when $\nu = 1$:

$$M_1(t) \;=\; E\{\exp(Y^2 t)\}$$

where $Y \sim N(\mu, 1)$ (dropping the subscript 1). Writing $Z = Y - \mu$ (standard normal), we have

$$M_1(t) \;=\; E[\exp\{(Z+\mu)^2 t\}] = \int \frac{1}{\sqrt{(2\pi)}} \exp\left[-\tfrac{1}{2}\{z^2 - 2(z+\mu)^2 t\}\right] dz.$$

Completing the square, the expression in parentheses may be written as

$$(1-2t)\left(z - \frac{2t}{1-2t}\mu\right)^2 \;-\; 2t\mu^2 \;-\; \frac{4t^2}{1-2t}\mu^2$$

for $t < \frac{1}{2}$. The last two terms can be combined and the resulting exponential factored out of the integral. The transformation from z to a new variable $u = \{z - 2t\mu/(1-2t)\}\sqrt{(1-2t)}$, with t assumed fixed, facilitates integration, yielding

$$M_1(t) \;=\; \frac{1}{\sqrt{(1-2t)}} \cdot \exp\left(\frac{t}{1-2t}\mu^2\right),$$

for $t < \frac{1}{2}$. Taking logs yields the cumulant generating function of Y^2; it is given by Equation (1.12) below with $\nu = 1$ and $\lambda = \mu$.

For $\nu > 1$, we need to convolve ν such Y^2's. The resulting cumulant generating function is the sum of the corresponding cumulant generating functions. With $\lambda^2 = \sum_{j=1}^{\nu} \mu_j^2$ this is

$$C_\nu(t) = \frac{t}{1-2t}\lambda^2 - \frac{\nu}{2}\log(1-2t), \tag{1.12}$$

again for $t < \frac{1}{2}$. As claimed earlier, the cumulant generating function, and hence also the distribution, depends on the μ_j's only through λ^2. Notice also that it collapses correctly to central chi-squared when $\lambda^2 = 0$. But we do not recognize the cumulant generating function in Equation (1.12) as that for a distribution considered earlier, so we must find its density by another method. We could use the inversion formula for the characteristic function to find the distribution function, but the following derivation is simpler.

Consider a pair (X, Y) of random variables with the conditional distribution of X given $Y = y$ (a non-negative integer) having a central chi-squared distribution with $2y + \nu$ degrees of freedom, and Y having a Poisson distribution with parameter $\frac{1}{2}\lambda^2$. The conditional moment generating function of X given $Y = y$ is

$$E\{\exp(tX)|Y = y\} = \left(\frac{1}{1-2t}\right)^{y+\nu/2}$$

The moment generating function of Y, evaluated at s, say, is

$$E\{\exp(sY)\} = \exp\left\{-\frac{1}{2}\lambda^2(1-e^s)\right\}. \tag{1.13}$$

The unconditional moment generating function of X is

$$E\{\exp(tX)\} = E\big[\{E\exp(tX)|Y\}\big] = \left(\frac{1}{1-2t}\right)^{\nu/2} E\{\exp(sY)\}$$

with $s = -\log(1-2t)$. Substituting $e^s = 1/(1-2t)$ in Equation (1.13) and simplifying gives

$$E\{\exp(tX)\} = \left(\frac{1}{1-2t}\right)^{\nu/2} \exp\left(\frac{\lambda^2 t}{1-2t}\right)$$

with cumulant generating function $C_X(t) = \log[E\{\exp(sX)\}]$ agreeing with (1.12). Since $C_X(t)$ exists in a neighborhood of zero, in fact for all $t < \frac{1}{2}$, the uniqueness theorem applies, so that X must itself have the postulated noncentral chi-squared distribution.

This result leads to a simple series expansion for the density, in terms of the central chi-squared densities $f(x; 2j+\nu)$ with $2j+\nu$ degrees freedom, namely

$$f(x;\nu,\lambda^2) = \sum_{j=0}^{\infty} e^{-\frac{1}{2}\lambda^2} \frac{(\frac{1}{2}\lambda^2)^j}{j!} \cdot f(x; 2j+\nu). \tag{1.14}$$

An alternative derivation and formula for this density are developed in *Exercises 15 and 16.*

The mean and variance of noncentral chi-squared can be determined in several ways: (i) directly from the definition, determining first the mean, second moment, and thence variance of $(Z+\mu)^2$ and then summing; (ii) expanding the cumulant generating function (1.12) and finding the coefficients of t and of $\frac{1}{2}t^2$, or (iii) from the alternative definition, using iterated expectation and variance formulas. We find, by whichever method,

$$E(X) = \nu + \lambda^2 \quad \text{and} \quad \text{var}(X) = 2\nu + 4\lambda^2.$$

To obtain the distribution function, we may integrate (1.14), using the expansion based on Equation (1.9) for the central chi-squared distribution function. Letting $k = i + j$ and reorganizing the double summation yields the following formula for the noncentral chi-squared distribution function:

$$
\begin{aligned}
F(x;\nu,\lambda^2) &= e^{-\frac{x}{2}}\sum_{k=0}^{\infty}\left(\frac{x}{2}\right)^{\frac{\nu}{2}+k}\Gamma\left(\frac{\nu}{2}+k+1\right)\cdot G\left(k,\tfrac{1}{2}\lambda^2\right) \\
&= e^{-y-\alpha}\sum_{k=0}^{\infty}\left\{\frac{y^{\mu+k}}{\Gamma(\mu+k+1)}\sum_{\ell=0}^{k}\frac{\alpha^\ell}{\ell!}\right\} \qquad (1.15)
\end{aligned}
$$

where $G(\cdot,\theta)$ is the distribution function of the Poisson distribution with mean θ and $y = x/2, \mu = \nu/2, \alpha = \lambda^2/2$. Formula (1.15) facilitates numerical computation of the distribution function; the sum converges quite fast.

The *t-distribution*, often called *Student's t*, is the distribution of the ratio of a standard normal random variable to the square root of an independent central chi-squared random variable divided by its degrees of freedom. Its *degrees of freedom* parameter is the degrees of freedom of the chi-squared random variable in the denominator. Its density is

$$
f(t) = \frac{1}{\sqrt{\nu}\mathrm{Be}(\frac{1}{2},\frac{\nu}{2})}\cdot\frac{1}{(1+\frac{t^2}{\nu})^{\frac{\nu+1}{2}}}
$$

with $\mathrm{Be}(\cdot,\cdot)$ the *beta function*, see the Appendix. The t-distribution arises as the distribution of a sample mean, divided by its estimated *standard error* (standard deviation of the distribution of the sample mean), when sampling from a normal distribution with mean zero. When $\nu = 1$, the denominator becomes the square root of a chi-squared distribution on one degree of freedom which reduces to a standard normal distribution and we obtain the density of the standard *Cauchy distribution* $f(x) = 1/\{\pi(1+x^2)\}$ (*Exercise 12*). The corresponding distribution function is $F(x) = \frac{1}{2} + (1/\pi)\arctan(x)$. The Cauchy distribution is the source of many counterexamples in probability because, despite the obvious symmetry of $f(x)$ about $x = 0$, the mean of the distribution does not exist since $\int |x|f(x)dx$ diverges. A more general form, with median ν and scale parameter τ, is

$$
f(x|\nu,\tau) = \frac{\tau}{\pi\{(x-\nu)^2+\tau^2\}},
$$

which integrates to $F(x) = \frac{1}{2} + (1/\pi)\arctan\{(x-\nu)/\tau\}$. Since $\arctan(1) = \pi/4$, the values $\nu \pm \tau$ are the upper and lower and quartiles of the density.

The *beta distribution*: The beta distribution is an absolutely continuous distribution on the unit interval, with two *shape parameters* α and β (both positive). The density is

$$
f(x) = \frac{1}{Be(\alpha,\beta)}x^{\alpha-1}(1-x)^{\beta-1}1_{(0,1)}(x)
$$

where the normalizing constant $Be(\cdot,\cdot)$ is the *beta function*. When $\alpha = \beta = 1$, it reduces to the *uniform distribution*. One derivation of the beta distribution is as the distribution of $V = X/(X+Y)$ where X and Y are independent gamma random variables with common scale parameters and with shapes α and β, respectively. This derivation also shows that $Be(\alpha,\beta) = \Gamma(\alpha)\Gamma(\beta)/\Gamma(\alpha+\beta)$. See the Appendix and *Exercise 11*. The beta distribution is also related to the *F-distribution*, in that F is proportional to $W = V/(1-V) = X/Y$, that is, to the ratio of two independent gamma random variables.

Other common absolutely continuous distributions are the *Laplace (double exponential) distribution*, the *logistic distribution*, the *Cauchy distribution*, the *Weibull distribution*, the *Pareto distribution* and the *lognormal distribution*. See *Exercise 10* for some properties of the latter. These distributions will be used for examples in later chapters.

Common discrete distributions (all on some set of nonnegative integers) are: the *Bernoulli* and *binomial distributions*, the *discrete uniform*, the *Poisson* and the *geometric* and *negative binomial distributions*. Again, some familiarity with their basic properties is recommended. The *multinomial distribution* is reviewed in Chapter 9.

See, for example, JOHNSON, KOTZ & BALAKRISHNAN (1994, 1997) and JOHNSON, KEMP & KOTZ (2008) for definitions and basic properties of these distributions.

1.6 The Multinormal Distribution

The joint density of $k > 1$ independent and identically distributed standard normal random variable's $Z_1, \ldots, Z_k$ is

$$f(\mathbf{z}) = \prod_{j=1}^{k} \varphi(z_j) = \frac{1}{(2\pi)^{\frac{k}{2}}} \exp(-\tfrac{1}{2}\mathbf{z}^T\mathbf{z}), \tag{1.16}$$

writing $\mathbf{z}$ for the column vector with coordinates $z_1, \ldots, z_k$ and T for transpose; hence, $\mathbf{z}^T\mathbf{z} = \sum_{j=1}^{k} z_j^2$. The mean vector is the zero vector $\mathbf{0}$ and the variance matrix is the *identity matrix* I. Such a k-vector $\mathbf{z}$ is said to have a *standard multinormal distribution* $N_k(\mathbf{0}, I)$. Its cumulant generating function is readily found to be

$$C_{\mathbf{z}}(\mathbf{t}) \equiv \log E_{\mathbf{z}}\{\exp(\mathbf{t}^T\mathbf{z})\} = \tfrac{1}{2}\textstyle\sum_{j=1}^{k} t_j^2 = \tfrac{1}{2}\mathbf{t}^T\mathbf{t} \quad \text{since} \quad \log E\{\exp(t_1 Z_1)\} = \tfrac{1}{2}t_1^2.$$

It can be shown that every symmetric positive definite matrix is expressible in the form C^TC for some invertible matrix C. Using this fact, we say that random k-vector $\mathbf{X}$ is *multinormally distributed* $N_k(\boldsymbol{\mu}, \Sigma)$ (of full rank k), for some k-vector $\boldsymbol{\mu}$ and some symmetric positive definite matrix Σ, if there exists $\mathbf{z}$ which is $N_k(\mathbf{0}, I)$ and

$$\mathbf{X} = \boldsymbol{\mu} + C^T\mathbf{z} \tag{1.17}$$

where C is a $k \times k$ matrix for which $C^TC = \Sigma$ that is, if $(C^T)^{-1}(\mathbf{X} - \boldsymbol{\mu})$ is $N_k(\mathbf{0}, I)$. Clearly, the mean of $\mathbf{X}$ is $\boldsymbol{\mu}$ and the variance matrix is $C^TIC = \Sigma$. The cumulant generating function of $\mathbf{X}$ is

$$\begin{aligned} C_{\mathbf{X}}(\mathbf{t}) &\equiv \log[E\{\exp(\mathbf{t}^T\mathbf{X})\}] = \log\left(E\left[\exp\left\{\mathbf{t}^T\boldsymbol{\mu} + (C\mathbf{t})^T\mathbf{z}\right\}\right]\right) \\ &= \mathbf{t}^T\boldsymbol{\mu} + C_{\mathbf{z}}(C\mathbf{t}) = \mathbf{t}^T\boldsymbol{\mu} + \tfrac{1}{2}\mathbf{t}^T\Sigma\mathbf{t}. \end{aligned} \tag{1.18}$$

See *Exercises 23 and 24* for the specialization to the bivariate case. Note also that, from Equations (1.17) and/or (1.18), if Σ is diagonal, so that all covariances and all correlation coefficients are zero, then C is diagonal with diagonal elements the square roots of those in Σ, and the coordinates of $\mathbf{X}$ are seen to be mutually independent. Hence, for jointly normal random variables, zero correlation or covariance is equivalent to independence.

What can be said about linear combinations of the coordinates of $\mathbf{X}$? We know the mean and variance, but how about the distribution? As in the Cramér-Wold device, we find the cumulant generating function of $Y \equiv \mathbf{a}^T\mathbf{X}$ to be

$$C_Y(s) = \log[E\{\exp(sY)\}] = \log[E\{\exp(s\mathbf{a}^T\mathbf{X})\}].$$

By Equation (1.18) this equals $s\mathbf{a}^T\boldsymbol{\mu} + \frac{1}{2}s\mathbf{a}^T\Sigma\mathbf{a}s = s\mu + \frac{1}{2}s^2\sigma^2$ where $\mu = \mathbf{a}^T\boldsymbol{\mu}$ and $\sigma^2 = \mathbf{a}^T\Sigma\mathbf{a}$, and this is recognized as the cumulant generating function of a $N(\mu, \sigma^2)$ random variable. So every linear combination of the coordinates of a multinormal variable is normally distributed. In fact, the argument can be reversed to conclude as an alternative useful characterization of multinormality:

$$\mathbf{X} \sim N_k(\boldsymbol{\mu}, \Sigma) \text{ if and only if } \mathbf{a}^T\mathbf{X} \sim N(\mathbf{a}^T\boldsymbol{\mu}, \mathbf{a}^T\Sigma\mathbf{a}) \text{ for each } \mathbf{a} \in \mathcal{R}^k. \tag{1.19}$$

(Unit vectors $\mathbf{a}$ are enough.)

From Equations (1.17) and (1.18), the multinormal density is found to be

$$f(\mathbf{X}; \boldsymbol{\mu}, \Sigma) = \frac{1}{(2\pi)^{\frac{k}{2}}\{\det(\Sigma)\}^{\frac{1}{2}}} \exp\left\{-\tfrac{1}{2}(\mathbf{X} - \boldsymbol{\mu})^T\Sigma^{-1}(\mathbf{X} - \boldsymbol{\mu})\right\}, \tag{1.20}$$

where $\det(\Sigma)$, the determinant of the covariance matrix Σ, arises from the Jacobean in the formula (1.6). For a binormal random variable (X, Y) with $EX = \mu$, $EY = \eta$, $\text{var}(X) = \sigma^2$, $\text{var}(Y) = \tau^2$, $\text{corr}(X, Y) = \rho$ the density becomes

$$\frac{1}{2\pi\sigma\tau\sqrt{(1-\rho^2)}} \exp\left[-\left\{\left(\frac{x-\mu}{\sigma}\right)^2 - \frac{2\rho(x-\mu)(y-\nu)}{\sigma\tau} + \left(\frac{y-\mu}{\tau}\right)^2\right\} \Big/ 2(1-\rho^2)\right]. \tag{1.21}$$

See *Exercise 23*. The form of this bivariate density will play an important role in the discussion of LeCam's third lemma in Chapter 4.

Conditional distributions of Y given X take a simple form for the bivariate normal distribution. We find that the conditional distribution of Y given $X = x$ is normal with mean $\nu + (\rho\tau/\sigma)(x - \mu)$ and variance $\tau^2(1-\rho^2)$, see *Exercise 24*. This facilitates computation of bivariate moments. For example, if both X and Y are standard normal, so that $\mu = \nu = 0$ and $\sigma = \tau = 1$, we find that $E(XY) = \rho$, $E(X^2Y) = E(XY^2) = 0$, $E(XY^3) = E(X^3Y) = 3\rho$ and $E(X^2Y^2) = 1 + 2\rho^2$. See *Exercise 26*.

More generally, if X and Y are both vectors and (X, Y) follows a multinormal distribution with mean vector $\mu = (\mu_x^T, \mu_y^T)^T$ and nonsingular covariance matrix Σ, with submatrices Σ_{xx}, $\Sigma_{xy} = \Sigma_{yx}^T$ and Σ_{yy} in an obvious notation, then the conditional distribution of Y given $X = x$ is multinormal with mean $\mu_y + \Sigma_{xy}\Sigma_{xx}^{-1}(x - \mu_x)$ and covariance matrix $\Sigma_{yy} - \Sigma_{yx}\Sigma_{xx}^{-1}\Sigma_{xy}$. This result can be proved using formulas for inverting partitioned matrices. See *Exercise 25*, and Appendix I of Chapter 6 for details of the matrix inversion formulas.

Now we turn to extensions of facts about squared normal random variables and the resulting chi-squared distributions. We proceed step by step, starting by assuming that the means are zero, and considering (i) the square of a single normally distributed X, then (ii) the sum of squares of several independent such X's and (iii) finally relaxing the independence. Then we repeat for the corresponding cases with means not necessarily zero.

i. Suppose that $X \sim N(0, \sigma^2)$. Then $Z \equiv X/\sigma \sim N(0,1)$ and $Z^2 = (X/\sigma)^2$ has a central chi-squared distribution with $\nu = 1$.

ii. Now let $X_1, \ldots X_k$ be mutually independent with $X_i \sim N(0, \sigma_i^2)$. Let $Z_i \equiv X_i/\sigma_i$ so that $\mathbf{Z}^T\mathbf{Z} = \sum_{i=1}^k Z_i^2$ has a central chi-squared distribution with $\nu = k$.

iii. Now remove the independence assumption and suppose instead that $\text{var}(\mathbf{X}) = \Sigma = C^TC$. Express $\mathbf{X} = C^T\mathbf{z}$ as in Equation (1.17) so that the components Z_i of the vector $\mathbf{z} = (C^T)^{-1}\mathbf{X}$ are mutually independent standard each with a standard normal distribution. Then $\mathbf{X}\Sigma^{-1}\mathbf{X} = \mathbf{Z}^T\mathbf{Z} = \sum_{i=1}^k Z_i^2$, again distributed as (central) chi-squared with $\nu = k$.

Now we introduce μ_i as the mean of X_i, and extend each of the above. In (i), we conclude that Z^2 now has a noncentral chi-squared distribution with $\lambda^2 = (\mu/\sigma)^2$, since $E(Z) = \mu/\sigma$. In (ii), we likewise have a noncentral chi-squared distribution with λ^2 now equal to $\sum_{i=1}^k(\mu_i/\sigma_i)^2$. Finally, for (iii), we similarly conclude that the *quadratic form* $Q \equiv \mathbf{X}^T\Sigma^{-1}\mathbf{X}$ is distributed as noncentral chi-squared with $\nu = k$ and $\lambda^2 = \boldsymbol{\mu}^T\Sigma^{-1}\boldsymbol{\mu}$, the sum of squares of the means of the Z_i's, defined by $\mathbf{z} = (C^T)^{-1}\mathbf{X}$, and noting that $\Sigma^{-1} = (C^TC)^{-1} = C^{-1}(C^T)^{-1}$. We summarize in

Proposition 1.1 *Suppose that* $\mathbf{X} \sim N_k(\boldsymbol{\mu}, \Sigma)$, *with* Σ *nonsingular. Then* $Q \equiv \mathbf{X}^T\Sigma^{-1}\mathbf{X}$ *is distributed as noncentral chi-squared with k degrees of freedom and noncentrality* $\lambda^2 \equiv \boldsymbol{\mu}^T\Sigma^{-1}\boldsymbol{\mu}$, *equal to zero if and only if* $\boldsymbol{\mu} = \mathbf{0}$.

For further reading, see RAO (1973) or textbooks on multivariate analysis. The *singular multinormal distribution* is considered in Chapter 9.

1.7 Stochastic Ordering

We divert from a review of basic facts about random variables and their distributions to introduce a useful *order relation* between random variables, or, more precisely, between their distributions.

Definition 1.1 X *is stochastically larger than* Y, *written* $X \succ Y$, *if* $\bar{F}(x) \geq \bar{G}(x)$ *for all* x, *where* $\bar{F}$ *and* $\bar{G}$ *are the survivor functions of* X *and* Y, *respectively.*

(A more exact terminology would be 'X is stochastically at least as large as Y', written $X \succeq Y$, but we will not make such distinctions.) Equivalently, we write $Y \prec X$ and say Y is *stochastically smaller than* X. If both $X \succ Y$ and $X \prec Y$, implying that $\bar{F}$ and $\bar{G}$ are identically equal we write $X =_d Y$, and say that X is *equal-in-distribution* to Y.

The concept is meant to reflect the idea that a typical value of X tends to be larger than a typical value of Y. See *Exercise 30*. More precisely, for any value c, there is greater probability that X exceeds c than that Y exceeds c. Stochastic ordering depends only on *marginal* distributions, whether or not any joint distribution is defined. Alternative notation for $X \succ Y$ is simply $\bar{F} \geq \bar{G}$ or equivalently, $F \leq G$ the phrase 'for all arguments x' being implicitly understood. No definite assurance can be given (typically) about the relationship between a realized x^* and a realized y^*. However, if a joint distribution *is* defined, and if $X \geq Y$ always (or with probability 1), then it certainly follows that $X \succ Y$, see *Exercise 32*.

Note that stochastic ordering is a *partial ordering* of distributions: Often neither $X \succ Y$ nor $X \prec Y$. Typically, a pair of distribution functions will cross somewhere, rather than remaining perfectly ordered.

As a first simple example, suppose that X and Y are each normally distributed with unit variances and respective means μ and ν, with $\mu < \nu$. Then it readily follows that $X \prec Y$ since the survival function of X, namely $\bar{\Phi}(x - \mu)$, is increasing in μ for each x (differentiate in μ). There is a *tendency* for X-values to be smaller than Y-values; however, no ordering of realized values is assured. However, we can construct X' so that $X' =_d X$ and $X' \leq Y$ (with certainty): simply let $X' = Y - \nu + \mu < Y$. This construction illustrates one application of this concept, namely that of facilitating proof of mathematical facts that relate only to marginal distributions. For example, with this same X and Y, suppose that $h(\cdot)$ is an increasing function, for example $h(x) = \exp(tx)$ for some positive t. Then we can use stochastic ordering to prove that $E\{h(X)\} < E\{h(Y)\}$. For, $h(X) =_d h(X')$ implies $E\{h(X)\} = E\{h(X')\} = E\{h(Y - \nu + \mu)\} < E\{h(Y)\}$, the last step following from the fact that $h(y - \nu + \mu) < h(y)$ for all y. And let us emphasize: the result holds whether X and Y are jointly bivariate normal, whatever the correlation coefficient, or even if no joint distribution is defined. We pursue this more generally below.

But before proceeding with extensions, applications and examples, here are two examples of non-ordered distributions:

i. X and Y are normally distributed with unequal variances; then no stochastic ordering holds, no matter whether the means are equal or not; the corresponding survivor functions (and distribution functions) cross. (See *Exercise 33.*)
ii. No pair of normal and Cauchy distributions are stochastically ordered, since the Cauchy distribution function approaches each of its asymptotes more slowly than does the normal distribution function, and hence the distribution functions must cross somewhere.

To extend beyond a pair of distributions to a real-parameter family, we define:

Definition 1.2 *A family (F_θ) of distribution functions, θ belonging to a real set Θ, is a stochastically ordered family ('increasing') if, for every $\theta' < \theta''$ $(\in \Theta)$, $\bar{F}_{\theta'} \leq \bar{F}_{\theta''}$ (and 'decreasing' if the ordering is reversed).*

Here are some examples of stochastically ordered families; verifications of (a) and (b) are trivial and omitted.

a. A *location family* is stochastically ordered. That is, suppose F is a distribution function on $\mathcal{R}$ and $F_\theta(x) = F(x - \theta)$ for all x; then $\{F_\theta\}$ is stochastically increasing.
b. A *scale family* of positive distributions is stochastically ordered. That is, suppose F is a distribution function on $\mathcal{R}^+$ and $F_\theta(x) = F(\theta x)$ for all $x > 0$ $(\theta > 0)$; then $\{F_\theta\}$ is stochastically decreasing.
c. Two densities $f_1(x)$ and $f_0(x)$ are said to have monotone likelihood ratio if the ratio $f_1(x)/f_0(x)$ is monotone in x, say, without loss of generality, increasing in x. The definition extends as above to a family of distributions $f_\theta(x)$. The concept is important in developing the Neyman-Pearson theory of hypothesis testing, see Chapter 7. It applies whenever $f_\theta(x)$ is a one-parameter exponential family. It is easily seen (*Exercise 34*) that this monotone likelihood-ratio property implies stochastic ordering of the corresponding distribution functions.

Stochastic ordering cannot be determined by a comparison of the density functions, except as in (c). But the implications for quantile functions are precise and useful:

Lemma 1.2 *Let Q_F and Q_G be the quantile functions corresponding to distribution function's F and G. Then $\bar{F} \leq \bar{G}$ if and only if $Q_F \leq Q_G$.*

As before, 'for all values of the argument' is implicit when we write inequalities between functions. This lemma simply states that, when looking at plots of two distribution functions, the ordering of the curves in the horizontal direction is equivalent to their ordering in the vertical direction.

Proof Suppose that $\bar{F} \leq \bar{G}$. Given $t \in (0,1)$, choose x so that $G(x) \geq t$. Since $\bar{F}(x) \leq \bar{G}(x)$, we have $F(x) \geq G(x)$ and conclude that $F(x) \geq t$, which, by Lemma 1.1, implies $Q_F(t) \leq x$. Now take the infimum of the right side of the last inequality over all such x, that is, x for which $G(x) \geq t$, obtaining $Q_G(t)$. Hence, $Q_F(t) \leq Q_G(t)$.

Proof of the converse (*Exercise 40*) is parallel to that above, using the formula $F(x) = \sup\{t | Q_F(t) \leq x\}$ proved in *Exercise 2* parallelling the defining formula for Q_F. □

As a corollary, quantile transforms corresponding to two stochastically ordered distributions, applied to the same uniform random variable, are ordered, not just stochastically ordered:

Corollary 1.1 *If $X \prec Y$, there exists (X', Y') satisfying*

$$X' =_d X, Y' =_d Y, \text{ and } X' \leq Y'.$$

Choose $X' = Q_F(U)$ and $Y' = Q_G(U)$ with U uniform on (0,1). □

This is the simplest version of a *Skorokhod construction.* These will be used to prove the *Continuous Mapping Theorem* in Chapter 2. As an application, it follows that if $X \prec Y$ and $h(\cdot)$ is increasing, then $E\{h(X)\} \leq E\{h(Y)\}$ whenever the expectations exist. A proof was indicated earlier.

Here is another application: Suppose that T is a test statistic, used to reject a null hypothesis when T is large. If the distribution of T is a member of a family that is stochastically ordered by a real parameter θ, then the power function of the test is monotone in θ, whatever the significance level of the test. To check this, note that the power function is the survival function of T at a *critical value* t_α.

Finally, the family of noncentral chi-squared distributions (with fixed degrees of freedom ν) is stochastically increasing in the noncentrality parameter (*Exercises 38 and 39*). This has implications for chi-squared tests, that is tests based on a statistic having a central chi-squared distribution under a null hypothesis and a noncentral chi-squared distribution under an alternative hypothesis. Specifically, the power function is monotone in the noncentrality parameter: The larger the noncentrality, the greater the power. This property will be important in Chapters 7 and 9.

1.8 The Change-of-Measure Identity

We consider a very simple identity, sometimes called a *likelihood-ratio identity*, that has proved useful in several different areas of statistical theory and practice.

Proposition 1.2 *Suppose that f and g are densities of a random variable or vector X with $g(x) = 0$ whenever $f(x) = 0$. Consider a function $h(x)$ and define*

$$H(x) \equiv h(x)g(x)/f(x) = h(x)L(x)$$

where $L(x) \equiv g(x)/f(x)$. Then

$$E_g\{h(X)\} = E_f\{H(X)\}. \tag{1.22}$$

In particular,

$$P_g(X \in A) = E_f\{\mathbf{1}[X \in A] \cdot L(X)\}. \tag{1.23}$$

The proof is trivial. Often, $L(x)$ is written as $\exp\{l(x)\}$ with $l(x) = \log\{g(x)/f(x)\}$, the *log likelihood-ratio.*

Here are four applications or variations:

1. Suppose we want to evaluate μ_g, the expectation of a function $h(X)$ of a random variable X whose distribution has density g, but cannot carry out the mathematical computation. A Monte Carlo evaluation can be done by generating multiple values of X from this distribution, yielding $x_1, \ldots, x_N$, say, and then estimating μ_g by the empirical average $N^{-1}\sum_{j=1}^{N} h(x_j)$. This estimate has variance $\sigma_g^2 = N^{-1}\text{var}_g\{h(X)\}$. Alternatively, we may generate X from the distribution with density f, yielding $x_1', \ldots, x_N'$, say, and estimate μ_g by $N^{-1}\sum_{j=1}^{N} H(x_j')$, using Equation (1.23). This estimate has variance $\sigma_f^2 = N^{-1}\text{var}_f\{H(X)\}$. Supposing that $L(x)$ is easy to compute, the latter estimate may be preferred if (a) it is easier to generate X from f than from g and/or (b) $\sigma_f^2 << \sigma_g^2$. The latter may happen if $H(x)$ is fairly flat, compared with $h(x)$; in particular, if h is everywhere positive, choosing $f(x) \propto h(x)g(x)$ would make $H(x)$ constant, and hence have zero variance! Of course it is not possible to choose such an f without having already determined the proportionality constant, namely $1/E_g\{h(X)\}$, but an approximating choice should lead to a small variance σ_f^2.

2. *LeCam's third lemma* discussed in Chapter 4 is a version of this identity applicable to distributions occurring in large-sample theory. This result plays an important role in our approach to the topic. The lemma connects the joint distribution function of some statistics of interest, along with so-called *score statistics*, under an 'alternative' parameter value θ_n, to the joint distribution of the same quantities under a 'null' parameter value θ. This will be clarified later.
3. An application in the theory of *rank statistics* is as follows. Consider the distribution of the ranks of observations associated with one of two independent random samples, based on a ranking of the pooled data from the two samples. A formula expressing this distribution, assuming the two samples come from different distributions, in terms of the expected value of a function of the order statistic when the two samples come from a specific common distribution, is due to HOEFFDING (1951). His formulas may be viewed as an example of the *Change-of-Measure Identity.* Our approach to the problem, in Section 8.8 of Chapter 8 makes direct use of LeCam's third lemma.
4. In *sequential analysis*, Wald's *Fundamental Identity* played a useful role historically (WALD, 1947). More recent versions are the *Likelihood-Ratio Identity* and its conditional analog, both being versions of Equation (1.23). See SIEGMUND (1985) for example.

1.9 Computing with R, Simulation Studies

We end the main part of this introductory chapter with a short discussion of some features of R, which we shall use for occasional detailed calculations and illustrations. The reader may well be familiar with many of the statistical and graphical features of R and of their use in data analysis. Our purpose here is to provide numerical illustrations of the mathematical results and techniques that we develop. Consider, for example a result that will be discussed in Chapter 3, that in large samples from a normal distribution the correlation coefficient between the sample mean and the sample median is approximately $\sqrt{(2/\pi)} = 0.7979$. The following short program provides an empirical confirmation of this result.

```
< Mean ← vector(“numeric”, 200)
< Median ← vector(“numeric”, 200)
< for (i in 1:200){
+ x ← rnorm(100)
+ Mean[i] ← mean(x)
```

```
+ Median[i] ← median(x)}
< plot(Mean, Median)
< sd(Mean)
< sd(Median)
< cor(Median,Mean)
```

This program generates 200 (pseudo)-independent random samples, each of size 100, of standard normal random variables, computes the mean and median of each sample, produces a scatter plot of the 200 pairs and calculates the sample standard deviations and correlation coefficient of the plotted points. Figure 1.2 shows a plot produced by running this program. As we would expect there is a strong positive relation between the two quantities, and their sample correlation coefficient of 0.811 is not far from the theoretical value.

It should not be surprising that the sample estimate does not agree exactly with the theoretical value. There are two reasons for this. First, the theoretical result applies to samples of infinite size, rather than samples of size $n = 100$.

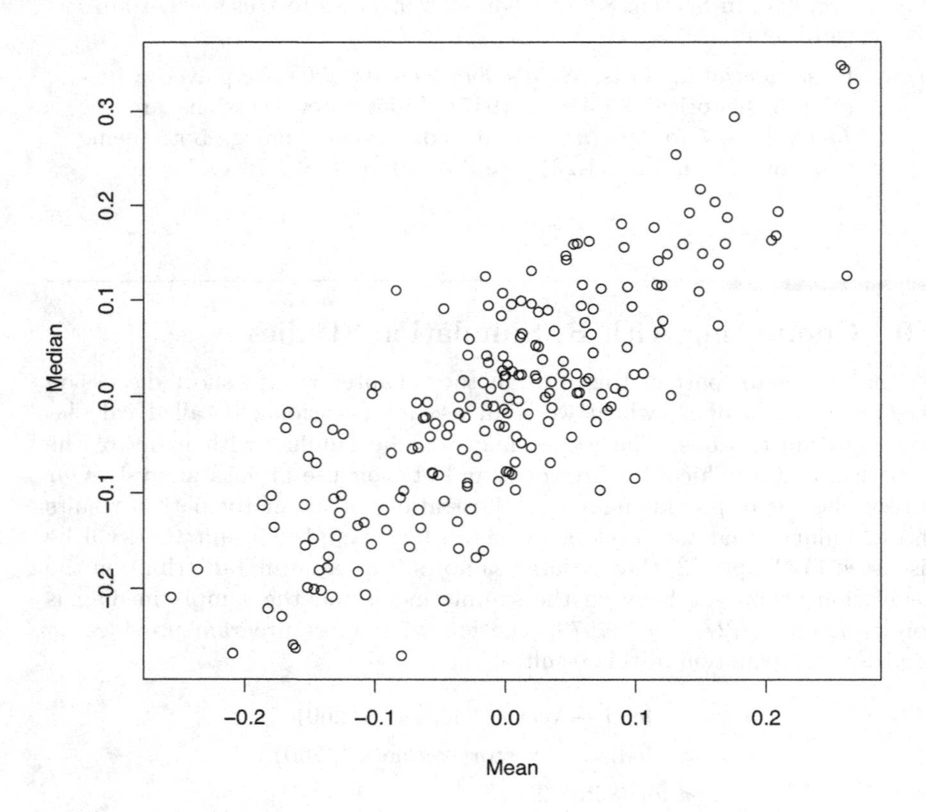

FIGURE 1.2
Medians and means, 200 simulations.

It is obvious for example (why?) that using the usual definition of a median for a sample of size $n = 2$ would yield a correlation coefficient of unity between the sample mean and sample median. The theoretical calculation for a sample of size $n = 3$ or greater is more difficult, involving integrals of the normal cumulative distribution function. A simpler but instructive example involves exponentially distributed variables, see *Exercise 46*. The second reason why the actual result may differ from the theoretical value is easier to address. If the number of simulations here (200) is judged too small to provide a reliable estimate, the problem can be solved by increasing this number. We can use standard statistical techniques to compute standard errors and confidence intervals to gauge the accuracy of our estimate and to plan the number of simulations needed to estimate the quantity of interest with specified precision. For example, the use of Fisher's z-transform for the correlation coefficient, to be discussed in *Exercise 34* of Chapter 3, yields a 95% confidence interval of (0.758, 0.854), for its true value based on the data shown in Figure 1.2. The 'n' to be included in the calculation of the confidence interval is the number of simulations, 200, the same as the number of points in the figure, not the sample size (100) on which each point is based.

R has simple commands for generating (pseudo)-random samples from all the distributions considered in this text (including the noncentral versions of the chi-squared and t-distributions, and for calculating their density, distribution and quantile functions. For example "rnorm(n,mu,sigma)" returns n independent random variables from a normal distribution with mean "mu" and standard deviation "sigma". Omitting the parameter values, as in the program above, defaults to the standard normal distribution. Also "dnorm(x, mu, sigma)", and "pnorm(x, mu, sigma)" calculate the corresponding density and distribution functions at the value x and "qnorm(u, mu, sigma) the u'th quantile. Similar naming conventions apply to other distributions. Also available are special functions including the log-gamma function and its derivatives. These appear frequently in examples.

Appendix: Some Special Functions

In this appendix, we present for future reference certain properties of the Gamma function and its logarithmic derivatives. These results are used later in calculations involving common distributions including the gamma, normal, beta and Weibull. Some of the derivations use the monotone convergence theorem and other properties of the Lebesgue integral - but the essence of the argument is clear without this background. Numerical values of these functions are readily available in R and other software packages.

A.1 The Gamma and Beta Functions

The gamma function arises in statistics as the normalizing constant in the gamma density

$$f(x;\kappa,\rho)=\frac{\rho^{\kappa}x^{\kappa-1}\exp(-\rho x)}{\Gamma(\kappa)}.$$

The substitution $u=\rho x$ shows that $\Gamma(\kappa)=\int u^{\kappa-1}\exp(-u)du$, where the integral is taken over $[0,\infty)$, and, as the notation implies, is free of ρ. Clearly $\Gamma(\kappa)$ is defined for all $\kappa>0$ and $\Gamma(1)=1$. The relation $\Gamma(\kappa)=(\kappa-1)\Gamma(\kappa-1)$ for $\kappa>1$ is easily shown using integration by parts, so that $\Gamma(n)=(n-1)!$ for any positive integer n. It is also possible to evaluate $\Gamma(n+\frac{1}{2})$ explicitly since

$$\Gamma(\tfrac{1}{2})=\textstyle\int_0^\infty x^{-1/2}\exp(-x)dx=\int_0^\infty\sqrt{2}\exp(-t^2/2)dt$$

by the substitution $x=t^2/2$. Since

$$\int_0^\infty\frac{1}{\sqrt{(2\pi)}}\exp(-x^2/2)dx=\tfrac{1}{2}$$

from the normal integral, we have $\Gamma(\frac{1}{2})=\sqrt{\pi}$. The celebrated formula due to Stirling,

$$\Gamma(\kappa)\approx\sqrt{(2\pi)}\exp(-\kappa)\kappa^{\kappa-\frac{1}{2}}$$

gives accurate approximations even for κ as small as 10. See, for example, FELLER (1968), Chapter 2.

If X and Y are independent with distributions $\Gamma(\kappa,1)$ and $\Gamma(\nu,1)$ respectively then the change in variables to $Z=(X+Y)$ and $U=X/(X+Y)$ makes U and Z independent with densities proportional to $u^{\kappa-1}(1-u)^{\nu-1}$ and $z^{\kappa+\nu-1}\exp(-z)$ respectively. The normalizing constant for the gamma density in z is $\Gamma(\kappa+\nu)$. So the density of U is

$$\frac{\Gamma(\kappa+\nu)}{\Gamma(\kappa)\Gamma(\nu)}u^{\kappa}(1-u)^{\nu}.$$

The reciprocal of the normalizing constant is the *beta function* denoted by

$$\mathrm{Be}(\kappa,\nu)=\int_0^1 u^{\kappa-1}(1-u)^{\nu-1}du=\frac{\Gamma(\kappa)\Gamma(\nu)}{\Gamma(\kappa+\nu)}.$$

See *Exercise 11* for these results.

A.2 The Digamma Function and Euler's Constant

The digamma function, denoted by $\psi(\kappa)$, is the derivative of the logarithm of the gamma function,

$$\psi(\kappa)=\frac{\Gamma'(\kappa)}{\Gamma(\kappa)}=\int\frac{(\log t)t^{\kappa-1}\exp(-t)}{\Gamma(\kappa)}dt.$$

Here and later differentiation under the integral can be justified by Fubini's theorem, since $\int t^{\kappa-1}|\log t|^n \exp(-t)dt$ is finite for all $\kappa > 0$ and $n \geq 0$. The relation $\Gamma(\kappa) = (\kappa - 1)\Gamma(\kappa - 1)$ gives, for $\kappa > 1$,

$$\psi(\kappa) = \frac{1}{\kappa - 1} + \psi(\kappa - 1). \tag{A.1}$$

We now show that the value of the digamma function at $\kappa = 1$ is $\psi(1) = -\gamma$, where $\gamma \approx 0.5772$ is Euler's constant, $\gamma = \lim_{n\to\infty} \gamma_n$ with

$$\gamma_n = \left(1 + \frac{1}{2} + \cdots + \frac{1}{n} - \log n\right).$$

The existence of this limit is easily proved by bounding the argument in the integral

$$\log n = \int_1^n \frac{1}{x} dx$$

by the floor and ceiling function applied to x, giving

$$\log n \leq \sum_1^n \frac{1}{j} \leq \log(n + 1),$$

and noting that the difference between the outside terms converges to zero. The connection of this limit with the digamma function is quoted in many standard references but a simple direct proof is quite hard to find. We follow WHITTAKER & WATSON (1927), Section 12.1. Since $\Gamma(1) = 1$ we must show that $\int(\log u) \exp(-u)du = -\gamma$. We first note that

$$1 + \frac{1}{2} + \cdots + \frac{1}{n} = \int_0^1 \frac{1 - (1 - t)^n}{t} dt.$$

This is easily shown by mathematical induction, since the result holds for $n = 1$ and the difference between the values of the right side evaluated at $n + 1$ and at n is

$$\int_0^1 \frac{(1 - t)^n \{(1 - (1 - t)\}}{t} dt = \int_0^1 (1 - t)^n dt = \frac{1}{n + 1}.$$

So

$$\gamma_n = \int_0^1 \frac{1 - (1 - t)^n}{t} dt - \int_1^n \frac{1}{t} dt.$$

Writing $u = nt$ in the first integral and replacing t by u in the second gives

$$\gamma_n = \int_0^n \frac{1 - (1 - u/n)^n}{u} du - \int_1^n \frac{1}{t} dt = \int_0^1 \frac{1 - (1 - u/n)^n}{u} du - \int_1^n \frac{(1 - u/n)^n}{u} du.$$

The second integral may be written

$$\int_1^\infty \frac{f_n(u)}{u} du,$$

where $f_n(u) = \max\{0, (1 - u/n)^n\}$. It is easily shown that for each u, $f_n(u)$ is increasing in n and that $f_n(u) \to \exp(-u)$ as $n \to \infty$. The monotone convergence theorem applies, so that

$$\lim_{n\to\infty} \gamma_n = \int_0^1 \frac{1-\exp(-t)}{t} dt - \int_1^\infty \frac{\exp(-t)}{t} dt.$$

Integration of each term by parts gives the result.

The recurrence relation (A.1) shows that, for integral $n > 1$, $\psi(n) = -\gamma + 1 + \frac{1}{2} + \cdots + 1/(n-1)$. It will be shown below that $\psi(\kappa)$ is monotone increasing in κ, for all $\kappa \in R^+$ so that $\psi(\kappa) \approx \log \kappa$ as $\kappa \to \infty$.

The relation (A.1) also provides to obtain a more explicit formula for $\psi(\kappa)$. We have, for $\kappa > 0$,

$$\psi(\kappa + n) = \frac{1}{\kappa + n - 1} + \cdots + \frac{1}{\kappa} + \psi(\kappa),$$

$$\psi(1 + n) = \frac{1}{n} + \cdots + 1 + \psi(1),$$

Subtracting gives

$$\psi(\kappa + n) - \psi(1 + n) = \psi(\kappa) - \psi(1) + \sum_{m=0}^{n-1} \left(\frac{1}{m+\kappa} - \frac{1}{m+1} \right).$$

For any fixed κ, $\psi(\kappa + n) - \psi(1 + n) \le \kappa/(\kappa + n) \to 0$ as $n \to \infty$. So, substituting for $\psi(1) = -\gamma$, we have

$$\psi(\kappa) = -\gamma - \frac{1}{\kappa} + \sum_{m=1}^{\infty} \left(\frac{1}{m} - \frac{1}{m+\kappa} \right). \tag{A.2}$$

An integral formula, obtained by putting together the previous results and using the formula $\log \kappa = \int \{(e^{-t} - e^{-\kappa t}/t)\} dt$, is

$$\psi(\kappa) = \int_0^\infty \left(\frac{e^{-t}}{t} - \frac{e^{-\kappa t}}{1 - e^{-t}} \right) dt.$$

A.3 The Trigamma Function

The second derivative of $\log \Gamma(\kappa)$ is called the trigamma function and denoted by $\psi'(\kappa)$. We have

$$\psi'(\kappa) = \frac{\Gamma''(\kappa)}{\Gamma(\kappa)} - \left\{ \frac{\Gamma'(\kappa)}{\Gamma(\kappa)} \right\}^2 = \frac{\Gamma''(\kappa)}{\Gamma(\kappa)}^2 - \psi(\kappa)^2.$$

The recurrence relation (A.1) gives

$$\psi'(\kappa) = \psi'(\kappa - 1) - \frac{1}{(\kappa - 1)^2}.$$

Using the fact that $\Gamma''(\kappa) = \int x^{\kappa-1}(\log x)^2 \exp(-x)dx$ we see that $\psi'(\kappa)$ equals the variance of $\log X$ when X follows a gamma distribution with index κ. So $\psi'(\kappa) > 0$ for all κ as was asserted above. Differentiation of the formula (A.2) shows that

$$\psi'(\kappa) = \sum_{n=0}^{\infty} \frac{1}{(n+\kappa)^2}.$$

A celebrated result of Euler, solving what had become known as the Basel problem, is that $\psi'(1) = \pi^2/6 \approx 1.645$. Many proofs of this result are now known. One recent example (*Exercise 13*) uses a simple calculation involving the distribution of the ratio of two independently folded Cauchy distributed variables.

The first and second derivatives $\Gamma'(\kappa)$ and $\Gamma''(\kappa)$ can easily be expressed in terms of $\Gamma(\kappa)$, $\psi(\kappa)$ and $\psi'(\kappa)$. Since $\psi(\kappa) = \Gamma'(\kappa)/\Gamma(\kappa)$ and $\psi'(\kappa) = \Gamma''(\kappa)/\Gamma(\kappa) - \psi(\kappa)^2$ we have $\Gamma'(\kappa) = \Gamma(\kappa)\psi(\kappa)$ and $\Gamma''(\kappa) = \Gamma(\kappa)\{\psi'(\kappa) + \psi(\kappa)^2\}$. so, for example $\Gamma'(1) = -\gamma$ and $\Gamma''(1) = \gamma^2 + \pi^2/6 \approx 1.978$. These results will be used in Chapter 4.

Exercises

The following exercises mostly focus on quantile functions, noncentral chi-squared distributions, the multinormal distribution, stochastic order and the change-of-measure identity.

1. Determine and sketch the quantile functions for the following distributions:

   (a) Standard exponential.

   (b) Binomial$(3, \frac{1}{2})$.

   (c) Standard exponential, censored at $x = 2$, that is, it is absolutely continuous on $(0, 2)$ with density given by the standard exponential form $f(x) = \exp(-x)$ and with a single mass point, namely $x = 2$, with jump $\exp(-2)$.

2. Verify Properties (i)–(iii) of Section 1.2, repeated below:

   (i) $Q\{F(x)\} \leq x$, [*Hint:* $x \in \{x' : F(x') \geq F(x)\}$];

   (ii) $F\{Q(u)\} \geq u$, [*Hint:* $F(\cdot)$ is right continuous, $Q(u) \in \{x' | F(x') \geq u\}$ and so $F\{Q(u)\} \geq u$];

   (iii) $Q(\cdot)$ is left-continuous, [*Hint:* $F(t) \geq u - \epsilon$ for all $\epsilon > 0$ if and only if $F(t) \geq u$].

   Show that equality in (i) holds unless x is contained in an interval on which F is flat and in (ii) if and only if u is in the range of F, that is there is some value of x for which $F(x) = u$.

3. Use Lemma 1.1 to show that $F(x) = \sup\{u | Q(u) \leq x\}$.

4. Prove the *iterated variance formula* of Section 1.3. [*Hint:* Use the *iterated expectation* formula.]

5. Verify that when X is a non-negative random variable with $E(X) < \infty$ then $E(X) = \int \{1 - F(x)\} dx$. Give separate proofs for the case of X continuous and X discrete. Show also that in both cases $E(X) = \int_0^1 Q(u) du$.

6. Derive the density function of the product of two independent random variables that are each uniformly distributed over $(0, 1)$.

7. Verify the exponential family representations claimed for the binomial, Poisson, normal and gamma densities discussed in Section 1.5. In each case, specify the functions $c(\theta)$, $t(x)$, $d(\theta)$ and $v(x)$ in Equation (1.10).

8. Suppose that X and Y are independent with density functions, each over the whole real line, proportional to $c^{-1} \exp(-\frac{1}{2}x^2)$ and $c^{-1} \exp(-\frac{1}{2}y^2)$ respectively. For now, the constant of proportionality given by $c = \int \exp(-\frac{1}{2}x^2)$ is assumed unknown.

   (a) Make the transformation $X = R\cos(U)$ and $Y = R\sin(U)$. Show that R and U are independent and that U is uniformly distributed over $(0, 2\pi)$. Find the density of R. Hence show that $c = \sqrt{(2\pi)}$.

   (b) Derive the same result by transforming to new variables (T, Z) given by $X = Z, Y = TZ$. (See also Exercise 12 below).

   (c) Show, by change of variables in the normal integral, that $\Gamma(\frac{1}{2}) = \sqrt{\pi}$.

9. (a) Verify that when Z has a standard normal distribution, $E(Z^2) = 1$, and $E(Z^4) = 3$.

   (b) By the substitution $x = u + t$ in the expression $M_x(t) = \int e^{tx} \varphi(x) dx$ show that $M_X(t) = \exp(\frac{1}{2}t^2)$, so $C_X(t) = \frac{1}{2}t^2$.

10. Right-skewed data, such as may arise in analyzing the concentrations of environmental toxins such as asbestos fibers, are often modeled by the *lognormal* distribution. Specifically, the distribution of a variable Y is said to be $LN(\mu, \sigma^2)$ if $\log Y \sim N(\mu, \sigma^2)$. Analysis can then proceed in the usual way after taking the logarithmic transform of the data. However exponentiating the arithmetic mean of the logarithms gives the geometric mean, not the arithmetic mean, of the original data and in many applications the arithmetic mean of the data is of more scientific relevance. Similar considerations apply to all nonlinear transformations. For the lognormal distribution, obtain the following explicit formulas:

(a) $E(Y) = \exp(\mu + \frac{1}{2}\sigma^2)$,

(b) $\text{var}(Y) = \exp(2\mu + \sigma^2)\{\exp(\sigma^2) - 1\}$.

Show also that $E(Y^n)$ is finite for all n but that the moment generating function $E\{\exp(tY)\}$ does not converge for any $t > 0$. [*Hint:* The moment generating function of $X = \log Y$ is $E\{\exp(tX)\} = \exp(t\mu + \frac{1}{2}t^2\sigma^2)$.]

11. If X and Y are independent with gamma densities $f_X(x) \propto \rho^\kappa x^{\kappa-1}\exp(-\rho x)$ and $f_Y(y) \propto \rho^\nu y^{\nu-1}\exp(-\rho y)$, show that $U = X/(X+Y)$ and $Z = X + Y$ are independent, that $f_Z(z) \propto \rho^{(\kappa+\nu)} z^{\kappa+\nu-1}\exp(-\rho z)$ and U has the *beta distribution*, with density $f_U(u) \propto u^{\kappa-1}(1-u)^{\nu-1}$, Use this argument to show that the constant of proportionality in the distribution of U is $\text{Be}(\kappa,\nu)^{-1}$, where
$$\text{Be}(\kappa,\nu) = \frac{\Gamma(\kappa)\Gamma(\nu)}{\Gamma(\kappa+\nu)}.$$

12. If X and Y are independently distributed over $(-\infty, \infty)$ with densities $f_X(x)$ and $f_Y(y)$ show that the ratio $U = X/Y$ has density
$$f_U(u) = \int y f_X(uy) f_Y(y) du.$$
Use this result to show that the ratio of two independent standard normal variables has the Cauchy density
$$f_U(u) = \frac{1}{\pi(1+u^2)}.$$
Show also that the distribution function can be evaluated explicitly as
$$F_U(u) = \tfrac{1}{2} + \tfrac{1}{\pi}\arctan(u).$$
and that the upper and lower quartiles of the distribution are at $u = \pm 1$.

13. (The "Basel Problem"). Let X_1 and X_2 be independently distributed each with the *folded Cauchy density* $f_X(x) = 2/\{\pi(1 + x^2)\}$ $(x > 0)$. Show that when $r < 1$ the density of the ratio $R = X/Y$ at r is
$$f_R(r) = \frac{-4\log r}{\pi^2(1-r^2)},$$
and find its form when $r > 1$. Argue by symmetry that $\text{pr}(R \leq 1) = \text{pr}(R \geq 1) = \frac{1}{2}$ and show that $\text{pr}(R \leq 1)$ can be expressed as
$$\frac{4}{\pi^2}\sum_{j=0}^{\infty}\frac{1}{(2j+1)^2}.$$

Hence show that

$$\sum_{j=0}^{\infty} \frac{1}{(2j+1)^2} = \frac{\pi^2}{8},$$

so that

$$\sum_{j=0}^{\infty} \frac{1}{(j+1)^2} = \frac{\pi^2}{6}.$$

This famous result is due to Euler. See PACE (2011) for this proof, further discussion and references.

14. The *quantile density function* for a continuous distribution is $g(u) = f\{Q(u)\}$ where $Q(u)$ is the quantile function. Show that the Laplace distribution, with density $f(x) = -\frac{1}{2}\exp(-|x|)$, has quantile density function $g(u) = \frac{1}{2} - |u - \frac{1}{2}|$. Show that both the Laplace distribution and the standard Cauchy distribution with density $f(x) = 1/\{\pi(1+x^2)\}$ have $g(0.25) = g(0.75) = \frac{1}{2}g(0.5)$.

15. This and the next exercise derive an alternative formula for the density $f(x;\nu,\lambda^2)$ of the noncentral chi-squared density. First we consider the case $\nu = 1$, that is the density of Y^2 with $Y \sim N(\lambda, 1)$. Transforming from standard normal Z in Equation (1.6) to $X = (Z+\lambda)^2$, with inverses $Z = \pm\sqrt{X} - \lambda$, yields the density for X. Adding over the two inverses, and noting that $|J^{-1}| = 1/\{2\sqrt{x}\}$ gives

$$f(x;1,\lambda^2) = \frac{1}{2\sqrt{x}}\{\varphi(\sqrt{x}-\lambda)+\varphi(-\sqrt{x}-\lambda)\} = f(x;1)e^{-\frac{1}{2}\lambda^2}\cosh(\lambda\sqrt{x}).$$

16. *Continuation.* To calculate $f(x;\nu,\lambda^2)$ convolve $f(x;1,\lambda^2)$ given above with the $f(x;n-1,0)$ of a central chi-squared with $\nu - 1$ degrees of freedom, as in Equation (1.10) but with ν reduced by 1:

$$f(x;\nu,\lambda^2) = \int_0^x f(x-u;1,\lambda^2)f(u;\nu-1)du.$$

Transforming from u to $z = \sqrt{(1-u/x)}$ in the integral, with x fixed, show that this yields

$$f(x;\nu,\lambda^2) = f(x;\nu)\cdot h(x;\nu,\lambda^2)$$

where the first factor is the central chi-squared density and the second factor is

$$h(x;\nu,\lambda^2) = e^{-\frac{1}{2}\lambda^2}\cdot\begin{cases}\cosh(\lambda\sqrt{x}) & \text{for } \nu = 1;\\ 2\int_0^1 (1-z^2)^{\frac{\nu-3}{2}}\cosh(z\lambda\sqrt{x})dz/\mathrm{Be}(\frac{1}{2},\frac{\nu-1}{2}) & \text{for } \nu > 1.\end{cases}$$

Here Be(u, v) is the beta function, see *Exercise 11* and the Appendix.

17. Show from Equation (1.12) that all cumulants of the noncentral chi-squared distribution increase with increasing ν and with increasing λ^2. How about standardized cumulants? Determine the third standardized cumulant (*skewness coefficient*), and show that it equals

$$\frac{3}{\lambda}\Big[1 - \frac{5}{12}\frac{\nu}{\lambda^2} + 0\Big(\frac{\nu^2}{\lambda^4}\Big)\Big].$$

Hence, for fixed ν, skewness decreases as λ increases. And, for $\nu << \lambda^2$, skewness decreases with increasing ν.

18. Show that the formula given in *Exercise 16* agrees with that in Equation (1.14). [*Hint:* Transform from z to $w = z^2$, and expand in a power series in λ.]

19. Verify the formula (1.15) for the cumulative distribution function of the noncentral chi-squared distribution.

20. Use the representation of the density of a non-central chi-squared distribution in Equation (1.14) to write a program in R to calculate the density of the noncentral chi-squared distribution using dchisq(x, df) and dpois(x, lambda) and check numerically (say when $x = 3.0$, $df = 2$) that this gives the same answer as the direct calculation from dchisq(x, df, lambda). Note that lambda does not have the same meaning in the two commands.

21. Verify the characterization (1.19), that $\mathbf{X}$ is multivariate normal if and only if $\mathbf{a}^T\mathbf{X}$ is univariate normal for every k-vector a, by using cumulant generating functions.

22. Verify the formula (1.20), by applying the transformation (1.17) to equation (1.16).

23. Use the general expression (1.20) to derive the density (1.21) of a bivariate normal random variable (X, Y).

24. Show that, if (X, Y) have the bivariate normal density (1.21) then the conditional distribution of Y given $X = x$ is normal with mean $\nu + (\rho\tau/\sigma)(x - \mu)$ and variance $\tau^2(1 - \rho^2)$. [*Hint:* The conditional density is $f_{Y|X}(y|x) = f_{XY}(x, y)/f_X(x)$. Equivalently, show that $f_{Y|X}(y|x)f_X(x) = f_{XY}(x, y)$, where $f_{Y|X}(y|x)$ is the hypothesized conditional density.]

25. Extend the result of the previous question, to the case where X and Y are themselves vector-valued, with means μ_x and μ_y and variance-covariance matrices Σ_{xx}, Σ_{yy} and Σ_{xy}. Note that if

$$\Sigma = \begin{pmatrix} \Sigma_{xx} & \Sigma_{xy} \\ \Sigma_{yx} & \Sigma_{yy} \end{pmatrix}$$

then

$$\Sigma^{-1} = \begin{pmatrix} \Sigma^{xx} & \Sigma^{xy} \\ \Sigma^{yx} & \Sigma^{yy} \end{pmatrix},$$

where

$$\Sigma^{xx} = (\Sigma_{xx} - \Sigma_{xy}\Sigma_{yy}^{-1}\Sigma_{yx})^{-1}$$

$$\Sigma^{yy} = (\Sigma_{yy} - \Sigma_{yx}\Sigma_{xx}^{-1}\Sigma_{xy})^{-1},$$

and

$$\Sigma^{xy} = -\Sigma^{xx}\Sigma_{xy}\Sigma_{yy}^{-1}, \ \Sigma^{yx} = -\Sigma^{yy}\Sigma_{yx}\Sigma_{xx}^{-1}.$$

See Appendix I in Chapter 6.

26. Verify the formulas $E(XY) = \rho$, $E(X^2Y) = E(XY^2) = 0$, $E(XY^3) = E(X^3Y) = 3\rho$ and $E(X^2Y^2) = 1 + 2\rho^2$ when (X, Y) have a bivariate normal distribution with standard normal marginals and correlation ρ.

27. Write out the statement of Proposition 1 for the case $k = 2$ but without any vector or matrix notation. Specifically, suppose that (X, Y) has a bivariate normal distribution as in Equation (1.21). What are the corresponding quadratic form $Q(X, Y)$ and noncentrality parameter λ^2?

28. Suppose that (X, Y) has a bivariate normal distribution with means zero, unit standard deviation and correlation coefficient ρ with $-1 < \rho < 1$. Show that

$$E(\mathbf{1}[X > 0, Y > 0]) = \text{pr}(X > 0, Y > 0) = \tfrac{1}{4} + \tfrac{1}{2\pi}\arcsin(\rho),$$

and that $E(X \cdot \mathbf{1}[Y > 0]) = \rho/\sqrt{(2\pi)}$. [*Hint:* Show that the transformation $X = U$, $Y = \rho U + \sqrt{(1 - \rho^2)}V$ makes U and V independent standard normals, so that their joint distribution has circular symmetry. Then transform to polar coordinates.]

29. *Extension.* Show that

$$E(X \cdot \mathbf{1}[X > 0, Y > 0]) = \frac{(1 + \rho)}{2\sqrt{(2\pi)}}$$

and that

$$E(XY \cdot \mathbf{1}[X > 0, Y > 0]) = \frac{\rho}{4} + \frac{\rho}{2\pi}\arcsin(\rho) + (1 - \rho^2)^{\frac{1}{2}}.$$

For the latter case, the calculations may be simplified slightly by first calculating $E\{Q(X, Y) \cdot \mathbf{1}[X > 0, Y > 0]\}$.

30. (a) Consider sampling from the population of American adults, and let X represent a measure of physical strength of the right arm and Y that of the left arm. Let F and G be the marginal distribution functions of X and Y. Argue that X and Y may be stochastically

ordered. Does this mean that everyone's right arm is stronger than their left arm? Discuss.

(b) Suppose that there are two different treatments for a serious medical condition, a surgical treatment which has a high initial risk of operative mortality, but offers the possibility of a long-term cure, and a medical treatment which does not have any immediate risk but offers no prospect of cure. Would the corresponding distributions of survival time be stochastically ordered? Why or why not? Illustrate with a sketch.

31. A *proportional hazards family* of distributions is defined by the survivor function

$$\bar{F}_\theta(x) \equiv \bar{F}(x)^\theta, \; (\theta > 0)$$

where $\bar{F}$ is a specified absolutely continuous survival function.

(a) Show that, for any two different values of θ, the corresponding hazard functions are proportional; hence, the name. (The *hazard function* $h(x)$ is defined as $f(x)/\bar{F}(x) = (d/dx)\{-\log \bar{F}(x)\}$, usually only when F has support in $\mathcal{R}^+$, but this restriction is not necessary.)

(b) Show that members of a proportional hazards family are stochastically ordered.

32. Show that, if the random vector (X, Y) satisfies $X \leq Y$, then $X \prec Y$. [*Hint:* The joint distribution is supported within the half-plane where $x \leq y$. For any number z, argue geometrically that $F_X(z) \geq F_Y(z)$.]

33. Show that, if $X \sim N(\mu, \sigma^2)$ and $Y \sim N(\nu, \tau^2)$, with $\tau \neq \sigma$, then X and Y are not stochastically ordered. [*Hint:* Assuming $\sigma < \tau$, show that for small enough c, $F_X(c) < F_Y(c)$ whereas, for large enough x, $F_X(c) > F_Y(c)$. The crossing point can be determined explicitly by expressing each probability in terms of the standard normal distribution. Notice that the corresponding density functions cross twice, the distribution functions only once.]

34. Show that, if the likelihood ratio $f_1(x)/f_0(x)$ is increasing in x then $\bar{F}_1(x) \geq \bar{F}_0(x)$ for all x, so that a monotone likelihood ratio implies stochastic ordering. [*Hint:* Suppose that $y > x$ so that $f_0(x)f_1(y) > f_1(x)f_0(y)$ say. Integrate this inequality (i) with respect to x over $-\infty \leq x \leq y$ with y fixed and then (ii) with respect to y over $x \leq y \leq \infty$ with x fixed.]

35. Prove the **Lemma**: *If X and U are independent, Y and V are independent, $X \prec Y$ and $U \prec V$, then $X + U \prec Y + V$.* Hence, stochastic ordering is preserved under convolution. [*Hint:* This can be done in two steps, first showing that $X + U \prec Y + U$].

36. Suppose that X has distribution function F. Then $V \equiv F(X)$ is called the *probability transform* of X. Let $U \sim U(0,1)$. Show that $V \prec U$. [*Hint:* Use *Exercise 3* and Lemma 1.1.] Moreover, F is continuous if and only if $V \sim U(0,1)$.

37. Prove the **Lemma:** *If* $X \prec Y$ *and* $U = h(X)$, $V = h(Y)$, *for some increasing function* $h(\cdot)$, *then* $U \prec V$. Hence, stochastic ordering is preserved under monotone transformation.

38. Show that the family of noncentral chi-squared distributions with one degree of freedom is a monotone likelihood ratio family, and hence (by *Exercise 32*) is stochastically increasing in the noncentrality parameter. [*Hint:* Use *Exercise 34*].

39. Show that the conclusion in the previous exercise extends to noncentral chi-squared distributions with arbitrary degrees of freedom; that is, for fixed degrees of freedom ν, the noncentral chi-square is stochastically increasing in the non-centrality parameter [*Hint:* Use *Exercise 35.*] (Alternatively, this conclusion follows from the stronger result that the noncentral chi-squared is a monotone likelihood ratio family, but this is more difficult to prove).

40. Prove the converse part of Lemma 1.2.

41. The *Tukey-λ family* of distributions, with real parameter λ, is defined by the quantile functions

$$Q_\lambda(u) = \frac{u^\lambda - (1-u)^\lambda}{\lambda}$$

for $\lambda \neq 0$ and $Q_0(u) = \log\{u/(1-u)\}$ to ensure by continuity at zero. Show that the two distributions with $\lambda = 1$ and $\lambda = 2$ are *not* stochastically ordered, and hence that the Tukey-λ family is not stochastically ordered in λ.

42. Suppose the distribution of X is supported on a finite interval I. Describe two methods of evaluating $E\{h(X)\}$ by Monte Carlo, one requiring generation of independent identically distributed X's and the other generation of independent identically distributed uniform random variable on I. Illustrate by implementing the two approaches by simulation routines in R to estimate $E(X^3)$ when X has density $f(x) \propto \exp(-x)$ over $[0,2]$. Note that we can generate a sample from this distribution by generating independent standard exponentials X_i over the entire real line, keeping the first N values of X that are less than 2 and discarding the others.

43. Suppose X has a beta distribution with parameters 2.2 and 10.8 and we need to evaluate $c = E(Y)$, where $Y = X^{-1}\exp(-X - X^2)$.

(a) Use the "rbeta" command in R to generate 10,000 realizations of X sampled from this distribution. Plot a histogram of the corresponding values of Y and comment on its shape. Calculate the sample mean and standard deviation of this distribution.

(b) Propose a change of measure to another beta distribution which will lead to a flatter distribution of Y on the unit interval, especially over the first half. Carry out a Monte Carlo evaluation of c as in (a) and estimate its standard error.

(c) Describe and compare the methods in (a) and/or (b) with numerical integration using Riemann sums.

(d) Make a guess as to why the author chose to use the numbers 2.2 and 10.8 for this example.

44. Extend the change-of-measure identity, eliminating the support assumption, and adding to the right side of Equation (1.22) the quantity $E_g\{h(X) \cdot \mathbf{1}[f(X) = 0]\}$. What is the effect on Equation (1.23)?

45. Suppose that two independent samples, each of size $n = 100$, are available say with $X_i \sim N(\mu, 1)$ and $Y_i \sim N(\tau, 1)$, and we want to test the null hypothesis $\mu = \tau = 0$, at significance level $\alpha = 5\%$. Consider the test statistic $W = n(\bar{X}_n^2 + \bar{Y}_n^2)$, with $\bar{X}_n$ and $\bar{Y}_n$ the two sample means, rejecting for large values of W_n.

(a) What is the null distribution function of W? Determine the critical value $w_{0.05}$, for which $\text{pr}(W > w_{0.05}) = 0.05$.

(b) What is the distribution of T for general values of μ and τ?

(c) Express the probability of T exceeding $t_{0.05}$ when the null hypothesis is in terms of the function $F(, x, \nu, \lambda^2)$ in Equation (1.15) with $\nu = 2$ and $\lambda^2 = n(\mu^2 + \tau^2)$. Use R to calculate the numerical value when $\lambda^2 = 5.0$ and $\lambda^2 = 9.0$.

(d) What is the smallest value of n needed to give 90% power to reject the null hypothesis when $\mu = \nu = 0.5$?

[*Note*: Similar computations provide large-sample approximations to the power for many popular statistical tests (Chapter 7); *Exercise 39* assures monotonicity of the power, or more precisely of the approximation to the power based on the chi-squared distribution]

46. This question uses special properties of random samples from the exponential distribution.

(a) Let observations $X_1, X_2, \ldots, X_{2n+1}$ be independent and exponentially distributed random variables with unit mean. Let $X_{(1)} < \cdots < X_{(2n+1)}$ be the same observations arranged in increasing order (the *order statistic*), and $Y_j = X_{(j)} - X_{(j-1)}$ with $j = 1, 2, \ldots 2n+1$ and $X_{(0)}$ set to zero by convention. Show that the Y_j are mutually independent and that Y_j has an exponential distribution with parameter $(2n+2-j)$. Show also that $\bar{X} = (2n+1)^{-1} \sum (2n+2-j) Y_j$ and that $\text{med}(X) = X_{(n+1)} = \sum_{j=1}^{n+1} Y_j$. Hence calculate the exact mean and variance of $\text{med}(X)$ and also $\text{corr}\{\bar{X}, \text{med}(X)\}$. What is the limiting behavior of these quantities as $n \to \infty$?

(b) Perform a simulation study like that described in Section 1.9 to verify these calculations numerically (i) when $n = 3$ and (ii) when $n = 101$.

2

Weak Convergence

We shall be concerned with a sequence $\{X_1, X_2, \ldots\}$ of random variables (or random vectors), often independent and identically distributed. The context is that $x_1^*, \ldots, x_n^*$ represent available data, modeled as a realization of random quantities $X_1, \ldots, X_n$. When the X_i are independent and identically distributed, this is a *random sample* from a *population* defined by their common distribution. We allow, conceptually, that n is yet to be chosen and that the chosen value will be 'large'. This leads us to consider the infinite sequence $\{X_1, X_2, \ldots\}$. As we are often interested in functions of the available data, *statistics*, we need likewise to consider a sequence $\{T_1, T_2, \ldots\}$ where $T_n = t_n(X_1, \ldots, X_n)$. Possibly after some normalization, there may be some stability in certain characteristic properties of T_n as n increases. The value of T_n evaluated at the actual data, namely $t_n^* = t_n(x_1^*, \ldots, x_n^*)$, is the statistic at hand. We interpret it according to the 'large-sample', or limiting, behavior of T_n.

We first focus on the large-sample behavior of sample averages: (i) To a first approximation, they act like constants, the corresponding population averages. (ii) To a finer approximation, they are approximately normally distributed. The first of these assertions is the *weak law of large numbers*, and the second is the *central limit theorem*. We often relate other necessary large-sample facts to these, expressing, to a satisfactory approximation, other random quantities as averages; see Chapter 3. Taylor expansions, through linear or quadratic terms (see Appendix I), often play a critical role.

Some of the material in this chapter is presented as an overview or simply a review. More details can be found in textbooks on probability. The key concepts and results, including the continuous mapping theorem and the "delta-method" are presented in Sections 2.1–2.8. Sections 2.9–2.11 discuss some applications. Sections 2.12–2.14 provide additional theoretical background. The Appendices review some results and notation from mathematical analysis.

2.1 Sequences of Random Variables; Weak Convergence

When the random variables $X_1, X_2, \ldots$ are independent and identically distributed, we can think of them as successive values sampled from some population, with distribution function F, density f, characteristic function ψ, and,

DOI: 10.1201/9780429160080-2

when these exist, moment generating function M, cumulant generating function C, mean μ, and standard deviation σ. The sample consists of the first n of these random variables, and since the sample size will often be tending to infinity in our discussions, we sometimes append it to the sample and to various sample statistics as a subscript; thus, the sample is now $\mathbf{X}_n = (X_1, \ldots, X_n)$, the sample mean is $\bar{X}_n$, the sample distribution function is F_n, etc. (However, we do not follow this convention for the sample standard deviation S_X, to avoid confusion with the score statistic S_n introduced in Chapter 4.) In general, $T_n = t_n(\mathbf{X}_n)$ is a statistic based on the first n sample values with its own distribution function denoted by G_n. When evaluating statistics at the actual observed sample values $\mathbf{x}_n^*$, we use lower case and sometimes append an asterisk: $t_n^* = t_n(\mathbf{x}_n^*)$. Note our use of bold font for vectors of length n.

Many statistics are nearly constant in large samples. This concept is formalized through the notion of *convergence in probability*: A sequence of statistics $\{T_1, T_2, \ldots\}$ is said to *converge in probability* to a constant c if the difference between T_n and c is small with high probability whenever n is large, that is:

$$\text{For every } \epsilon > 0, \ \text{pr}(|T_n - c| > \epsilon) \to 0 \quad \text{as} \ \ n \to \infty$$

We then write $T_n \to_p c$. Equivalently, $T_n - c \to_p 0$. Alternative notation is: $T_n = c + o_p(1)$, where $o_p(1)$ is read as 'little-oh-p one' and stands for a quantity that converges in probability to zero. See Section 2.2.

For a specified $\epsilon > 0$, the probabilites $\text{pr}(|T_n - c| > \epsilon)$ form a sequence of real numbers $\{a_1, a_2, \ldots\}$ say, and the requirement is that $a_n \to 0$ in the ordinary sense. That is, for every $\delta > 0$, there must an exist an N such that $n \geq N$ implies $|a_n| < \delta$. Putting this together with the previous statement gives the following restatement of the definition:

Definition 2.1 *For every pair of positive real numbers ϵ and δ there exists an integer N such that*

$$n \geq N \Rightarrow \text{pr}(|T_n - c| < \epsilon) > 1 - \delta.$$

Alternative phraseology is that the asserted inequality holds "for all sufficiently large n". Sometimes we write $N = N(\epsilon, \delta)$ to emphasize that the value of N depends on both ϵ and δ.

The most familiar example of convergence in probability is the *weak law of large numbers.* In the case of independent and identically distributed observations, it asserts that $\bar{X}_n$ *converges in probability to* μ, so long as μ is finite. The sample mean converges in probability to the population mean. Consequently, $n^{-1} \sum_{j=1}^{n} h(X_j)$ converges in probability to $E\{h(X)\}$ whenever the latter is defined and finite, as is seen by writing $Y_j = h(X_j)$ and $Y = h(X)$. In particular, sample proportions converge (in probability) to the corresponding probability: *the proportion of successes in n Bernoulli trials is arbitrarily close to the success probability π with probability close to 1 if n is sufficiently large.* If σ is assumed finite, an elementary proof follows from the *Chebyshev inequality*: $\text{pr}(|\bar{X}_n - \mu| \geq \epsilon) \leq \sigma^2/(n\epsilon^2)$, since for fixed $\epsilon > 0$, the right side tends to zero as $n \to \infty$. In view of the central importance of the result, we present a proof in the general case in Section 2.14. This proof is based on that in FELLER (1968, Chapter X).

As noted, convergence in probability allows the approximation of various statistics by constants, for example the approximation of $\bar{X}_n$ by μ, of a sample proportion by a probability, or of the sample standard deviation S_X by the population standard deviation σ (see below). However, a more detailed approximation is often possible, magnifying the difference $\bar{X}_n - \mu$ by a factor $\sqrt{n}$. This leads to the concept of *convergence in distribution*, also known as *weak convergence*.

A sequence of statistics $\{T_1, T_2, \ldots\}$ is said to *converge in distribution* to a random variable T if the sequence $\{G_n\}$ of distribution functions of the T_n converges to the distribution function G of T at every point where G is continuous. We write $T_n \to_d T$ and $G_n \to_w G$. This provides a basis for approximating the distribution function of T_n by that of T when n is large. Note that even when T is discrete, its distribution function $G(t) = \text{pr}(T \leq t)$ is defined over the whole real line and is discontinuous only at the atoms of T.

Convergence in distribution, also called *weak convergence* or *convergence in law*, includes convergence in probability as a special case when the limiting random variable T is a constant (*Exercise 1*). The reason for term "weak" is that the statement involves only the (marginal) distributions of the T_n in the sequence and the distribution of the limit random variable (in the case of convergence in distribution): there is no requirement that the values of the statistics T_n themselves converge. Indeed there is no requirement that the joint distribution of collections of the T_n even be defined. Such convergence, involving the joint behavior of the whole sequence and its limit, is termed "strong" and is not of primary interest here. However, we will see in Section 2.4 that strong convergence can be a useful tool in proving weak convergence. Summarizing, weak convergence concerns approximation of the *distribution* of T_n by some limit *distribution*; in the special case of convergence in probability, the limit distribution is degenerate.

A related concept is *boundedness in probability*: A sequence $\{T_1, T_2, \ldots\}$ is bounded in probability if, given $\epsilon > 0$, there exists some finite interval I_ϵ for which $\text{pr}(T_n \in I_\epsilon) > 1 - \epsilon$ for all sufficiently large n. In other words, 'no probability escapes to infinity'. We write: $T_n = O_p(1)$ and say 'T_n is big-oh-p one'. The next section describes O_p/o_p notation in more detail. *Exercise 4* gives a simple example of a sequence that is unbounded in probability. It is clear that any sequence converging in distribution is bounded in probability; choose an I_ϵ for which $\text{pr}(T \in I_\epsilon) > 1 - \frac{1}{2}\epsilon$, but the converse is not true; see *Exercise 3*.

Our primary need for the concept of convergence in distribution is to provide a basis for normal approximations and associated chi-squared approximations. For normal approximations, some other terminology and notation is suggestive: we say T_n is *asymptotically normal*, $T_n \sim AN(\nu_n, \tau_n^2)$, if the standardized variables $Z_n \equiv (T_n - \nu_n)/\tau_n$ converge in distribution to a standard normal variable Z. Then $T_n \sim AN(\nu_n, \tau_n^2)$ whenever $(T_n - \nu_n)/\tau_n \rightarrow_d N(0,1)$. We need to be careful: the sequence $\{T_1, T_2, \ldots\}$ itself need not converge in distribution; we only state the convergence holds for an appropriate standardization of the T_n. Although the precise statement is 'asymptotically normal', the operational meaning is 'approximately normal', and such approximation is preserved under linear transformation: It is correct, as well as intuitive, to say that if $T_n \sim AN(\nu_n, \tau_n^2)$ then T_n is approximately normally distributed with mean ν_n and variance τ_n^2. The former statement describes limiting behavior as $n \rightarrow \infty$, the latter is a consequence, valid to a greater or lesser degree of approximation, for finite n.

And although ν_n and τ_n^2 are play roles similar to those of the mean and variance of T_n, they need not be exactly the mean and variance.

The primary example of convergence in distribution, and of asymptotic normality, is provided by the *central limit theorem.* In the case of independent and identically distributed case random variables, this asserts that $\bar{X}_n \sim AN(\mu, \sigma^2/n)$, so long as $\sigma < \infty$. This implies that sample averages are approximately normally distributed whenever the sample size is sufficiently large and the population standard deviation is finite; here, the roles of ν_n and τ_n^2 are the actual mean μ and variance σ^2/n of $\bar{X}_n$. We say 'the large-sample distribution of $\bar{X}_n$' is $N(\mu, \sigma^2/n)$. See below for a proof.

The quality of the approximation in the central limit theorem depends on how close to normal the population distribution is, as well as on how large n is. Specifically, for exactly normal populations, $\bar{X}_n$ is exactly normally distributed for every n, while for highly skewed populations, or population distributions with multiple modes or other 'non-normal' features, a large n, perhaps 100 or more, may be required; however, $n = 25$ is often sufficient for a fairly crude approximation.

When σ is finite, we can exhibit both the weak law of large numbers and the central limit theorem in the following expression defining Z_n:

$$\bar{X}_n = \mu + \frac{\sigma}{\sqrt{n}} Z_n. \tag{2.1}$$

According to the weak law of large numbers, the second term on the right is $o_p(1)$, that is, it converges in probability to zero. Subtracting μ from each side and multiplying through by the 'magnification factor' $\sqrt{n}/\sigma$ gives $Z_n = \sqrt{n}(\bar{X}_n - \mu)/\sigma$. The central limit theorem asserts that $Z_n \rightarrow_d N(0,1)$ and is consequently bounded in probability. It follows that $Z_n/\sqrt{n} \rightarrow_p 0$, so that the weak law of large numbers is a corollary to the central limit theorem.

Another mode of convergence of a sequence of random variables is *convergence in mean-square*: T_n converges in mean-square to c if $E\,(T_n - c)^2 \rightarrow 0$. Its only use for us is as a convenient method for verifying convergence in probability, which it implies by the *Chebyshev Inequality*:

$$\mathrm{pr}\big(|T_n - c| > \epsilon\big) \;\leq\; \frac{E(T_n - c)^2}{\epsilon^2}$$

mentioned above. Indeed, it is sufficient that $E|T_n - c|^p \to 0$ for any positive p, but $p = 2$ is usually the most convenient choice.

2.2 The O_p/o_p Notation

We now formalize the O_p/o_p notation used in the previous section. It is an extension to random variables of the O/o notation used in mathematical analysis which is reviewed in Appendix II. We consider a sequence $\{Y_n\}$ of random variables, usually statistics $Y_n = y_n(\mathbf{X}_n)$, say. We express the 'order in probability' of the sequence, possibly relative to a sequence $\{b_n\}$ of real numbers, as follows.

Definition 2.2 $Y_n = O_p(1)$ *if* Y_n *is bounded in probability.*

That is, given any positive δ, there is a number $K = K(\delta)$ and an integer $N = N(\delta)$ for which $\text{pr}(|Y_n| \leq K) > 1 - \delta$ for every $n \geq N$;

$$Y_n = O_p(b_n) \text{ if } Y_n/b_n = O_p(1).$$

We say: 'Y_n is of order b_n in probability' or 'Y_n is "big-Oh-p" b_n'.

Note the structure of this definition. We are permitted to make an arbitrary choice of $\delta > 0$, and given this choice, there must exist numbers K and N such that the required inequality holds for all $n \geq N$. Of course the choice of K and N is not unique.

Definition 2.3 $Y_n = o_p(1)$ *if* $Y_n \to_p 0$ *as* $n \to \infty$.

That is, given positive numbers ϵ and δ, there is an integer $N = N(\epsilon, \delta)$ for which $\text{pr}(|Y_n| \leq \epsilon) > 1 - \delta$ for every $n \geq N$. By extension,

$$Y_n = o_p(b_n) \text{ if } Y_n/b_n = o_p(1).$$

We say: 'Y_n is of smaller order than b_n in probability' or 'Y_n is 'little-oh-p' b_n'.

Here we are allowed to make arbitrarily small choices both of a positive δ and a positive ϵ; given these choices there must exist an N such that the required inequality holds for all $n \geq N$. Again the choice of N is not unique.

If $Y_n = O_p(b_n)$ and $b_n \to 0$, then $Y_n = o_p(1)$, just as for non-stochastic order. Also, algebraic relations such as $O_p(b_n) + o_p(b_n) = O_p(b_n)$ and $O_p(b_n)o_p(b_n) = o_p(b_n)$ hold as just as for their non-stochastic counterparts, $O(b_n) + o(b_n) = O(b_n)$, $O(b_n)o(b_n) = O(b_n)$. If $Y_n \to_d Y$, then we can find a K for which virtually all of the distribution of Y is within $\pm K$, say $\text{pr}(|Y| < K) > 1 = \frac{1}{2}\epsilon$. If also N is such that $\text{pr}(|Y_n - Y| > \delta) < \frac{1}{2}\epsilon$ for $n > N$ then also $\text{pr}(|Y_n| \leq K + \delta)$ for $n > N$ so Y_n is $O_p(1)$. So convergence in distribution implies boundedness in probability.

We can extend the definition to the comparison of two stochastic sequences, $Y_n = o_p(X_n)$ if $Y_n/X_n = o_p(1)$.

Here are two simple results we will need later.

Lemma 2.1 *Suppose that $\{X_n\}$ is a sequence of random variables with $X_n \to_p 0$ and that $R_n = g(X_n)$ for some function g such that $g(x) = o(x)$ as $x \to 0$. Then $R_n = o_p(X_n)$.*

Proof We must show that for any $\epsilon > 0$ and $\delta > 0$, there exists an N such that $n \geq N$ implies $\text{pr}(|R_n/X_n| > \epsilon) < \delta$. This is straightforward. First choose η such that $|x| < \eta$ implies $|r/x| < \epsilon$. Then choose N such that $\text{pr}(|X_n| > \eta) < \delta$ when $n \geq N$. □

Lemma 2.2 *Suppose that $T_n \to_d T$ and $Y_n \to_p 0$. Then $T_n + Y_n \to_d T$.*

Proof Let t be a continuity point of the distribution function $F(t)$ of T. Then $F_n(t) = \text{pr}(T_n \leq t) \to F(t)$ and we must show that $G_n(t) = \text{pr}(T_n + Y_n \leq t) \to F(t)$. Let $\delta > 0$ be given. Then, by the continuity of $F(t)$ at t we can find an $\epsilon > 0$ such that t, $t - \epsilon$ and $t + \epsilon$ are all continuity points of $F(t)$ and such that $|F(x) - F(t)| < \frac{1}{3}\delta$ if $|x - t| \leq \epsilon$. We can then find an N such that, if $n \geq N$, $\text{pr}(|Y_n| > \epsilon) < \frac{1}{3}\delta$ and $|F_n(x) - F(x)| < \frac{1}{3}\delta$ for $x = t$ and for $x = t \pm \epsilon$. Now if $T_n + Y_n \leq t$ then either $T_n \leq t + \epsilon$ or $|Y_n| > \epsilon$ or both. That is, in a set-theoretic sense, the first event is included in the union of the second and third events. So for $n \geq N$,

$$\begin{aligned} G_n(t) = \text{pr}(T_n + Y_n \leq t) &\leq \text{pr}(T_n \leq t + \epsilon) + \text{pr}(|Y_n| > \epsilon) \\ &\leq F_n(t + \epsilon) + \tfrac{1}{3}\delta \\ &\leq F(t + \epsilon) + \tfrac{2}{3}\delta \leq F(t) + \delta. \end{aligned}$$

Similarly, if $T_n + Y_n > t$ then either $T_n > t - \epsilon$ or $|Y_n| \geq \epsilon$ or both, so that $1 - G_n(t) \leq 1 - F_n(t - \epsilon) + \frac{1}{3}\delta$. Rearranging this expression gives

$$G_n(t) \geq F_n(t - \epsilon) - \tfrac{1}{3}\delta \geq F(t - \epsilon) - \tfrac{2}{3}\delta \geq F(t) - \delta.$$

Combining the two results we have that for $n \geq N$,

$$|G_n(t) - F(t)| \leq \delta.$$

Since δ is arbitrary, this shows that $G_n(t) \to F(t)$. Since this property holds for every continuity point t of F, we have shown that $G_n \to_w F$. □

This result is a simple case of the continuous mapping theorem, discussed in more detail below. It is very useful in calculations of asymptotic distributions as it allows us to ignore remainder terms that converge to zero in probability. There is no requirement on the form of the joint distribution of T_n and Y_n.

2.3 Other Criteria for Weak Convergence

Recall that the distribution of a random variable may be specified in a variety of ways: by the distribution function, density, quantile function, cumulant generating function, etc. The question naturally arises: Having defined convergence in distribution as convergence of distribution functions (at all continuity points of the limit distribution function), how about convergence of other corresponding ways of describing a distribution? There is no universal answer such as: 'If the distribution function of T_n is well approximated by the distribution function of T, then any other description of the distribution of T_n is well approximated by that of T.' For example, a discrete distribution function can be well approximated by a continuous one, but the corresponding densities have different meanings! And distributions with no finite moments can converge to distributions with finite moments (and vice versa), so moments need not converge. However, the following sufficient conditions for convergence in distribution may be proved, and these are enough for our purposes. Brief sketches and references to proofs are given below.

Moment Generating Functions and *Characteristic Functions*: If the moment generating functions $M_{T_n}(t)$ converge to a moment generating function $M_T(t)$ for every t in a neighborhood of zero, then the distribution functions converge, that is $T_n \to_d T$. The same is true for characteristic functions, and any limit continuous at zero is of necessity a characteristic function. The converse holds also, so long as all moment generating functions exist, including that for the limit distribution. See texts on probability theory, for example Breiman (1968) or Shorack (2000).

Density Functions: If density functions $g_{T_n}(x)$ converge to a density function $g_T(x)$ for every x, then the distribution functions converge. This is known as *Scheffé's Theorem.* It is essential to verify that the limit function *is* a proper density (which integrates or sums to unity) however, and not just a subdensity. See Pratt (1960). For theorems on the convergence of densities of $\sqrt{n}\,(\bar{X}_n - \mu)/\sigma$, so-called *local limit theorems*, see Gnedenko (1962).

Quantile Functions: If the quantile functions $Q_{T_n}(p)$ converge to a quantile function $Q_T(p)$ for every $p \in (0,1)$ at which $Q_T(p)$ is continuous, then the distribution functions converge (at continuity points of F), and conversely. If F is continuous and strictly increasing, the proof is immediate. The general case is more complicated, the proof given in Appendix III follows that in Section 1.5.6 of Serfling (1980).

Moments: If, for *every* integer r, the rth moment (or cumulant) of T_n converges to the rth moment (or cumulant) of the standard normal distribution, then $T_n \to_d N(0,1)$. See the discussion in Section 1.5.1 of Serfling (1980), for example. Convergence of the first few moments may suggest, but cannot confirm, asymptotic normality.

The first of these results enables a proof of the basic central limit theorem: With $Z_n \equiv \sqrt{n}(\bar{X}_n - \mu)/\sigma$, its cumulant generating function is

$$C_{Z_n}(t) = nC_X\left(\frac{s}{n}\right)\bigg|_{s=\sqrt{n}\frac{t}{\sigma}} - \sqrt{n}\frac{\mu}{\sigma}t \tag{2.2}$$

assuming the cumulant generating function of X exists. See *Exercise 5*. Expanding C_X as $C_X(t) = \sum_{r=1}^{\infty} \kappa_r t^r/r!$ for t near zero, we see each term in Equation (2.2) except that in t^2 tends to zero, and that the term in t^2 is $\frac{1}{2}t^2$. Hence, the limiting cumulant generating function is the standard normal cumulant generating function, namely $\frac{1}{2}t^2$. (The order of operations can be interchanged due to absolute convergence.) A general proof requires characteristic functions, which always exist; the proof sketched here presumes the existence of moment generating functions, but the method is essentially the same.

Another example shows that the Poisson distribution is approximately normal when its mean λ is large. Here, the sequence is of Poisson random variables X_n with associated means λ_n increasing to infinity. Consider the cumulant generating function of the standardized values $(X_n - \lambda_n)/\sqrt{\lambda_n}$. Dropping the subscript n, this is $\lambda\{\exp(t/\sqrt{\lambda}) - 1\} - \sqrt{\lambda}t$. After expanding the exponential, this simplifies to $\frac{1}{2}t^2 + O(\lambda^{-\frac{1}{2}})$ as $\lambda \to \infty$, and hence asymptotic normality obtains. But, of course, the Poisson density does not converge to a normal density.

As an example of density convergence, the density of the t-distribution may be shown to converge to a standard normal density as the degrees of freedom tend to infinity, and hence the distribution functions also converge. And as an alternative proof of the central limit theorem (again assuming that the cumulant generating function exists), recall that the rth cumulant of $\bar{X}_n$ is κ_r/n^{r-1}; from this it follows [why?] that the rth cumulant of $\sqrt{n}(\bar{X}_n - \mu)/\sigma$ is $\kappa_r/(\sigma^r n^{\frac{r}{2}-1})$ for $r > 1$. This is zero for $r = 1$ and unity for $r = 2$. For $r > 2$ the expression converges to zero, implying asymptotic standard normality. The normal approximation to the Poisson distribution may be proved in the same manner. Note, however, that it is necessary to consider cumulants of *every* order, not just the first three or four.

2.4 The Continuous Mapping Theorem

Another broadly useful result is that convergence in probability and convergence in distribution are preserved under continuous transformation, just as is ordinary convergence of a sequence of real numbers. Thus, if h is a continuous function (on the support of T) and $T_n \to_d T$, then $h(T_n) \to_d h(T)$; the same holds for $T_n \to_p c$ (continuity of h at c now being sufficient). For example, supposing as before that $\bar{X}_n$ is the sample mean of n independent and identically distributed random variables with finite mean μ and variance σ^2, $\log \bar{X}_n \to_p \log \mu$ (assuming that $\mu > 0$ so that $\log \mu$ is defined), and $n(\bar{X}_n - \mu)^2/\sigma^2 \to_d \chi_1^2$ (chi-squared with one degree of freedom, the square of a standard normal). A useful extension of these, with additional arguments converging in probability and others non-random, is:

Theorem 2.1 *(Continuous Mapping Theorem): Suppose that $U_n \to_d U$, $V_n \to_p b$, and (non-random) $w_n \to c$, and that the function $h(u,v,w)$ is continuous at (u,b,c) for every u (in the support of U). Then $h(U_n, V_n, w_n) \to_d h(U,b,c)$.*

Indeed, there can be several V_n's and w_n's though only a single U_n, which can however be multidimensional, see Section 2.8 below.

In particular, $U_n + V_n + w_n \to_d U + b + c$ and $U_n V_n w_n \to_d Ubc$. These latter facts, a slight extension of Lemma 2.2 above, are often called *Slutsky's Theorem.* A proof of the general theorem is best done through a Skorokhod construction, to be sketched below. Bickel & Doksum (2015) give an alternative proof of the simpler Slutsky's Theorem along the lines of that of Lemma 2.2.

One implication of this theorem is that, if we have an algebraic expression that converges in distribution, we can replace any constants in the expression by random quantities that converge in probability to these constants or by non-random quantities that converge to them. As an example, consider the t-statistic $T_n = \sqrt{n}(\bar{X}_n - \mu)/S_X$, where S_X is the sample standard deviation. If S_X were replaced by σ, we would have $T_n \sim AN(0,1)$; so we only need verify that $S_X \to_p \sigma$. To this end, we apply the continuous mapping theorem (without a U_n component but with two V_n components): $S_X^2 \equiv \sum(X_i - \bar{X}_n)^2/(n-1) = \{n/(n-1)\}\sum(X_i - \mu)^2 - (\bar{X}_n - \mu)^2\}/n$ and $w_n = n/(n-1) \to 1$, while both $\sum(X_i - \mu)^2/n \to_p \sigma^2$ and $\bar{X}_n - \mu \to_p 0$ by the weak law of large numbers. Application of the continuous mapping theorem, with $V_n = \sum(X_i - \mu)^2/n$, $V_n' = \bar{X}_n - \mu$, $w_n = n/(n-1)$, and $h(v, v', w) = \sqrt{w(v - v'^2)}$, completes the proof.

The statement of Theorem 2.1 implies that for each n, U_n and V_n are defined on the same probability space, so that, conceptually, their joint distribution needs to be defined. However, because it is assumed that $V_n \to_p b$, this joint distribution is irrelevant to the limiting behavior. In our proof, we restrict attention to a single scalar random variable, here relabeled T_n, to avoid a notational clash with the uniformly distributed random variable U introduced in the proof. The construction described below can easily be extended to include the elements V_n and w_n. Here is the restatement:

Let $T_1, T_2, \ldots$ be a sequence of random variables converging in distribution to a random variable T and let $h(t)$ be a function that is continuous at t for all t in the support of T. Then $h(T)$ converges in distribution to T.

Proof A *Skorokhod construction* creates versions of $T_1, T_2, \ldots$ and of T as functions of a single uniformly distributed random variable U, using *quantile transforms.* Specifically, with G_n and G the distribution functions of T_n and T, respectively, let $Q_n(u) = G_n^{-1}(u)$ and $Q(u) = G^{-1}(u)$ be the quantile function corresponding to G_n and G respectively, and define

$$T_n = Q_n(U) \text{ and } T = Q(U).$$

Then by Lemma 2.1 in Chapter 1, $T'_n =_d T_n$ and $T' =_d T$. Unlike the T_n, the T'_n are all defined on the same probability space, because they are all functions of the same random variable U. Moreover, as shown in Appendix III, if $T_n \rightarrow_d T$, then the quantile functions also converge weakly. That is, $Q_n(u) \rightarrow Q(u)$ for every $u \in (0,1)$ at which $Q(\cdot)$ is continuous. This implies that $T'_n \rightarrow T'$, except possibly for U taking values in the countable set (if any) where $Q(u)$ has jumps, a set which has probability zero under the uniform distribution. So $T'_n \rightarrow T'$ *with probability 1*. This is a much stronger statement than weak convergence, being a simultaneous statement about the whole sequence of random variables $\{T'_n\}$ and the limit random variable T', all being functions of the single U. Now, from ordinary analysis, $t_n \rightarrow t$ and h continuous at t imply that $h(t_n) \rightarrow h(t)$. This in essence is the definition of continuity of the function h at t. So, here, $h(T'_n) \rightarrow h(T')$ with probability one. The weaker conclusion that $h(T'_n) \rightarrow_d h(T')$ then follows by Lemma 2.3 below. This shows that, for any sequence of random variables $\{Y_n\}$, $Y_n \rightarrow 0$ with probability one implies $Y_n \rightarrow_p 0$. Our result follows by setting $Y_n = h(T'_n) - h(T')$ and using Lemma 2.2 above. But this conclusion is a statement about the distribution functions of $h(T'_n)$ and of $h(T')$, and these are the same as the distribution functions of $h(T_n)$ and $h(T)$. Hence, $h(T_n) \rightarrow_d h(T)$. This completes the proof of the basic continuous mapping theorem. □

Lemma 2.3 *If the sequence of random variables* $\{Y_1, Y_2, \ldots\}$, *all defined on the same probability space, is such that* $\text{pr}(Y_n \rightarrow 0) = 1$, *then* $Y_n \rightarrow_p 0$.

Proof It is helpful here to recall that random variables are (measurable) functions on a probability space Ω, so that $Y_n = Y_n(\omega)$ say, where $\omega \in \Omega$. Writing E for the event that $Y_n(\omega) \rightarrow 0$, we see that

$$E = \bigcap_{\epsilon>0} \bigcup_{N\geq 1} \bigcap_{n\geq N} \{|Y_n(\omega)| < \epsilon\}.$$

This equation is a restatement in the terminology of set theory of the definition of convergence in analysis, that for every $\epsilon > 0$ there exists an N such that for all n greater than or equal to N, $|Y_n(\omega)| < \epsilon$. So strong convergence implies $\text{pr}(E) = 1$. Since E is an intersection of sets $\bigcup_N \bigcap_{n\geq N}\{|Y_n(\omega)| < \epsilon\} = E_\epsilon$ say, it is contained in each of these sets and so $\text{pr}(E_\epsilon) = 1$ for each $\epsilon > 0$. In turn, $E_\epsilon = \bigcup_N E_{\epsilon,N}$, where $E_{\epsilon,N} = \bigcap_{n\geq N}\{|Y_n(\omega)| < \epsilon|\}$. Clearly $E_{\epsilon,N} \subseteq E_{\epsilon,N+1}$ for all N, since the right side of the inequality is an intersection over a smaller class of sets than the left side. So, for each $\epsilon > 0$, the $E_{\epsilon,N}$ form an increasing class of sets, and their union has probability one. It follows that for any $\delta > 0$

there must be some N for which $\text{pr}(E_{\epsilon,N}) > 1 - \delta$. Writing $F_{\epsilon,n}$ for the set $\{|Y_m(\omega)| \leq \epsilon\}$, so that $E_{\epsilon,N} = \bigcap_{n \geq N} F_{\epsilon,n}$ we see that for all $n \geq N$, $\text{pr}(F_{\epsilon,n}) \geq \text{pr}(E_{\epsilon,n}) > 1 - \delta$. So we have shown that for all $\epsilon > 0$ and $\delta > 0$ there exists an $N = N(\delta, \epsilon)$ such that $n \geq N$ implies $\text{pr}(|Y_n(\omega)| < \epsilon) > 1 - \delta$, which is precisely the definition of convergence in probability to zero. □

2.5 Convergence of Expectations

When random variables converge in distribution, do their expectations converge to the corresponding expectation? That difficulties that can arise are evident from the following examples: First, suppose $G_n(x) = (1 - 1/n)\Phi(x) + (1/n)F(x - n)$ for some distribution function F. This sequence converges to the limit distribution function $\Phi(x)$. If F has mean zero, then $F(\cdot - n)$ has mean n so that G_n has mean one for every n; yet the limit distribution function has mean zero. If instead F is Cauchy, G_n has no existing mean for any n; yet the limit has mean zero. Next, consider a standard Cauchy distribution truncated to the interval $(-n, n)$. This distribution, which has mean zero for every n, converges to a standard Cauchy distribution, for which the mean does not exist.

There are two very useful results regarding convergence of expectations. First, consider the expectation of a bounded continuous function of the random variable X_n, say $E(Y_n) = E\{h(X_n)\}$, with the function $h(x)$ bounded and continuous. The resulting distribution of Y_n is confined to a fixed finite interval, and the difficulties exhibited above cannot arise. And indeed weak convergence of X_n to X is sufficient for the expectations of Y_n to converge to the expectation of $Y = h(X)$. This is a theorem attributed to Helly-Bray (or sometimes called *Helly's Second Theorem*). The proof is outlined in Section 2.12. In fact, this provides a useful equivalent definition of weak convergence: X_n *converges weakly to* X if, for every bounded continuous function h, $E\{h(X_n)\} \to E\{h(X)\}$. This turns out to be a more useful definition when X_n takes values in an infinite-dimensional space, for example for a *stochastic process* $\{X_n(\cdot)\}$ but this need not concern us here.

If the function h is *not* bounded, a necessary and sufficient condition is that $Y_n = h(X_n)$ be *uniformly integrable*, that is, the convergence of the approximating sums defining each integral be uniform in n; see SERFLING (1980) for more details. A useful sufficient condition is that $E|Y_n|^2$ (or even $E|Y_n|^{1+\epsilon}$ for $\epsilon > 0$) be bounded in n.

2.6 The "Delta Method"

As a final general result on weak convergence, we combine convergence in distribution with a Taylor expansion. Theorem 2.1 led to conclusions regarding the approximate distribution of a function of a statistic that converges in distribution. The next theorem concerns the approximate distribution of a function of a statistic that converges in probability, and, when suitably normalized, also converges in distribution. In particular, look back at expression (2.1), and consider a function h of $\bar{X}_n$, which is then expanded in a Taylor series (through first derivatives only) around μ; we would obtain $h(\bar{X}_n) = h(\mu) + h'(\mu)\sigma n^{-1/2}Z_n +$ terms of smaller order. See Appendix I. As a consequence, $h(\bar{X}_n)$, being approximately linear in the approximately normally distributed random variable Z_n, is also approximately normally distributed.

More generally, if $T_n \sim AN(\nu, \tau^2/n)$, then $T_n = \nu + \epsilon_n$ with $\epsilon_n \sim AN(0, \tau^2/n)$ and hence 'small' (with high probability). Therefore, $h(T_n) = h(\nu + \epsilon_n) \approx h(\nu) + h'(\nu)\epsilon_n$ and hence also asymptotically normal. Formally, the resulting theorem is

Theorem 2.2 *(Delta Method): If $T_n \sim AN(\nu, \tau^2/n)$, and if h is a function with non-zero derivative h' at ν, then $U_n \equiv h(T_n) \sim AN\big[h(\nu), \{h'(\nu)\tau\}^2/n\big]$.*

The statement remains true even if $h'(\nu) = 0$, but then the limit distribution is degenerate.

Proof The existence of the derivative $h'(\nu)$ at ν implies that

$$h(t) - h(\nu) = h'(\nu)(t - \nu) + o(t - \nu).$$

Write $x = t - \nu$, $X_n = T_n - \nu$, and $g(x) = h(x + \nu) - h(\nu) - xh'(\nu)$. Then $g(x)/x = o(1)$ and so, since $X_n \to_p 0$, Lemma 2.1 implies $g(X_n)/X_n = o_p(1)$ and $g(X_n) = o_p(X_n)$. Therefore

$$h(T_n) - h(\nu) = h'(\nu)(T_n - \nu) + o_p(T_n - \nu),$$

and

$$\sqrt{n}\{h(T_n) - h(\nu)\} = h'(\nu)Y_n + o_p(Y_n),$$

where $Y_n = \sqrt{n}(T_n - \nu)$. By hypothesis $Y_n \to_d N(0, \tau^2)$ and so $o_p(Y_n) \to_p 0$ from which the result follows. □

For two simple examples, if $\mu > 0$ then $\log \bar{X}_n \sim AN\{\log \mu, \sigma^2/(n\mu^2)\}$, and for $\mu \neq 0$, $\bar{X}_n^2 \sim AN\{\mu^2, (2\mu\sigma)^2/n\}$. Regarding the latter, we could instead use the fact that the square of an normal variable has a noncentral chi-squared distribution. There is no contradiction, since its noncentrality parameter tends to infinity with n, and such a noncentral chi-squared may be shown to be asymptotically normal (*Exercise 6*). The result also holds, trivially, when $\mu = 0$, bacause $\sqrt{n}\bar{X}_n^2 \to_p 0$ in this case.

Let us contrast once more the difference between the basic continuous mapping theorem and the delta method. Each is a statement about $h(T_n)$. In the former, T_n converges in distribution, to T say, taking the distribution function of $h(T_n)$ to the distribution function of $h(T)$. In the latter, T_n converges in probability to a constant ν, but, when the difference is magnified by $\sqrt{n}$, converges in distribution to normality. Since T_n is close to ν, a linear approximation is validated, and asymptotic normality transferred to $h(T_n)$. In the continuous mapping theorem, the T_n need *not* be close to T; only their distributions must be close. In particular, for the mean $\bar{X}_n$ of a random sample, the continous mapping theorem concerns the large-sample behavior of $h\{\sqrt{n}\,(\bar{X}_n - \mu)\}$ whereas the delta method concerns the large-sample behavior of $h(\bar{X}_n)$, or more specifically that of $\sqrt{n}\{h(\bar{X}_n) - h(\mu)\}$.

2.7 Variance-Stabilizing Transformations

For distributions involving a single unknown parameter, for example the Poisson with mean λ, the binomial with mean p (assuming the index k is known) or the gamma with scale parameter ρ, assuming the index parameter κ is known, the variance of an observation X will be a function of its mean. For the Poisson distribution with mean λ, $\text{var}(X) = \lambda$. When $X \sim B(k,p)$ the estimate $\hat{p} = X/k$ has $\text{var}(\hat{p}) = p(1-p)/k$. For the gamma distribution with index κ and scale parameter ρ, $\text{var}(X) = \kappa/\rho^2 = \{E(X)\}^2/\kappa$. Each of these distributions converges to the normal under the appropriate limiting operations: $\lambda \to \infty$; $k \to \infty$ with p fixed; $\kappa \to \infty$ with ρ fixed. For example for the Poisson distribution, if, for $n = 1, 2, \ldots$, X_n has a Poisson distribution with mean λ_n with $\lambda_n \to \infty$, then $(X_n - \lambda_n)/\sqrt{\lambda_n} \to_d N(0,1)$. However, the width of the confidence interval for λ will depend on λ. It is often convenient to examine such data on a transformed scale, in which the asymptotic variance is (approximately) free of the parameter.

The delta method provides a simple approach to finding an appropriate transformation. Suppose that $h(\cdot)$ is a differentiable function over the range of possible parameter values and let $Y_n = h(X_n)$. The delta method shows that $Y_n - h(\lambda_n) \sim AN(0, \sigma_n^2)$, where

$$\sigma_n^2 = h'(\lambda_n)^2 \text{var}(X) = h'(\lambda_n)^2 \lambda_n.$$

We can choose the function $h(\cdot)$ to make this λ-free by setting $h'(\lambda)^2\lambda =$ constant. This is a first-order differential equation in $h(\cdot)$ which is easily solved. We have

$$h'(\lambda) \propto \frac{1}{\lambda^{1/2}}, \quad h(\lambda) \propto \lambda^{1/2} + \text{ constant},$$

Taking the multiplicative and additive constants as one and zero respectively gives simply $h(\lambda) = \sqrt{\lambda}$, so that $Y_n = \sqrt{X_n}$. We then have

$$\sqrt{n}(Y_n - \lambda^{1/2}) \to_d N(0, \tfrac{1}{4}).$$

as $\lambda \to \infty$. Various authors have suggested higher-order improvements to this formula, for example $h(X) = \sqrt{(X + 3/8)}$ due to Anscombe (1948) or $h(X) = \sqrt{X} + \sqrt{(X+1)}$ due to Freeman & Tukey (1950).

Examples of variance-stabilizing transformations for other distributions are presented in *Exercises 15–17*. See also *Exercise 34* in Chapter 3.

2.8 Weak Convergence of Sequences of Random Vectors

We now consider k statistics simultaneously, say $T_n = (T_{1n}, \ldots, T_{kn})^T$ with $T_{jn} = t_{jn}(X_1, \ldots, X_n)$ and the superscript T indicating vector transpose. What is the large-sample (n large) joint behavior of these k statistics?

Convergence in probability and in distribution can be defined for such multidimensional random variables. For 'in probability', it is sufficient to define convergence component-wise. An alternative definition would be in terms of a *norm* of the vector T_n, for example the Euclidean norm: T_n converges in probability to a constant vector c if the norm of $T_n - c$, a real-valued random variable, converges in probability to 0; that is, $\| T_n - c \| = o_p(1)$. These definitions can be shown to be equivalent, and so we may use the simpler definition that $T_n \to_p c$ if and only if each coordinate of T_n converges in probability to the corresponding coordinate of c.

Similar results hold for a vector version of 'bounded in probability'. We can define it by requiring a norm of T_n to be $O_p(1)$, or equivalently by requiring each component to be $O_p(1)$ (*Exercise 19*).

Convergence *in distribution* may be defined using joint distribution functions, requiring convergence at each point in $\mathcal{R}^k$ at which the limit distribution function is continuous. In particular, we say $T_n \sim AN_k\{\nu, (1/n)\Sigma\}$ if $\sqrt{n}(T_n - \nu) \to_d N_k(\nu, \Sigma)$, and interpret 'asymptotically multinormal' (AN) as 'approximately multinormal' in distribution. In contrast to the previous two concepts, joint convergence in distribution is *not* the same as marginal (coordinate-wise) convergence (*Exercise 20*). However, the simplest way to deal with joint convergence is through the *Cramér-Wold device*: A two-dimensional random variable (X_n, Y_n) converges in distribution to (X, Y) if and only if every linear combination $aX_n + bY_n$ converges in distribution to the corresponding univariate random variable $aX + bY$; similarly for higher dimensional random variables. Just as in the argument used in Chapter 1, this follows from the continuity theorem for characteristic functions: $a^T T_n$ converges in distribution to $a^T T$ for every vector $a \in \mathcal{R}^k$ if and only if T_n converges in distribution to T. This result is especially convenient for convergence in distribution to multivariate normal, since linear combinations of multivariate normal variables are themselves normal. Hence, we only need to deal with univariate normal distributions. Moreover, we know the means and variances of the components of the limiting multivariate normal distribution from the convergence of marginal distributions: $T_{jn} \to_d T_j$, and we can determine the correlation coefficients by considering the T_{jn} in pairs (j and j', say). As a consequence, we have no need to be specific about the vector a. An example (see below) will help.

Also, the continuous mapping theorem and the delta-method continue to hold in higher dimensions. The statement of the former needs no change. But importantly, in contrast to the 'converging in probability' or simply 'converging' terms, *joint* convergence in distribution is required. A statement of the extended delta method is:

Theorem 2.3 *(Multivariate Delta Method): Suppose that* $T_n \sim AN_k(\nu, (1/n)\Sigma)$ *and* $h(t)$ *has a row vector* $\dot{h}(t)$ *of partial derivatives, continuous in a neighborhood of* $t = \nu$*. Then*

$$h(T_n) \sim AN\Big\{h(\nu), \frac{1}{n}\dot{h}(\nu)\Sigma\dot{h}(\nu)^T\Big\}.$$

The proof is left to the reader (*Exercise 22*).

To extend the continuous mapping theorem to higher dimensions we first extend Skorokhod constructions to random vectors by using k independent and identically distributed uniform random variables, successively considering the conditional distribution function of each coordinate given the previous ones and inverting; details are omitted. Then the proof proceeds as before for $h(T_n)$, and similarly for $h(T_n, V_n, w_n)$. The extended delta method simply uses the multivariate mean-value theorem (Appendix I).

Here is an example of each of these: Suppose that (X, Y) has mean (μ, ν), variances σ^2 and τ^2 and correlation coefficient ρ, and $(X_1, Y_1), \ldots, (X_n, Y_n)$ is a random sample; let Σ represent the variance matrix. Then the sample averages $\bar{X}_n$ and $\bar{Y}_n$ have means μ and ν and variance matrix $(1/n)\Sigma$, and hence the same correlation coefficient ρ. They jointly converge in distribution to bivariate normal by the univariate central limit theorem and the Cramér-Wold device (since $a\bar{X}_n + b\bar{Y}_n$ is the average of n independent and identically distributed Z_i's with $Z_i \equiv aX_i + bY_i$). The appropriate quadratic form, namely $n(\bar{X}_n - \mu, \bar{Y}_n - \nu)\Sigma^{-1}(\bar{X}_n - \mu, \bar{Y}_n - \nu)^T$, therefore converges in distribution to chi-squared with two degrees of freedom (by Theorem 2.1). If $\mu \neq 0$, the ratio $\bar{Y}_n/\bar{X}_n$ converges in probability to ν/μ (by Theorem 2.1, or Slutsky's theorem), and furthermore $\bar{Y}_n/\bar{X}_n \sim AN\{\nu/\mu, (\nu^2\sigma^2 - 2\mu\nu\rho\sigma\tau + \mu^2\tau^2)/(n\mu^4)\}$, using a Taylor expansion of $h(\bar{x}_n, \bar{y}_n) = \bar{y}_n/\bar{x}_n$ around (μ, ν) and Theorem 2.3. See *Exercise 23.* We will have many more examples in Chapter 3.

The first example above extends to give the *multivariate central limit theorem:* If $\mathbf{X}_1$, $\mathbf{X}_2$, ..., are independent and identically distributed with mean vector $\boldsymbol{\mu}$ and finite variance matrix Σ, then

$$\mathbf{Y}_n \;\equiv\; \sqrt{n}\,(\bar{X}_{1n} - \mu_1, \ldots, \bar{X}_{kn} - \mu_k)^T \;\to_d\; N_k(\mathbf{0}, \Sigma)$$

that is, $\bar{\mathbf{X}}_n$ is *asymptotically multinormal* $AN_k\{\boldsymbol{\mu}, (1/n)\Sigma\}$. And a corollary, with $\mathbf{V}_n \equiv (\bar{X}_{1n} - \mu_1, \ldots, \bar{X}_{kn} - \mu_k)^T$, is that

$$n\mathbf{V_n}^T\Sigma^{-1}\mathbf{V}_n \to_d \text{ chi-squared with } k \text{ degrees of freedom,}$$

assuming that Σ is nonsingular. (For the singular case, see Chapter 9.) The same conclusion holds if Σ is replaced by an analog estimate, for example the sample variance matrix, since the quadratic form is continuous in the elements of Σ and the elements of the sample covariance matrix each converge in probability to the corresponding element of Σ.

2.9 Example: Estimation for the Lognormal Distribution

As mentioned in Chapter 1, the lognormal distribution is often used to model highly skewed data, as arises, for example, in the modeling of data on environmental contaminants. Recall that the original data X are such that $Y = \log X$ follows a normal distribution with mean ν and variance η, both typically unknown. We showed that $E(X) = \exp(\nu + \frac{1}{2}\eta) = \mu$ say and that $\text{var}(X) = \exp(2\nu + \eta)\{\exp(\eta) - 1\} = \sigma^2$ say. These results suggest that μ be estimated by substituting the sample mean $\bar{Y}$ and sample variance S_Y^2 of the log-transformed data Y for their population values, giving

$$\hat{\mu} = \exp(\bar{Y} + \tfrac{1}{2}S_Y^2).$$

This has the form $\hat{\mu} = h(\bar{Y}, S_Y^2)$ where $h(u, v) = \exp(u + \frac{1}{2}v)$. The partial derivatives of h are

$$\frac{\partial h}{\partial u} = h(u, v), \quad \frac{\partial h}{\partial v} = \tfrac{1}{2}h(u, v).$$

From previous work, we know that

$$\sqrt{n}\begin{pmatrix} \bar{Y} - \nu \\ S_Y^2 - \eta \end{pmatrix} \to_d N\left\{\begin{pmatrix} 0 \\ 0 \end{pmatrix}, \begin{pmatrix} \eta & 0 \\ 0 & 2\eta^2 \end{pmatrix}\right\}.$$

The multivariate delta method shows that $\sqrt{n}(\hat{\mu} - \mu) \to N(0, \tilde{\sigma}^2)$, where

$$\tilde{\sigma}^2 = \{\exp(\nu + \tfrac{1}{2}\eta)\}^2\eta + \{\tfrac{1}{2}\exp(\nu + \tfrac{1}{2}\eta)\}^2 2\eta^2.$$

It is interesting to compare this with the limiting distribution of the sample mean of the original X_i, which has $\sqrt{n}(\bar{X} - \mu) \to_d N(0, \sigma^2)$. The ratio of the asymptotic variances is

$$\frac{\tilde{\sigma}^2}{\sigma^2} = \frac{\eta + \frac{1}{2}\eta^2}{\exp(\eta) - 1}.$$

This ratio is free of ν. It is always less than unity, as we might expect, since $\hat{\mu}$ is based on strong parametric assumptions, while $\bar{X}$ requires only that the variance of X be finite. As $\eta \to 0$ the ratio is $1 - O(\eta^{-2})$, so the gain from using the parametric estimator is small, whereas as $\eta \to \infty$ the ratio tends to

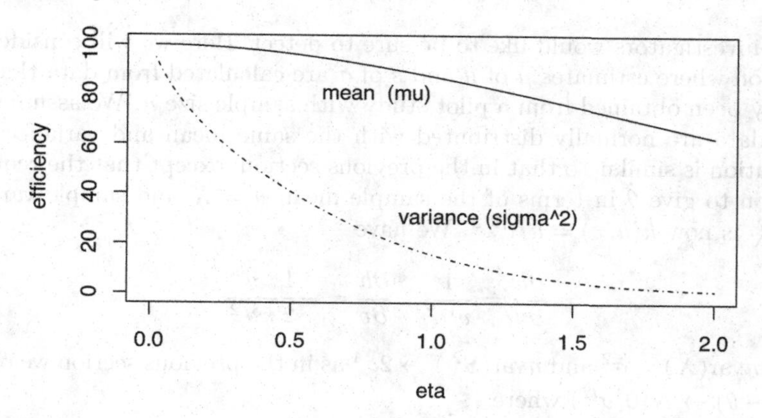

FIGURE 2.1
Estimation efficiencies for lognormal parameters.

zero. A similar but lengthier calculation can be made concerning the estimate of the variance σ^2. Here the ratio of the asymptotic variances of the parametric to the nonparametric estimator decays much more rapidly as η increases. See *Exercise 29.* Figure 8.1 shows the ratio of the asymptotic variances of the two estimators of μ and of σ^2 as a function of the lognormal variance η.

2.10 Example: Estimation of Effect Size for Power Calculations

Power calculations are a major part of statistical practice. We will consider the simple case of testing whether the mean of a normally distributed sample is zero. It is well known that the required sample size is

$$N = (z_\alpha + z_\beta)^2 \frac{\sigma^2}{\mu^2},$$

where z_α and z_β are the standard normal deviates corresponding to the required α and β levels of the test. Often $z_\alpha = 1.96$, corresponding to a two-sided α level of 0.05, and $z_\beta = 0.84$ or 1.28 to give power of 0.8 or 0.9 respectively. The required sample size N, or alternatively the power $1 - \beta$ achieved by a specified N, depends on μ and σ through the ratio $\theta = |\mu|/\sigma$, which is often called the *effect size.* When $\mu > 0$, which we will assume here, it is the reciprocal of the coefficient of variation. When μ and σ are both known or specified the calculation is straightforward. Often μ is specified from practical or scientific considerations such as the smallest value distinct from zero which investigators would like to be sure to detect. Here we will consider the situation where estimates $\hat{\mu}$ of μ and s of σ are calculated from data that has already been obtained from a pilot study with sample size n. We assume that these data are normally distributed with the same mean and variance. The calculation is similar to that in the previous section, except that the requisite function to give $\hat{\theta}$ in terms of the sample mean $u = \bar{X}$ and sample variance $v = S_X^2$ is now $h(u, v) = u/v^{1/2}$. We have

$$\frac{\partial h}{\partial u} = \frac{1}{v^{1/2}}, \quad \frac{\partial h}{\partial v} = -\frac{1}{2}\frac{u}{v^{3/2}}$$

Using $n\text{var}(\bar{X}) = \sigma^2$ and $n\text{var}(S_X^2) \to 2\sigma^4$ as in the previous section we obtain $\sqrt{n}(\hat{\theta} - \theta) \to N(0, \sigma'^2)$ where

$$\sigma'^2 = \frac{1}{\sigma^2}\sigma^2 + \frac{1}{4}\frac{\mu^2}{\sigma^6}2\sigma^4 = 1 + \frac{1}{2}\frac{\mu^2}{\sigma^2} = 1 + \tfrac{1}{2}\theta^2.$$

The precision with which the effect size can be estimated depends on its value. Usually we are interested in effect sizes in the range 0.2–0.8. In this case, the first term dominates the sum, indicating that the imprecision in the estimate of μ is more important than the imprecision in the estimate of σ^2. We will return to this example several times in future chapters.

The explicit result here assumes that the estimates of $\bar{X}$ of μ and S_X^2 of σ^2 are both based on a single sample of size n. The results are different for testing the hypothesis $\mu_1 = \mu_2$ when independent samples of sizes n_1 and n_2 are available from distributions $N(\mu_1, \sigma^2)$ and $N(\mu_2, \sigma^2)$. Assuming, for simplicity that $n_1 = n_2 = n$ say, the estimated effect size is

$$h(u_1, u_2, v) = \frac{u_1 - u_2}{\sqrt{v}},$$

where $u_1 = \bar{X}_1$, $u_2 = \bar{X}_2$ and $v = S_p^2$, the pooled estimate of variance from the two samples. See *Exercise 24* for the result.

2.11 Example: Combining Estimators, Weighted Means

Suppose that we have independent estimators $\hat{\mu}_1, \ldots, \hat{\mu}_k$ of the same parameter μ. Suppose further that each $\hat{\mu}_j$ is unbiased (that is $E(\hat{\mu}_j) = \mu$) and has finite variance σ_j^2. Then for any selection of weights w_j such that $\sum w_j = 1$, the linear combination $\tilde{\mu} = \sum w_j \hat{\mu}_j$ is also unbiased, since $E(\sum w_j \hat{\mu}_j) = \sum w_j E(\hat{\mu}_j) = \sum w_j \mu = \mu \sum w_j = \mu$. The variance of $\tilde{\mu}$ is

$$\text{var}(\tilde{\mu}) = \sum w_j^2 \sigma_j^2.$$

The best choice of the weights w_j in the sense of minimizing $\text{var}(\tilde{\mu})$ is to make $w_j \propto 1/\sigma_j^2$. and the variance of the resulting estimator $\hat{\mu}$ say, is the proportionality constant

$$\text{var}(\hat{\mu}) = \left(\sum \frac{1}{\sigma_j^2}\right)^{-1},$$

the reciprocal of the sum of the reciprocals of the individual variances.

This result can be proved by calculus or algebra. See *Exercise 25* for a simple proof using the fact that a variance cannot be negative. Chapter 6 discusses the closely related Cauchy-Schwarz inequality and its matrix extensions.

The preceding analysis extends easily to the large-sample setting. Suppose we have independent estimators $\hat{\mu}_j$ of a parameter μ based on sample sizes n_j with $\sum n_j = n \to \infty$ in such a way that

$$\sqrt{n}(\hat{\mu}_j - \mu) \sim AN(0, \sigma_j^2),\ (j = 1, \ldots, k),$$

where the σ_j^2 are known. For example, we may take $n_j = \pi_j n$ where the π_j are fixed. Then, for any choice of weights w_j with $\sum w_j = 1$, $\tilde{\mu} = \sum w_j \hat{\mu}_j$ satisfies

$$\sqrt{n}(\tilde{\mu}_j - \mu) \sim AN(0, \tilde{\sigma}^2),$$

where $\tilde{\sigma}^2 = \sum w_j \sigma_j^2$. As before, the optimal choice of weights w_j to minimize $\tilde{\sigma}_j^2$ is to choose $w_j \propto 1/\sigma_j^2$.

However in practice, the σ_j^2 may be unknown and will often be estimated from the same data that are used to estimate the μ_j, for example by the sample variances S_j^2. There will then be some additional variability in $\hat{\mu}$. We know that $S_j^2 - \sigma_j^2 = O_p(n_j^{-1/2})$ and that, in a single sample, estimation of σ does not affect the asymptotic distribution of $\sqrt{n}(\hat{\mu} - \mu)$, although it does affect the finite-sample distribution. What happens to the asymptotic distribution of $\hat{\mu}$ when the σ_j^2 are replaced by their estimates S_j^2?

We consider the more general case of estimating a specified weighted average $\sum w_j \mu_j$ of k distinct parameters μ_j by an estimate $\sum \hat{w}_j \hat{\mu}_j$ (with $\sum w_j = 1$), where $\sqrt{n}(\hat{\mu}_j - \mu_j) \to_d N(0, \sigma_j^2)$. We will assume that that $\sum \hat{w}_j = 1$ and that $\hat{w}_j - w_j = O_p(n^{-1/2})$, where n is the total sample size. The latter can be ensured by having the individual sample sizes $n_j \to \infty$ in fixed proportions as stated above.

We make no assumption about the form of the distribution of the $\hat{w}_i$, in some applications the vectors $(\hat{w}_1, \ldots, \hat{w}_k)$ and $(\hat{\mu}_1, \ldots, \hat{\mu}_k)$ may not be independent, even asymptotically. We have

$$\begin{aligned}\sum \hat{w}_j \hat{\mu}_j - \sum w_j \mu_j &= \sum \hat{w}_j (\hat{\mu}_j - \mu_j) + \sum (\hat{w}_j - w_j)\mu_j \\ &= \sum \{w_j + O_p(n^{-1/2})\}(\hat{\mu}_j - \mu_j) + \sum (\hat{w}_j - w_j)\mu_j.\end{aligned}$$

In the first term, the $O_p(n^{-1/2})$ in parentheses will not affect the asymptotic distribution of $\hat{\mu}$. The second term will in general be of order $O_p(n^{-1/2})$ and so will not be negligible asymptotically. However, when the true values μ_j are all equal ($\mu_j = \mu$), usually the case of primary interest, this term reduces to $\mu(\sum \hat{w}_j - \sum w_j) = 0$, since $\sum \hat{w}_j = \sum w_j = 1$. We may conclude that in this case $\sqrt{n}(\sum \hat{w}_j \hat{\mu}_j - \mu)$ has the same asymptotic distribution as $\sqrt{n}(\sum w_j \hat{\mu}_j - \mu)$.

The result extends to estimators that are not independent. A curiosity here is that the optimal combination of two unbiased, but correlated, estimators may not be convex, see *Exercise 26*.

2.12 Further Theorems on Weak Convergence

We present two standard theorems regarding weak convergence of distribution functions. Since we are primarily interested in convergence to normality, we confine attention to the special case of a limit distribution function which is continuous, allowing minor simplifications in the proofs. General versions appear in textbooks on probability. We also confine attention to distribution functions on the real line; extensions to higher dimensions are straightforward. Throughout, we consider a sequence of random variables X_n, with distribution functions F_n, converging weakly to a random variable X with continuous distribution function F.

The first theorem asserts that weak convergence to a continuous distribution is *uniform*.

Theorem 2.4 *(Uniformity of Weak Convergence): Weak convergence of a sequence $\{F_n\}$ to a continuous distribution F is uniform in its argument x; that is,*

$$F_n \to_w F \Rightarrow \sup_{x \in \mathcal{R}} \left| F_n(x) - F(x) \right| \to 0 \text{ as } n \to \infty.$$

Proof Since F is continuous, for each integer m, let $x_{m,0} = -\infty$, $x_{m,i} = F^{-1}(i/m)$ $(i = 1, \ldots, m-1)$ and $x_{m,m} = +\infty$. Let N_m be such that

$$|F_n(x_{m,i}) - F(x_{m,i})| < \frac{1}{m} \text{ for every } i = 1, \ldots, m-1 \text{ and all } n > N_m. \quad (2.3)$$

We can always find such an N_m since $m - 1$ is finite and $F_n(x) \to F(x)$ pointwise by the continuity of F amd the assumption of weak convergence of F_n to F.

Given x, choose an $i = i(x)$ for which $x \in [x_{mi}, x_{m,i+1}]$. Then by the monotonicity of F and F_n

$$\frac{i}{m} \leq F(x) \leq \frac{i+1}{m}, \quad (2.4)$$

$$F_n(x_{mi} \leq F_n(x) \leq F_n(x_{m,i+1}). \quad (2.5)$$

But for $n > N_m$, $F_n(x_{m,i}) > F(x_{m,i}) - 1/m$ and $F_n(x_{m,i+1}) < F(x_{m,i+1}) + 1/m$, so that $F(x)$ and $F_n(x)$ lie within the same interval $(F(x_{m,i}) - 1/m, F(x_{m,i+1}) + 1/m)$, of length less than $3/m$. Since m is arbitrary, the result follows. □

Note The theorem as stated is clearly false if $F(x)$ is not continuous, for example if $X_n \sim U(0, 1/n)$ then $X_n \to_d X$, where $F(x) = \mathbf{1}[x \geq 0]$ and $\sup |F_n(x) - F(x)| = 1$ for all n.

We next prove convergence of expectations:

Theorem 2.5 *(Helly-Bray): Suppose that $g(\cdot)$ is bounded and continuous on $\mathcal{R}$. Then*

$$E\{g(X_n)\} \to E\{g(X)\}, \text{ that is } \int g(x)\, dF_n(x) \to \int g(x)\, dF(x). \tag{2.6}$$

Proof Let $B = \sup |g(x)|$. For given $a < b$, both finite,

$$\begin{aligned}
&\left| \int g dF_n - \int g dF \right| \\
&= \left| [\int g dF_n - \int_a^b g dF_n] + [\int_a^b g dF_n - \int_a^b g dF] + [\int_a^b g\, dF - \int g\, dF] \right| \\
&\leq \left| \int g dF_n - \int_a^b g dF_n \right| + \left| \int_a^b g\, dF_n - \int_a^b g\, dF \right| + \left| \int_a^b g dF - \int g\, dF \right|.
\end{aligned} \tag{2.7}$$

Given $\epsilon > 0$, choose $a = a_\epsilon$ and $b = b_\epsilon$ so that $F(a) = \epsilon/B$ and $F(b) = 1 - \epsilon/B$. And choose N_ϵ so that

$$\sup_x |F_n(x) - F(x)| < \frac{\epsilon}{B} \text{ for } n > N_\epsilon, \tag{2.8}$$

possible by the previous theorem. Then, for such n,

$$\left| \int_{-\infty}^a g dF_n \right| \leq \int_{-\infty}^a |g| dF_n \leq B \cdot F_n(a) < B \cdot F(a) + \epsilon = 2\epsilon.$$

Similarly, $| \int_{-\infty}^a g dF| \leq \epsilon$, $| \int_b^\infty g dF_n| < 2\epsilon$, and $| \int_b^\infty g dF| \leq \epsilon$. Therefore

$$\left| \int g dF_n - \int_a^b g dF_n \right| = \left| \int_{-\infty}^a g dF_n + \int_b^\infty g dF_n \right| < 4\epsilon,$$

while

$$\left| \int_a^b g dF - \int g dF \right| \leq 2\epsilon.$$

Now consider the middle term on the right in Equation (2.7). First, partition $[a, b]$ into m intervals of equal length with right endpoints $x_{mi} = a + (i/m)$

$(b-a)$ and let $x_{m0} = a$. Let $g_m(x)$ be the step function on $(a, b]$ with $g_m(x) = g(x_{mi})$ for $x \in (x_{m,i-1}, x_{mi}]$ $(i = 1, \ldots, m)$. And let $\delta_m = \sup_{x \in [a,b]} |g(x) - g_m(x)|$, which $\to 0$ as $m \to \infty$ by the continuity of g. Here we use the fact that a continuous function on a closed interval is uniformly continuous. The approximation of the function $g(x)$ by $g_m(x)$ is essentially the process used in defining a Riemann integral.

Adding and subtracting g_m in the middle term on the right in Equation (2.7) yields

$$\left|\int_a^b g dF_n - \int_a^b g dF\right| \leq \left|\int_a^b (g - g_m) dF_n\right| + \left|\int_a^b g_m \, dF_n - \int_a^b g_m \, dF\right| + \left|\int_a^b (g_m - g) dF\right|. \tag{2.9}$$

Now $|\int_a^b (g - g_m) dF_n| \leq \int_a^b |g - g_m| dF_n \leq \delta_m$ and similarly with F_n replaced by F. The middle term on the right in Equation (2.9) equals $|\sum_{i=1}^m g(x_{mi}) \cdot (\delta_{nmi} - \delta_{mi})|$ where $\delta_{nmi} = F_n(x_{mi}) - F_n(x_{m,i-1})$ and $\delta_{mi} = F(x_{mi}) - F(x_{m,i-1})$; and this is $< m \cdot B \cdot 2\,[\delta_m/(mB)] = 2\delta_m$ for $n > N_\epsilon$ with $\epsilon = \delta_m/m$ by Equation (2.8).

Therefore, the left side in Equation (2.9) $< 6\epsilon + 4\delta_m \leq 10\delta_m$ for $n > N_\epsilon$ with $\epsilon = \delta_m/m$. Hence, by choosing m large enough for δ_m to be sufficiently small, the left side in Equation (2.7) is small for all n sufficiently large. □

This theorem can be extended to g and to F with discontinuities, so long as g is continuous at the discontinuities of F. And there is a converse: If (2.6) holds for *every* bounded continuous $g(\cdot)$, then $X_n \to_d X$. This latter fact provides an alternative definition of weak convergence, a definition that extends easily to higher dimensions and to function-valued random variables (stochastic processes), but is not needed here.

2.13 Variables That Are Independent but Not Identically Distributed

Here we present forms of the weak law of large numbers and the central limit theorem when the component random variables are independent but not identically distributed. Let μ_j and σ_j^2 be the mean and variance of X_j, assumed finite, and write

$$\bar{\mu}_n \equiv \frac{1}{n}\sum_{j=1}^n \mu_j \text{ and } \bar{\sigma}_n^2 \equiv \frac{1}{n}\sum_{j=1}^n \sigma_j^2.$$

Then a form of the weak law of large numbers (based on *Chebyshev's Inequality*) is

$$\bar{X}_n - \bar{\mu}_n = o_p(1) \quad \text{if} \quad \bar{\sigma}_n^2 = o(n).$$

and *Lyapounov*'s form of the Central Limit Theorem (not proved here) is:

$$\bar{X}_n \sim AN(\bar{\mu}_n, \frac{1}{n}\bar{\sigma}_n^2) \quad \text{if for some real } \nu > 2,$$

$$\frac{1}{n\bar{\sigma}_n^{\nu}} \sum_{j=1}^{n} E\left|X_j - \mu_j\right|^{\nu} = o\left(n^{\frac{\nu}{2}-1}\right).$$

For example, for $\nu = 3$ and $\bar{\sigma}_n$ bounded between $b > 0$ and $B < \infty$, the requirement is that the average third absolute central moment grows more slowly than $\sqrt{n}$.

A multivariate form of *Lyapounov's Theorem*, applicable to a sequence of k-dimensional vectors $X_n = (X_{1n}, \ldots X_{kn})^T$, may be derived using the the Cramér-Wold device with the help of *Minkowski's Inequality.* We find that

$$\bar{X}_n \text{ is } AN_k(\bar{\mu}_n, \frac{1}{n}\bar{\Sigma}_n) \text{ if } \bar{\Sigma}_n \equiv \frac{1}{n}\sum_{j=1}^{n} \text{var}(X_j) \to \Sigma$$

and, for some $\nu > 2$,

$$\frac{1}{n}\sum_{j=1}^{n} E\left|X_{ij} - \mu_{ij}\right|^{\nu} = o(n^{\frac{\nu}{2}-1}) \text{ for every } i = 1, \ldots, k.$$

More general forms of these theorems are discussed in SERFLING (1980), and other books on probability theory.

2.14 The Weak Law of Large Numbers - Proof by Truncation

We conclude this chapter with a proof of the weak law of large numbers for independent and identically distributed random variables with no condition on their common distribution function other than the existence of the mean. This result is used frequently in the following chapters. The proof, based on that of FELLER (1968) Chapter X, is an example of a technique known as truncation. It builds on the proof by Chebychev's theorem which requires the finiteness of the common variance. As in Section 2.1, we suppose that $X_1, X_2, \ldots$ is a sequence of independent identically distributed random variables with finite mean μ and write $\bar{X}_n = (1/n)\sum_{i=1}^{n} X_i$, the mean of the first n terms. We write $\eta = E(|X_i|)$ which must be finite for μ to exist. If $\eta = 0$ then the X_i

are almost surely zero, so the result is trivial, so we assume $\eta > 0$. We are required to show that for any pair $(\epsilon > 0, \delta > 0)$ we can find an N such that $n > N$ implies that

$$\text{pr}(|\bar{X}_n - \mu| > \epsilon) < \delta. \tag{2.10}$$

For each n, and $i = 1, 2, \ldots, n$ we write

$$X'_{ni} = \begin{cases} X_i, & \text{if } |X_i| \leq n\nu; \\ 0, & \text{otherwise.} \end{cases}$$

and $X''_{ni} = X_i - X'_{ni}$. Here ν is a number to be specified later. For each n, the random variables $X'_{ni}, i = 1, \ldots, n$ form a finite set of independent and identically distributed random variables. Their common distribution depends on n. For obvious reasons, a set of random variables like this is often called a *triangular array.* Write $\bar{X}'_n = (1/n)\sum_{i=1}^n X'_{ni}$ and $\mu'_n = E(X'_{ni}) = E(\bar{X}'_n)$. The proof proceeds by applying Chebychev's inequality to show that for a suitably small choice of ν and all sufficiently large n, $\text{pr}(|\bar{X}'_n - \mu| < \epsilon)$ exceeds $1 - \frac{1}{2}\delta$ and showing that the probability that any of the X''_{ni} is non-zero may be bounded by $\frac{1}{2}\delta$, again for all sufficiently large n.

We need several Lemmas:

Lemma 2.4 *Under the stated condition, there exists an N_1, depending on ν, such that $|\mu'_n - \mu| < \frac{1}{2}\epsilon$ for all $n > N_1$.*

Proof It suffices to show that $E(|X''_{n1}|) \to 0$ as $n \to \infty$. This is immediate, since $\int_{x \geq n\nu} |x| dF(x) \to 0$ as $n \to \infty$, since μ is finite. □

Lemma 2.5 $\text{var}(X'_{ni}) \leq n\nu\eta$.

Note Obviously $\text{var}(X'_{ni}) \leq E(X'^2_{ni}) \leq n^2\nu^2$. However, we need a stronger result for the sequel, giving a bound proportional to n, not to n^2.

Proof Since $|X'_{ni}| \leq |X_i|$, $E|X'_{ni}| \leq E(|X_i|) = \eta$ and

$$\text{var}(X'_{ni}) \leq E(X'^2_{ni}) = E(|X'_{ni}|^2) = E(|X'_{ni}||X'_{ni}|) \leq n\nu E(|X'_{ni}|) \leq n\nu\eta. \quad \square$$

We can use this bound to show that $\text{var}(\bar{X}'_n)$ is $O(1)$ so that $\bar{X}'_n = O_p(1)$.

Lemma 2.6 $\text{var}(\bar{X}'_n) \leq \nu\eta$.

Proof

$$\text{var}(\bar{X}'_n) = \frac{1}{n^2}\sum_{i=1}^n \text{var}(X_{ni}) \leq \frac{1}{n^2} n(n\nu\eta) = \nu\eta. \quad \square$$

We now turn to the X''_{ni}.

Lemma 2.7 *The finiteness of μ implies that $t\{1 - F(t)\} \to 0$ as $t \to \infty$ and $tF(t) \to 0$ as $t \to -\infty$.*

Proof Since $\eta = E(|X_i|)$ is finite so are

$$\int_0^\infty x dF(x) \text{ and } \int_\infty^0 |x| dF(x).$$

These results imply respectively

$$\int_t^\infty x dF(x) \to 0 \text{ as } t \to \infty \text{ and } \int_{-\infty}^t x dF(x) \to 0 \text{ as } t \to -\infty.$$

But

$$t\{1-F(t)\} = \int_t^\infty t dF(x) \le \int_t^\infty x dF(x) \text{ and } |t|F(t) = \int_\infty^t |t| dF(x) \le \int_t^\infty |x| dF(x).$$

This completes the proof. □

Corollary 2.1 *For any $\nu > 0$, $\text{pr}(\sum_{i=1}^n X''_{ni} \neq 0)$ may be made arbitrarily small for sufficiently large n.*

Proof

$$\begin{aligned}\text{pr}(\sum_{i=1}^n X''_{ni} \neq 0) &\le \sum_{i=1}^n \text{pr}(X''_{ni} \neq 0)\\ &= n \cdot \text{pr}(X''_{n1} \neq 0) \le \frac{1}{\nu}(n\nu) \cdot \text{pr}(|X_i| > n\nu).\end{aligned}$$

For any fixed $\nu > 0$ this converges to zero as $n \to \infty$ by Lemma 2.7. □

We now collect these results together to prove

Theorem 2.6 *(Weak Law of Large Numbers): Let $X_1, \ldots, X_n$, be a sequence of independent and identically distributed random variables with finite mean μ. Then $\bar{X}_n = (1/n)(X_1 + \cdots + X_n) \to_p \mu$ as $n \to \infty$.*

Proof We are required to show that for any $\epsilon > 0$ and $\delta > 0$ we can find an N such that $n > N$ implies that (2.10) holds. In the following argument ϵ and δ are kept fixed, the resulting N will depend on these quantities.

We are at liberty to first choose ν, and then, depending on the choice of ν, N. We show that we can choose ν to make $\text{pr}(|\bar{X}'_n - \mu'_n| > \frac{1}{2}\epsilon) \le \frac{1}{2}\delta$ for all n. By Lemma 2.6 and Chebychev's inequality, we have, for all n,

$$\text{pr}(|\bar{X}'_n - \mu'_n| > \tfrac{1}{2}\epsilon) \le \nu\eta/(\tfrac{1}{2}\epsilon)^2.$$

This is less than $\frac{1}{2}\delta$ if

$$\nu < \frac{\epsilon^2\delta}{8\eta}.$$

By Lemma 2.4, this choice of ν ensures that, for all $n > N_1$, $\text{pr}(|\bar{X}'_n - \mu| > \epsilon/2) < 1 - \frac{1}{2}\delta$.

To complete the proof, we must find an $N \geq N_1$ such that

$$\text{pr}\Big(\sum_{i=1}^{n} X''_{in} \neq 0\Big) < \tfrac{1}{2}\delta, \tag{2.11}$$

for all $n > N$. From Corollary 2.1.

$$\text{pr}\Big(\sum_{i=1}^{n} X''_{in} \neq 0\Big) \leq \frac{1}{\nu}\{n\nu \cdot \text{pr}(|X| \geq n\mu)\}.$$

With ν fixed, the term in parentheses tends to zero as $n \to \infty$ by Lemma 2.7. If N is chosen so that this term is less than $\frac{1}{2}\delta\nu$ for all $n \geq N_1$ then (2.11) and therefore (2.10) will hold. □

Appendix I: Taylor Series Expansions

Recall the *Mean-Value Theorem* of differential calculus of functions of a real variable. It states that $f(x) = f(x_0) + f'(z)(x - x_0)$ for some z between x and x_0, so long as f is continuous on the interval from x to x_0 and has a derivative everywhere in between. When x is close to x_0, this suggests the linear approximation

$$f(x) \;\approx\; f(x_0) \;+\; f'(x_0)(x - x_0)$$

Indeed, the error may be shown to be of order $o(x - x_0)$. (See Appendix II for $o(\cdot)$ notation.)

Taylor's Theorem is a generalization; we state two versions of it. For most purposes, we need to use the second form.

Theorem A - 1 *(Taylor's Theorem, Lagrange's form): Suppose that the function $f(x)$ has finite derivative $f^{(m)}$ of order m at each x between x_0 and $x_0 + \delta$ and that the $(m-1)$th-order derivative is continuous throughout the closed interval. Then there exists a $\vartheta \in (0, 1)$ for which*

$$f(x_0 + \delta) = f(x_0) + \sum_{j=1}^{m-1} \frac{f^{(j)}(x_0)}{j!}\delta^j + \frac{f^{(m)}(x_0 + \vartheta\delta)}{m!}\delta^m. \tag{A.1}$$

(See Apostol, 1957, for example.) If $f^{(m)}(x)$ is continuous at x_0, then $f^{(m)}(x_0 + \vartheta\delta) = f^{(m)}(x_0) + o(1)$ as $\delta \to 0$, and the last term in Equation (A.1) equals $f^{(m)}(x_0)\delta^m/m! + o(\delta^m)$. This latter form of Equation (A.1) holds with weaker assumptions:

Theorem A - 2 *(Taylor's Theorem, Young's form): Suppose that f has a finite derivative $f^{(m)}$ of order m at $x = x_0$. Then*

$$f(x_0 + \delta) \;=\; f(x_0) + \sum_{j=1}^{m} \frac{f^{(j)}(x_0)}{j!}\delta^j + o(\delta^m) \quad \text{as } \delta \to 0.$$

(See HARDY, *1952, pp. 289–290.)*

These theorems extend to higher dimensions, to functions of k variables. The assumptions appear to be slightly stronger, but the continuity of partial derivatives in the coordinate variables assures their existence in other directions, a potential problem not encountered in the one-dimensional case. (See GELBAUM & OLMSTED (1964), pp. 115–117, for example.) We state them only for $m = 1$ or 2, the only cases we use:

Theorem A - 3 *(Multidimensional Taylor Theorem): Suppose that the function f of a k-dimensional vector x has continuous partial derivatives of order m, with $m = 1$ or 2, at each point in a neighborhood of the point x_0; write $\dot{f} = (f_1, \ldots, f_k)$ for the row vector of first partial derivatives and $\ddot{f} = (f_{ij})$ for the $k \times k$ matrix of second-order partial derivatives. And suppose $x_0 + \delta$ is another point in this neighborhood. Then:*

(A) There exist $\vartheta \in (0,1)$ and $\vartheta' \in (0,1)$ for which

$$f(x_0 + \delta) = f(x_0) + \dot{f}(x_0 + \vartheta\delta)\delta = f(x_0) + \sum_{j=1}^{k} [f_j(x_0) + \vartheta\delta)\delta_j] \qquad \text{(A.2)}$$

and, if $m = 2$,

$$\begin{aligned} f(x_0 + \delta) &= f(x_0) + \dot{f}(x_0)\delta + \tfrac{1}{2}\delta^T \ddot{f}(x_0 + \vartheta'\delta)\delta \\ &= f(x_0) + \sum_{j=1}^{k} f_j(x_0)\delta_j + \tfrac{1}{2}\textstyle\sum_{i=1}^{k}\sum_{j=1}^{k} f_{ij}(x_0 + \vartheta'\boldsymbol{\delta})\delta_i\delta_j. \end{aligned}$$

(B) Also (with $\| \delta \|^2 \equiv \delta^T\delta$)

$$f(x_0 + \delta) = f(x_0) + \dot{f}(x_0)\delta + o(\| \delta \|) \quad \text{as } \delta \to 0$$

and, if $m = 2$,

$$f(x_0 + \delta) = f(x_0) + \dot{f}(\mathbf{x}_0)\,\boldsymbol{\delta} \;+\; \tfrac{1}{2}\,\delta^T \ddot{f}(\mathbf{x}_0)\,\boldsymbol{\delta} \;+\; o(\| \delta \|^2) \quad \text{as } \delta \to 0.$$

Formula (A.2) is the *Multivariate Mean-Value Theorem.* Notice that $x_0 + \vartheta\delta$ is on the line segment connecting x_0 and $x_0 + \delta$, and not just a point in the neighborhood.

Appendix II: The Order (O/o) Notation in Analysis

The O/o ('big-Oh' and 'Little-oh') notation is a concise way to express the limiting behavior of one sequence $\{a_n\}$ of real numbers relative to that of another sequence $\{b_n\}$. It forms the basis for the O_p/o_p notation introduced in Section 2.3. We start by taking the b_n to be constant.

Definition A - 1 $a_n = O(1)$ *if* $\{a_n\}$ *is bounded, that is, there is a number* K *for which* $|a_n| \leq K$ *for every* n. *Then*

$$a_n = O(b_n) \quad \text{if} \quad a_n/b_n = O(1).$$

We say: 'a_n *is of order* b_n*' or '*a_n *is "big-Oh"* b_n*'.*

Often (but not always), b_n is a simple sequence such as a power of n. The idea is to grade sequences a_n according to the rate at which they increase or decrease. This enables us to pick out the dominant sequence among several sequences added together, or to readily determine the effect of multiplying sequences together. Thus, $a_n = O(1/n)$ means the sequence a_n behaves like a bounded quantity divided by n; hence, it dominates $b_n = O(1/n^2)$ which behaves like a bounded quantity divided by n^2. The sum $a_n + b_n$ is then $O(1/n)$ while the product $a_n b_n$ is 'big-Oh' n^{-3}, a bounded quantity divided by n^3.

Definition A - 2 $a_n = o(1)$ *if* $a_n \to 0$ *as* $n \to \infty$, *that is, for every positive number* ϵ, *there is an integer* $N = N(\epsilon)$ *for which* $|a_n| \leq \epsilon$ *for every* $n > N$;

$$a_n = o(b_n) \quad \text{if} \quad a_n/b_n = o(1).$$

We say: 'a_n *is of smaller order than* b_n*' or '*a_n *is "little-oh"* b_n*'.*

Thus, the sequence c/n is $o(1)$; a more precise statement is that it is $O(1/n)$. It is also $o(1/\sqrt{n})$ (since dividing by $1/\sqrt{n}$, equivalently multiplying by $\sqrt{n}$, would lead to a quantity that still tends to 0). The sequence $c/(\log n)$ is also $o(1)$ but not $O(1/n)$ (since $n/\log n$ tends to infinity); it is $O(1/\log n)$, however.

Here are some more examples:

$$2 + \frac{1}{n} = O(1) \;\; [\text{take } K = 3]; \quad \frac{1}{1+n} = o(1) \quad \text{and} \quad = O\Big(\frac{1}{n}\Big).$$

For the latter, multiply by n and note that it is then bounded by $K = 1$.

$$\Big[2 + \frac{1}{n} - (1 + \frac{3}{n})\sqrt{n}\Big]^2 = O(n); \;\; \frac{1}{n} - \frac{1}{n+1} = O\Big(\frac{1}{n^2}\Big).$$

For the latter, cross-multiply, multiply by n^2, and bound by $K = 1$.

Similar order statements may be defined for functions of a real variable x (rather than an integer n), with some understanding as to whether x is tending to infinity, or to zero, or whatever. For example, with $x \to 0$, we may write $e^{-x} = O(1)$ (i.e., bounded for all x near 0) and with $x \to \infty$, we may write $e^{-x} = o(1)$ (i.e., tending to 0). This enables a simple expression for the error term in a Taylor expansion; for example

$$f(x) = f(0) + \frac{f'(0)}{1!}x + \frac{f''(0)}{2!}x^2 + o(x^2) \quad \text{as} \quad x \to 0.$$

This means that the 'error term', when divided by x^2, tends to zero as x tends to zero; hence, it is of smaller order than the previous term, which is $O(x^2)$ bounded when divided by x^2 but not tending to zero (unless $f''(0)$ happens to be zero).

This has implications for sequences too; here is an example:

$$\exp(c/\sqrt{n}) \;=\; 1 + \frac{c}{\sqrt{n}} + o\Big(\frac{1}{\sqrt{n}}\Big) \quad \text{[Taylor]}$$

and it also is $1 + O(1/\sqrt{n})$.

The basic *multiplicative facts* for order relations are:

$$O(a_n) \cdot O(b_n) = O(a_n b_n), \;\; O(a_n) \cdot o(b_n) = o(a_n b_n), \;\; o(a_n) \cdot o(b_n) = o(a_n b_n).$$

For example, $O(n)O(1) = O(n)$, $O(n)o(1) = o(n)$. The basic *additive fact* for order relations is that a (finite) sum of O and o expressions has the order of the largest, with O dominating o. For example, $o(1)+O(n^{-\frac{1}{2}})+O(n^{-1}) = o(1)$ and $o(n^{-1}) + O(n^{-1}) = O(n^{-1})$. Order relations ignore signs, so that subtraction is just like addition; and since order relations do not necessarily give the 'exact order' of things, division by these little-oh and big-Oh expressions is not defined.

Appendix III: Weak Convergence of Sample Quantile Functions

As discussed in Chapter 1 the quantile function $Q(\cdot)$ corresponding to the distribution function $F(\cdot)$ is the function whose graph $x = Q(u)$ is (almost) the reflection of the graph of $u = F(x)$ in the 45 degree line. To ensure that a unique value of $Q(u)$ is obtained for each value of u in $(0, 1)$, allowing for both flat spots and jumps in $F(\cdot)$ we define

$$Q(u) = F^{-1}(u) = \inf\{x' | F(x') \geq u\}.$$

We now prove an important result concerning weak convergence of quantile functions. Although the proof is surprisingly intricate it involves only simple notions in analysis, not probability. This lemma is used in the Skorohod construction. The proof is based on that in SERFLING (1980). RESNICK (1999) gives a similar treatment. Recall from Section 1.2 of Chapter 1 that $Q(\cdot)$ is left-continuous and satisfies the inequalities (i) $Q\{F(x)\} \leq x$ and (ii) $F\{Q(u)\} \geq u$.

Lemma A - 1 *Suppose that Q_n and Q are the quantile functions corresponding to distribution functions F_n and F. Suppose that $F_n \to F$. Then $Q_n(u) \to Q(u)$ for all points u such that $Q(u)$ is outside flat intervals of F.*

Proof We suppose that u is such that $Q_n(u) \not\to Q(u)$. Then there exists an $\epsilon > 0$ such that either $Q_n(u) < Q(u) - \epsilon$ for infinitely many n, or $Q_n(u) > Q(u) + \epsilon$ for infinitely many n, or possibly both. We may choose ϵ to be such that $Q(u) + \epsilon$ and $Q(u) - \epsilon$ are continuity points of $F(\cdot)$.

Suppose first that $Q_n(u) < Q(u) - \epsilon$ infinitely often. Then $F_n\{Q_n(u)\} \leq F_n\{Q(u) - \epsilon\}$ infinitely often by the monotonicity of F_n. But $F_n\{Q_n(u)\} \geq u$ by Property (ii) above. So $u \leq F_n\{Q_n(u)\} \leq F_n\{Q(u) - \epsilon\}$ infinitely often. Letting $n \to \infty$ shows that $u \leq F\{Q(u) - \epsilon\}$. Hence $Q(u) \leq Q[F\{Q(u) - \epsilon\}]$. Since $Q\{F(x)\} \leq x$ for all x (Property (i) above) this gives $Q(u) \leq Q(u) - \epsilon$, a contradiction.

Suppose now that $Q_n(u) > Q(u) + \epsilon$ infinitely often. By Lemma 1.1 of Chapter 1, (replacing $\leq$ by $>$ throughout), we have $u > F_n\{Q(u)+\epsilon\}$ infinitely often, so $u \geq F\{Q(u) + \epsilon\}$ by the convergence of F_n at $Q(u) + \epsilon$. Since $F\{Q(u)\} \geq u$ (Property (i)) we must have $F\{Q(u)\} = F\{Q(u)\} + \epsilon$, so $F(t)$ is constant on the interval $[Q(u), Q(u) + \epsilon]$. □

Corollary A - 1 *The set of points u for which $Q_n(u) \not\to Q(u)$ is countable.*

Proof If $Q_n(u) \not\to Q(u)$ then there exists an ϵ such that $F(\cdot)$ is constant over the interval $[Q(u), Q(u) + \epsilon]$ and so $Q(u)$ has a positive jump of at least ϵ at u. However, a monotone function can have at most countably many jumps. □

Exercises

1. Show directly that convergence in distribution to the degenerate distribution with distribution function $F(x) = \mathbf{1}_{[c,\infty)}(x)$ is equivalent to convergence in probability to c.

2. The proportion p of successes in n Bernoulli(π) trials, being an average, has $p \sim AN\{\pi, \pi(1-\pi)/n\}$. The normal approximation to the probability that $p \geq s/n$ for some integer s can be improved by using a *continuity correction*, namely replacing s by $s - \frac{1}{2}$ before applying the normal approximation. Compare this normal approximation with exact values obtained from the binomial distribution when $n = 25$, $\pi = \frac{1}{2}$ and $s = 13, 14, \ldots, 25$. Repeat for $\pi = 0.8$.

3. Consider a sequence of random variables $T_1, T_2, \ldots$. Suppose that, for every $m \geq 1$, T_{2m} is $N(0, 1)$ and T_{2m-1} is $U(0, 1)$. Show that T_n is bounded in probability but does not converge in distribution.
4. Let $H(x)$ be the distribution function of $U(0, 1)$, and suppose that T_n has distribution function $G_n(x) = (1 - p)\Phi(x) + pH(x - n)$ for some $p \in (0, 1)$.

   (a) Show that $G_n(x)$ has a limit for every x (as $n \to \infty$), but the limit is not a distribution function (but a 'sub-distribution function').

   (b) Show that T_n is *not* bounded in probability. [A fraction of the probability 'escapes to infinity'.]
5. Verify Equation (2.2) in Section 2.3, and the claim that for $r > 2$, the r'th cumulant of Z_n tends to zero.
6. Show that a sequence of noncentral chi-squared random variables $\{X_n\}$, each with the same degrees of freedom ν but with X_n having non-centrality $\lambda_n^2 \to \infty$, has $X_n \sim AN(\lambda_n^2, 4\lambda_n^2)$. [*Hint:* Use Equation (1.12) in Chapter 1.]
7. Show that a sequence of noncentral chi-squared random variables $\{X_\nu\}$, $\nu = 1, 2, \ldots$, with degrees of freedom $\nu \to \infty$ but with common noncentrality parameter λ^2 has $X_n \sim AN(\nu, 2\nu)$. [*Hint:* Use Equation (1.12) in Chapter 1.]
8. Show that, if F_n converges weakly to F, then the characteristic functions also converge. [*Hint:* See the *Helly-Bray theorem* and the remark at the end of Section 2.12]
9. Show that, if T_n has a t-distribution with n degrees of freedom, it converges in distribution to $N(0, 1)$. Use the *Scheffé Theorem* of Section 2.3. Why cannot the moment method, or the moment generating function method, be used?
10. Suppose that $X_1, X_2, \ldots$ are mutually independent random variables (and are therefore defined on the same probability space) with $\text{pr}(X_n = 1) = 1/n$ and $\text{pr}(X_n = 0) = 1 - 1/n$. Show that $X_n \to_p 0$ but $\text{pr}(X_n \to 0) = 0$. [*Hint:* show that, for any n, $\text{pr}(X_{n+1} = X_{n+2} = \cdots = X_{2n} = 0) = \frac{1}{2}$.] What happens if we replace $1/n$ by $1/n^2$ in the above? In either case, does it matter whether or not the X_n are mutually independent?
11. Suppose that $X_1, X_2, \ldots$ are independent and identically distributed with mean μ and variance $\sigma^2 > 0$.

    (a) What is the large-sample distribution of $Y_n = \bar{X}_n / \sqrt{(1 + \bar{X}_n^2)}$?

    (b) Show that $E\left\{\sqrt{n}\bar{X}_n \middle/ \sqrt{(1 + n\bar{X}_n^2)}\right\} = o(1)$.
12. (Continuation) Describe the large-sample behavior of $\sqrt{n}(\bar{X}_n^2 - \mu^2)$ and of $n(\bar{X}_n - \mu)^2$. Consider the cases $\mu \neq 0$ and $\mu = 0$ separately.

13. What is the large-sample distribution of $1/\bar{X}_n$? Be explicit about any needed assumptions.

14. (a) Find the asymptotic distribution of $\log S_X^2$, where S_X^2 is the sample variance of n independent normally distributed random variables each with mean μ and variance σ^2.

    (b) Use (a) to estimate how large n must be for the standard error of the estimate of σ^2 (or of σ) to be less than 10% of its true value. [*Hint:* $\log(1+a) \approx a$ for a small].

15. (a) The *arcsine transformation* of a sample proportion $\hat{p} = X/n$ say, where $X \sim Binomial(n, p)$ and n is large, is defined by $h(\hat{p}) \equiv 2 \cdot \arcsin^{-1}(\sqrt{\hat{p}})$. Show that the large-sample variance of $h(\hat{p})$ does not depend on p so that this transformation is variance stabilizing in the sense of Section 2.7.

    (b) Use this to approximate the probability that $p \geq s/n$ for the cases considered in *Exercise 2.* Again, use the continuity correction.

16. Suppose that $\text{var}(X) \propto E(X)^2$, as happens for the gamma distribution with known index and unknown scale parameter. Show that the appropriate variance-stabilizing transformation is $h(X) = \log X$.

17. For estimates $\hat{p}$ of continuous proportions p it is sometimes found that the variance $\text{var}(\hat{p}) \propto p^2(1-p)^2$, which approaches zero more rapidly than the usual binomial variance as p approaches the extremes of its range $(0, 1)$. See WEDDERBURN (1974) for discussion of data on the percentage of leaf area of barley infected with leaf blotch. Show that the *logistic* transform $h(\hat{p}) = \log\{\hat{p}/(1-\hat{p})\}$ is variance stabilizing here.

18. Show that if the logistic transformation is applied to binomially distributed data the variance of the transformed data approaches infinity as p approaches either end of the range $(0, 1)$. [*Note*: To avoid infinite values it is common to take $\hat{p} = (X + \frac{1}{2})/(n+1)$ here instead of X/n. This does not affect the large sample distribution of $\hat{p}$ or $h(\hat{p})$.] The logistic transform has many advantages for modeling complex categorical data, but care is needed in interpreting estimates corresponding to probabilities close to zero or unity.

19. Show that a random vector T_n (dimension d) is bounded in probability if and only if each coordinate is bounded in probability. [*Hint:* Use the 'coordinatewise' norm $|t| \equiv \max_j |t_j|$, and show that the joint probability exceeds $1 - d\delta$ whenever each probability exceeds $1 - \delta$; the converse is simpler.]

20. (a) Consider a sequence of pairs (X_n, Y_n) of random variables, and suppose $X_n \to_d X$ and $Y_n \to_p c$. Show that $(X_n, Y_n) \to_d (X, c)$.

(b) Show that '$\to_p c$' in (a) cannot be replaced by '$\to_d Y$' when Y is not constant (with probability one) by constructing a counterexample; that is, demonstrate that marginal convergence in distribution does not imply joint convergence in distribution. [*Hint:* Consider $X_n \equiv Z$ and $Y_n \equiv (-1)^n Z$ with Z standard normal.]

(c) Show that joint convergence in distribution implies marginal convergence in distribution.

21. Suppose that $X_1, X_2, \ldots$ are independent and identically distributed, $a_{rn} = (1/n)\sum_{j=1}^{n} X_j^r$, and $\alpha_r = E(X^r)$ finite for $r = 1$, 2, 3, 4. Show that a_{1n} and a_{2n} are jointly approximately bivariate normal in distribution when n is large. What are the parameters?

22. Prove Theorem 2.3 [*Hint:* Use the multidimensional form of Taylor's theorem in Appendix 1].

23. Prove that if $(X_1, Y_1), \ldots, (X_n, Y_n)$ is a random sample from a bivariate normal distribution with means μ, ν, variances σ^2 and τ^2 and correlation coefficient ρ then $\bar{Y}/\bar{X} \sim AN\{\nu/\mu, (\nu^2\sigma^2 - 2\mu\nu\rho\sigma\tau + \mu^2\tau^2)/(n\mu^4)\}$

24. Complete the calculation outlined at the end of Section 2.10 of the asymptotic distribution of the estimated effect size $(\bar{X}_1 - \bar{X}_2)/S_p$, using data from independent samples of size n from the distributions $N(\mu_1, \sigma^2)$ and $N(\mu_2, \sigma^2)$ and the pooled variance estimate S_p^2 for σ^2.

25. Consider the problem of choosing $w_1, \ldots, w_k$, subject to $\sum w_j = 1$, to minimize $\sum w_j^2 \sigma_j^2$, where the σ_j^2 are specified positive quantities. For candidates $w_1, \ldots w_k$, let V be a discrete random variable with $\mathrm{pr}(V = v_j) = \theta_j$, where

$$v_j = w_j \sigma_j^2, \quad \text{and} \quad \theta_j = \frac{1}{\sigma_j^2}\left(\sum \frac{1}{\sigma_j^2}\right)^{-1}.$$

Calculate $E(V)$ and $E(V^2)$. Conclude that

$$\sum w_j^2 \sigma_j^2 \geq \left(\sum \frac{1}{\sigma_j^2}\right)^{-1},$$

with equality if and only if $w_j \propto 1/\sigma_j^2$.

26. Suppose that $\hat{\mu}_1$ and $\hat{\mu}_2$ are each unbiased estimators of a parameter μ but are dependent, with standard deviations σ_1 and σ_2 and correlation coefficient ρ. Find the value of λ that minimizes $\mathrm{var}\{\lambda\hat{\mu}_1 + (1 - \lambda)\hat{\mu}_2\}$ and show that if $\rho\sigma_1 > \sigma_2$ this value is negative. Many authors have noticed this possibility and some have

recommended that the minimum variance estimator not be used when this occurs. Show also that the minimum variance is

$$\frac{\sigma_1^2\sigma_2^2(1-\rho^2)}{\sigma_1^2+\sigma_2^2-2\rho\sigma_1\sigma_2},$$

so that it is advantageous to have $\rho < 0$.

27. (Illustration). Suppose that X_i, Y_i, Z_i for $i = 1, \ldots, n$ have independent Poisson distributions with means $\mu_X = \mu\exp(-\beta)$, $\mu_Y = \mu$ and $\mu_Z = \mu\exp(\beta)$, where both $\mu > 0$ and β are unknown, and let $\bar{X}_n$, $\bar{Y}_n$ and $\bar{Z}_n$ denote the sample means of the X_i, Y_i and Z_i respectively. Show that $U_n = \log(\bar{Y}_n/\bar{X}_n)$ and $V_n = \log(\bar{Z}_n/\bar{Y}_n)$ are each asymptotically normally distributed with mean β and find their asymptotic joint distribution. Use the previous exercise to calculate an asymptotically optimal weighted average of U_n and V_n for estimating β. Compare the asymptotic variance of this estimate with that of $\frac{1}{2}\log(\bar{Z}_n/\bar{X}_n) = \frac{1}{2}(U_n + V_n)$. Consider the cases where $\beta = 0$ and $\beta \neq 0$.

28. (Extension). Suppose that $\hat{\mu}_1, \ldots, \hat{\mu}_k$ are unbiased estimators of the same (scalar) parameter μ but with (nonsingular) covariance matrix Σ.

    (a) Show that the weighted average $\sum w_j\hat{\mu}_j = w^T\hat{\mu}$ has variance $w^T\Sigma w$. We wish to choose w to minimize this variance subject to the condition $\sum w_j = e^T w = 1$ where e is a vector of units. Note that we do not require the w_j to be nonnegative.

    (b) Recall the *Cauchy-Schwarz* inequality: $(x^Ty)^2 \leq \|x\|^2\|y\|^2$ for any two k-vectors x and y. Let C denote a matrix such that $C^TC = \Sigma$ (see Chapter 1, Section 1.6). Then C is also nonsingular. Letting $x = Cw$ and $y = (C^T)^{-1}e$, show that $x^Tx = w^T\Sigma w$, $y^Ty = e^T\Sigma^{-1}e$ and $x^Ty = w^Te = 1$. Deduce that

    $$w^T\Sigma w \geq \frac{1}{e^T\Sigma^{-1}e}.$$

    (c) Show that $w = \Sigma^{-1}e/e^T\Sigma^{-1}e$ satisfies the constraint $e^Tw = 1$ and gives equality in (b), solving the minimization problem.

    See Chapter 6 for more extensions of the Cauchy-Schwarz inequality.

29. (Finney, 1941). Extend the discussion of Section 2.9 to the estimation of the variance σ^2 of the lognormal distribution. Compare the variances of the parametric and nonparametric estimators of σ^2. You may use the result, to be shown in Chapter 3, that whatever the common distribution of the X_j, provided it has finite fourth moment, the asymptotic distribution of $\sqrt{n}(S_X^2 - \sigma^2)$ is normal with mean zero and variance $m_4 - m_2^2$, where $m_j = E[\{X - E(X)\}^j]$.

Finney gives the ratio of the asymptotic variances of the two estimators of σ^2 as

$$\frac{4\eta(\theta-1)^2+2\eta^2(2\theta-1)^2}{(\theta-1)^2(\theta^4+2\theta^3+3\theta^2-4)},$$

where $\theta=\exp(\eta)$, correcting a typographical error in his formula, where the term $2\theta^3$ is incorrectly given a negative sign. Finney's plot of this function, similar to our Figure 2.1, is correctly drawn.

30. Prove the weak law of large numbers for non-identically distributed data given in Section 2.13.

31. Suppose that $X_1, X_2, \ldots$ are independent Bernoulli random variables with $\text{pr}(X_j=1)=1-\text{pr}(X_j=0)=p_j$, where the sequence of parameters p_j converges to a limit $p\in[0,1]$,

    (a) Show that, for any value of p, including 0 and 1, the weak law of large numbers holds.

    (b) Show that if $0<p<1$ the central limit theorem also holds.

32. Prove the multivariate version of Lyapounov's theorem given in Section 2.13.

33. Consider the function $h(x,y)=x/y$ with $(y>0)$, and let $(x_0,y_0)=(1,1)$.

    (a) What is the linear approximation to h near (x_0,y_0)? Evaluate it at $(x,y)=(1.1,0.9)$, and compare it with $h(x,y)$ there (numerically).

    (b) Determine ϑ so that equality holds in the Lagrange form of the bivariate Taylor theorem of order 1 (that is, the bivariate mean-value theorem), with $(x_0,y_0)=(1,1)$ and the two coordinates of δ equal to 0.1 and -0.1.

    (c) Repeat (a) and (b) for a second-order approximation to h.

34. Suppose that $X_1, X_2, \ldots$ are independent and identically distributed with common density

$$f(x)=\frac{1}{(1+|x|)^3},\quad(-\infty<x<\infty).$$

which is symmetric with mean zero, but has infinite variance. Let $\bar{X}_n$ denote the mean of $X_1,\ldots,X_n$. Use the proof of Theorem 2.6 to find a numerical value of N such that $n>N$ implies that $\text{pr}(|\bar{X}_n|>0.1)\leq 0.2$.

3

Asymptotic Linearity of Statistics

In this chapter, we show that many statistics based on random samples behave, to a first approximation, like sample averages. As a consequence, all such statistics are approximately normally distributed in large samples; moreover, pairs of such statistics are approximately bivariate normal in distribution, and several such statistics are together approximately multinormal. These facts underlie all general methods of large-sample inference. Sections 3.1 and 3.2 outline the general results and present some examples. Sections 3.3, 3.4, and 3.5 go into more depth regarding sample quantiles, the important class of U-statistics, and Gâteaux differentiation, respectively. Section 3.6 gives some further examples, and Section 3.7 gives some concluding comments.

3.1 Asymptotic Linearity

Consider a sequence of independent and identically distributed random variables (or vectors) $X_1, X_2, \ldots$; we refer to a typical one as X with support $\mathcal{X}$ and write $\mathbf{X}_n$ for the vector with components $(X_1, \ldots, X_n)$, the first n elements. The common distribution function is F and the density is f (in absolutely continuous, discrete or mixed cases).

A typical sample *statistic*, such as the sample mean, sample variance, sample median, sample correlation coefficient and sample proportion of positive values, is written as $T_n = t_n(\mathbf{X}_n)$. In this chapter, t_n is usually real-valued, that is, we consider one statistic at a time. When considering a higher-dimensional statistic, for example, the sample mean and sample standard deviation simultaneously, we write $t_n = t_n(\mathbf{X}_n) = \{t_1(\mathbf{X}_n), \ldots, t_k(\mathbf{X}_n)\}$, a vector of length k, with each coordinate $t_j(\mathbf{X}_n)$ a real-valued statistic. We do not use boldface for vectors whose length does not change with the sample size n.

Some statistics, for example, sample raw moments and sample proportions, are themselves averages. Many other statistics act very much like averages, to a first order of approximation. With this in mind, we propose the following

Definition 3.1 *A statistic T_n is asymptotically linear $T_n \sim AL\{\nu, k(\cdot), \tau^2\}$ if there exists a kernel $k: \mathcal{X} \to \mathcal{R}$ with expectation $E\{k(X)\} = 0$ and a finite*

DOI: 10.1201/9780429160080-3

and positive variance $\tau^2 \equiv \text{var}\{k(X)\}$, *and*

$$T_n = \nu + \frac{1}{n}\sum_{j=1}^{n} k(X_j) + o_p\Big(\frac{1}{\sqrt{n}}\Big). \tag{3.1}$$

The requirement that the variance be finite and positive is not essential, but simplifies the development. Of course, ν, $k(\cdot)$ and τ all depend on F, or on θ if $F \equiv F_\theta$ is defined parametrically.

The remainder or error term $o_p(1/\sqrt{n})$ is such that, when multiplied by $\sqrt{n}$, it converges in probability to zero (as $n \to \infty$). Writing $Y = \nu + k(X)$ and $\bar{Y}_n = \sum_{j=1}^{n} Y_j/n$ for the average of n such Y's, we have $T_n = \bar{Y}_n + o_p(1/\sqrt{n})$ that is, T_n differs from an average by an error term which is $o_p(1\sqrt{n})$.

The simplest examples are true averages; for example, the sample mean has $\bar{X}_n \sim AL(\mu, k, \sigma^2)$ with $k(x) = x - \mu$ and with μ and σ^2 the mean and variance (the latter assumed finite and positive) of X. Similarly, with $a_{rn} \equiv \sum_{j=1}^{n} X_j^r/n$ and $\alpha_r \equiv E(X^r)$, the rth raw sample moment a_{rn} has $a_{rn} \sim AL(\alpha_r, k, \alpha_{2r} - \alpha_r^2)$ with $k(x) = x^r - \alpha_r$, assuming $\alpha_r^2 < \alpha_{2r} < \infty$. Also, a sample proportion of values in a set A being an average, say $\bar{p}_n = \sum_{j=1}^{n} \mathbf{1}[X_j \in A]/n$, this has $\bar{p}_n \sim AL\{\pi, \mathbf{1}[\cdot \in A] - \pi, \pi(1-\pi)\}$ with $\pi = \text{pr}(X \in A)$.

We shall indicate shortly that sample central moments, sample cumulants, sample quantiles and many other common statistics are also asymptotically linear. In particular, the sample variance $S_X^2 = \sum_{j=1}^{n}(X_j - \bar{X}_n)^2/(n-1)$ has $S_X^2 \sim AL\{\sigma^2, (x-\mu)^2 - \sigma^2, \mu_4 - \sigma^4\}$ where $\mu_r = E(X-\mu)^r$, the rth central moment (shown in Section 3.2 below). Alternatively, we can write $\mu_4 - \sigma^4 = \sigma^4 \cdot (KUR + 2)$ where KUR is the *excess kurtosis coefficient* κ_4/κ_2^2 and κ_r is the rth *cumulant*. For definitions of cumulants, and of skewness and excess kurtosis coefficients SK and KUR, see Chapter 1.

Before presenting and developing more examples, we note three important reasons for identifying the asymptotically linear structure of statistics:

(A) This structure displays their large-sample behavior. Since averages converge in probability to their expectations, an asymptotically linear statistic T_n converges in probability to ν. Moreover, by the standard central limit theorem, averages are asymptotically normal, and the 'error term' in the asymptotically linear expansion is sufficiently small (by the continuous mapping theorem or Slutsky's theorem) so that

$$T_n \sim AN(\nu, \frac{1}{n}\tau^2); \tag{3.2}$$

that is,

$$\sqrt{n}(T_n - \nu) \to_d N(0, \tau^2)$$

Hence, asymptotically statistics are approximately normally distributed (3.2), with 'large-sample variance' τ^2/n. That is, τ^2/n is the variance of the approximating normal distribution.

(B) The structure also displays large-sample *joint* behavior. If two statistics T_{1n} and T_{2n} are each asymptotically linear, $T_{jn} \sim AL(\nu_j, k_j, \tau_j^2)$ for $j = 1, 2$, then they are jointly asymptotically bivariate normal with asymptotic correlation coefficient

$$\rho = \text{corr}\{k_1(X), k_2(X)\} \tag{3.3}$$

and similarly several asymptotically linear statistics are jointly asymptotically multinormal (see below). Thus, besides their marginal approximate normality (3.2), their joint behavior is also readily quantified by correlation coefficients as in Equation (3.3). This enables the

Definition 3.2 *A vector-valued statistic T_n is asymptotically linear whenever its components are asymptotically linear.*

The joint asymptotic linearity also specifies all pairwise covariances or correlations, but joint and marginal asymptotic linearity properties are otherwise identical. We write: $T_n \sim AL\{\nu, k(\cdot), \Sigma\}$ where Σ is the variance matrix of the vector function $k(\mathbf{X})$ of the sample $\mathbf{X}$. Proof of multinormality is from the multivariate central limit theorem, since $\sqrt{n}(T_n - \nu) = \sqrt{n}\sum_{i=1}^n k(X_i)/n + o_p(1)$, or more directly from the Cramér-Wold device and the univariate central limit theorem: $\sqrt{n}a^T(T_n - \nu) = \sqrt{n}\sum_{i=1}^n a^T k(X_i)/n + o_p(1)$; the $o_p(1)$ terms can be ignored in the asymptotics (by the continuous mapping theorem).

(C) The first-order influence of each component X_j on T_n is made explicit. The kernel $k(\cdot)$ is often called the *influence function* of the statistic T_n. Ignoring the error term, $k(X_j)/n$ is the amount that X_j pulls T_n away from its limiting value ν. Characteristics of the influence function identify *robustness* features of a statistic. In particular, a statistic is said to be *robust* if its influence function is bounded; then a single maverick observation has a bounded influence on the statistic. Section 3.5 gives some examples of influence functions.

A fourth role for asymptotically linear expansions will be developed in later chapters: In the local large-sample theory of estimation and testing, we shall find that the asymptotically linear structure of a statistic identifies its large-sample behavior under local alternative models as well as under the model F.

3.2 Deriving Asymptotically Linear Expansions

We identify five ways of deriving asymptotically linear expansions, here labelled *algebraic, Taylor expansion, Bahadur representation, Hájek projection* and *Gâteaux differentiation.* The last three of these are amplified in later sections.

[1] <u>algebraic</u>: This method is best described by example, rather than through formal definition. We do so for the sample variance S_X^2. First consider the sample second central moment, $m_{2n} \equiv \sum(X_j - \bar{X}_n)^2/n = \{(n-1)/n\}S_X^2$. After adding and subtracting μ inside the summand, we readily find that

$$m_{2n} = \sigma^2 + \frac{1}{n}\sum_{j=1}^{n}\{(X_j - \mu)^2 - \sigma^2\} - (\bar{X}_n - \mu)^2.$$

Now

$$(\bar{X}_n - \mu)^2 = \sqrt{n}(\bar{X}_n - \mu) \cdot (\bar{X}_n - \mu) \cdot \frac{1}{\sqrt{n}}. \tag{3.4}$$

In Equation (3.4) the term $\sqrt{n}(\bar{X}_n - \mu) \sim AN(0, \sigma^2)$, and so is $O_p(1)$, while the remaining terms converge in probability to zero, and hence the whole expression in Equation (3.4) is $O_p(1) \cdot o_p(1/\sqrt{n}) = o_p(1/\sqrt{n})$. Writing $k(X) = (X - \mu)^2 - \sigma^2$, with mean zero and variance $\tau^2 \equiv \mu_4 - \sigma^4$, we see that $m_{2n} \sim AL\{\sigma^2, k(\cdot), \tau^2\}$. Finally, $S_X^2 = \{1 + 1/(n-1)\}m_{2n}$, and since $m_{2n}/(n-1)$ is $O_p(1/n)$, we conclude that S_X^2 is likewise asymptotically linear, with the same expansion as m_{2n}.

This verification for the sample variance was somewhat delicate and lengthy; shorter ones appear below. Other sample central moments may be expanded in similar fashion (*Exercise 1*). For the third and higher central moments, replacing the population mean by the sample mean changes the asymptotic variance, not always in the direction one might expect.

[2] <u>*Taylor expansion*</u>: Here, we determine the asymptotically linear expansion of a statistic by Taylor expansion in other known asymptotically linear statistics. We express this in two propositions.

Proposition 3.1 *Suppose that a statistic T_n has $T_n \sim AL\{\nu, k(\cdot), \tau^2\}$ and $U_n = h(T_n)$ where h has a non-zero continuous derivative h' at ν. Then $U_n \sim AL\{h(\nu), l(\cdot), \omega^2\}$ where $l(\cdot) = h'(\nu)k(\cdot)$ and $\omega^2 = \{h'(\nu)\tau\}^2$.*

Proof By Taylor's theorem,

$$u = h(t) = h(\nu) + h'(\nu)(t - \nu) + o(t - \nu).$$

So, by Lemma 2.1 in Chapter 2, since $T_n - \nu \rightarrow_p 0$

$$U_n = h(T_n) = h(\mu) + h'(\mu)(T_n - \mu) + o_p(T_n - \mu),$$

and

$$\sqrt{n}\{U_n - h(T_n)\} = h'(\nu)\sqrt{n}(T_n - \nu) + o_p\{\sqrt{n}(T_n - \nu)\}.$$

However, $\sqrt{n}(T_n - \nu) = O_p(1)$ by the asymptotic linearity of T_n, so the last term on the right side is $o_p(1)$ as $n \to \infty$ and only the first term is relevant. This proves the result. □

As an example, the sample standard deviation S_X (the square root of the sample variance) has $S_X \sim AL\{\sigma, l(\cdot), \omega^2\}$ with $l(x) = \{(x-\mu)^2 - \sigma^2\}/(2\sigma)$ and $\omega^2 = (\mu_4 - \sigma^4)/(4\sigma^2) = \sigma^2 \cdot (KUR+2)/4$, and its logarithm has $\log S_X \sim AL\{\log\sigma, (1/\sigma)l(\cdot), (1/4)(KUR+2)\}$.

A similar result holds if U_n is a function of several asymptotically linear statistics; we state it for two rather than several. The proof (omitted) is again by Taylor's Theorem.

Proposition 3.2 *Suppose that* $T_{jn} \sim AL\{\nu_j, k_j(\cdot), \tau_j^2\}$ *for each* $j = 1, 2$ *and* $U_n = h(T_{1n}, T_{2n})$ *for some* h *with continuous partial derivatives* h_1 *and* h_2 *at* (ν_1, ν_2). *Then, writing* $\tau^2 = h_1^2\tau_1^2 + 2h_1h_2\tau_{12} + h_2^2\tau_2^2 > 0$ *(assumed positive) where* $\tau_{12} = \mathrm{cov}\{k_1(X), k_2(X)\}$, $U_n \sim AL\{h(\nu_1, \nu_2), l(\cdot), \tau^2\}$ *with* $l(x) = h_1(\nu_1, \nu_2)k_1(x) + h_2(\nu_1, \nu_2)k_2(x)$.

As an example, consider again m_{2n} and write it as $a_{2n} - \bar{X}_n^2$. Both a_{2n} and $\bar{X}_n$ are asymptotically linear, and hence so is m_{2n} by Proposition 3.2. Then the asymptotic linearity of the sample variance, already established by the algebraic method, readily follows. It is easily checked that the kernel and variance from Proposition 3.2 agree with those found earlier (*Exercise 2*). Other examples include various sample cumulants and skewness and kurtosis coefficients (*Exercises 3, 4*).

[3] Bahadur representation: The *Bahadur Representation Theorem*, Bahadur (1966) and Kiefer (1967, 1970) discussed in detail in Section 3.3, asserts that sample quantiles are asymptotically linear. Specifically, for a specified p in (0,1), let Q_p be the quantile function at p corresponding to the distribution function F: $Q_p \equiv \inf\{x | F(x) \geq p\}$, and suppose that F is differentiable at Q_p with derivative $f_p \equiv F'(Q_p) > 0$. Let F_n be the sample empirical distribution function, namely $F_n(x) \equiv \sum_{j=1}^n \mathbf{1}[X_j \leq x]/n$, the function whose value at x is the proportion of the first n X_j's which are at most x. Let $Q_{n,p}$ be any version of the pth sample quantile, that is, allowing any choice of inverse of F_n when there is an interval of inverse values. Then the *Bahadur representation* of the sample quantile is

$$Q_{n,p} = Q_p - \frac{F_n(Q_p) - p}{f_p} + o_p\left(\frac{1}{\sqrt{n}}\right). \tag{3.5}$$

Since $F_n(x)$ is a proportion, and a proportion is an average, (3.5) is really an asymptotically linear representation: $Q_{n,p} \sim AL(Q_p, k, \tau^2)$ with

$$k(x) = -\frac{\mathbf{1}[x \leq Q_p] - p}{f_p} = \frac{\mathbf{1}[x > Q_p] - (1-p)}{f_p}$$

and $\tau^2 = p(1-p)/f_p^2$.

Note that discrete distributions are excluded, since, for them, wherever F is differentiable the derivative is zero. But distributions of mixed type are not excluded, so long as they are differentiable at Q_p.

As an important special case, write *med* for the sample median and *MED* for the population median (of F). Then $med \sim AL(MED, k, \tau^2)$ with $k(x) = \{\mathbf{1}[x > MED] - \frac{1}{2}\}/f(MED)$ and $\tau = 1/\{2f(MED)\}$, assuming only that F has a positive derivative at *MED*.

The implication of this asymptotic linearity is that a sample quantile is approximately normally distributed around the corresponding population quantile:

$$Q_{n,p} \sim AN\left\{Q_p, \frac{p(1-p)}{nf_p^2}\right\} \quad \text{if } f_p > 0. \tag{3.6}$$

Moreover, consider two sample quantiles $Q_{n,p}$ and $Q_{n,q}$ with $p > q$, say, and suppose that both f_p and f_q are positive. An easy calculation (*Exercise 9*) shows that the covariance between the sample proportions $F_n(Q_p)$ and $F_n(Q_q)$ is $p(1-q)/n$ and that their correlation coefficient is $\sqrt{[p(1-q)/\{(1-p)q\}]}$. Then the joint asymptotic linearity of $Q_{n,p}$ and $Q_{n,q}$ implies their asymptotic bivariate normality with asymptotic covariance $p(1-q)/(nf_pf_q)$ and with the same asymptotic correlation coefficient as that between $F_n(Q_p)$ and $F_n(Q_q)$. A so-called *central order statistic* X_{n,k_n} has the same representation (3.5) and the same large-sample behavior (3.6); here, $X_{n,k}$ is the kth order statistic in $\mathbf{X_n}$ and $k_n = [np]$. Section 3.4 gives more detail and interpretive comments about Equation (3.5).

[4] Hájek projection: Statistics may be 'projected' onto the space spanned by averages of terms, each term depending only on a single X_j; such a projection may provide the basis for an asymptotically linear expansion. In brief, the candidate kernel is $nE(T_n - \nu|X_1 = x)$, if n-free. The most important application of this technique is to the class of *U-statistics*, discussed in Section 3.5. In particular, Chapter 8 discusses the *Wilcoxon signed-rank statistic*, defined as the proportion, p say, of positive pairwise averages among n sample values. We show there that if the underlying distribution function is F, then $p \sim AL(\pi, k, \tau^2)$ with $\pi = \text{pr}(X_1 + X_2 > 0)$, $k(x) = 2\{1 - \pi - F(-x)\}$ and $\tau^2 = 4\{\text{pr}(X_1 + X_2 > 0, X_1 + X_3 > 0) - \pi^2\}$. For more examples and discussion, see HETTMANSPERGER (1984) and SERFLING (1980).

[5] Gâteaux differentiation: Many statistics may be represented as functionals of the sample distribution function F_n, and only of F_n (not even depending further on n). Such a statistic is the *sample analog* of the corresponding function of F, the distribution function from which the sample came. Thus, if $\nu_F = \nu(F)$ represents the mean of the distribution function F, then $\nu(F_n)$ is the sample mean. Well-behaved functionals $\nu(F)$ have *Gâteaux derivatives* $\nu'(F)(x)$, a point function calculated by differentiating $\nu\{(1-\epsilon)F + \epsilon\delta_x\}$ in ϵ and letting $\epsilon \to 0$ through positive values. Here $\delta_x(\cdot)$ is the distribution with distribution function $F(y) = \mathbf{1}[y \geq x]$ assigning probability 1 to the point x. It turns out, typically, that $E_F\{\nu'(F)(X)\} = 0$, a perturbation of the distribution in the average direction of itself does not change the value of $\nu(F)$. Let $\tau^2 = \text{var}_F\{\nu'(F)(X)\}$, assumed finite and positive. Further, write

$$\nu(F_n) = \nu(F) + \frac{1}{n}\sum_{j=1}^{n} \nu'_F(x_j) + R_n, \tag{3.7}$$

where R_n is a remainder term. If R_n can be shown to be $o_p(1/\sqrt{n})$, then the statistic $\nu(F_n)$ is $AL\{\nu, \nu'_F(\cdot), \tau^2\}$. Section 3.5 illustrates the approach with applications to certain *robust* statistics such as *trimmed means*. See also SERFLING (1980).

Methods [1] and [3–5], and extensions to other statistics by method [2], lead to the asymptotic linearity of many, even most, statistics of interest (based on a single random sample). This includes most estimates (Chapter 5) and most (one-sided) test statistics under null hypotheses (Chapter 7). Then, what statistics are *not* asymptotically linear? Not many of the common ones, but here are several categories:

a. Statistics that do not treat all n observations equally, that is, that depend explicitly on the subscript labels: For example, the mean of the first half of the sample values. It is asymptotically linear in half the values, but this does not satisfy the proper definition of asymptotically linearity in the full sample. Such statistics are not functions of the sufficient statistic, the collection of sample values without regard to their labels, and hence are of little interest.

b. Some statistics have distributions that are not at all well-approximated by normal distributions. In particular, some are *asymptotically quadratic*, for example the square of any asymptotically linear statistic for which $\nu = 0$ (*Exercise 25*). Others include extreme order statistics (the maximum and minimum of a sample), Kolmogorov's goodness-of-fit statistic, and various density estimates; see *Exercise 26* for the first of these.

c. Certain 'irregular' statistics, defined by modifying 'regular' statistics in certain neighborhoods to achieve 'superefficiency,' such as *Hodges's estimate* of the mean of a normal distribution and *Stein's estimate* of a multinormal mean; these will be discussed in Chapters 4 and 5, as will the concept of *regularity* of estimates.

3.3 The Bahadur Representation of a Sample Quantile

We now give a detailed proof of the asymptotic linearity of a sample quantile. In essence, we follow the paper by GHOSH (1971). We assume that we have a random sample $X_1, \ldots, X_n$ from a distribution function $F(x) =$

$\text{pr}(X \le x)$, with corresponding quantile function $Q(p) = \inf\{x : F(x) \ge p\}$, see Chapter 1.

We focus on a particular $p \in (0,1)$, with $F(x)$ continuous at $x = Q_p \equiv Q(p)$, so that $F(Q_p) = p$. We also assume that $f(x) = F'(x)$ exists and is positive at $x = Q_p$ and write $f(Q_p) = f_p$.

Let $F_n(x)$ be the sample distribution function, and let $Q_{n,p}$ be any sample p-quantile: $Q_{n,p} \in \{x | F_n(x-) \le p \le F_n(x)\}$ (not necessarily the smallest such number). The *Bahadur representation* of $Q_{n,p}$ is

$$Q_{n,p} = Q_p - \frac{F_n(Q_p) - p}{f_p} + \frac{1}{\sqrt{n}} R_n. \tag{3.8}$$

We now show that $R_n \to_p 0$.

Theorem 3.1 *(Ghosh): Suppose that $F(x)$ is differentiable at $x = Q_p$ with derivative $f_p > 0$. Then Equation (3.8) holds with $R_n = o_p(1)$.*

Before giving the proof, we provide two partial insights into Equation (3.8), ignoring the remainder term R_n.

i. $Q_{n,p} > Q_p \Leftrightarrow F_n(Q_p) < p$; so that, after subtracting Q_p, both sides have the same sign. In particular, if the sample median *med* exceeds the population median *MED*, then fewer than half of the observations are less than *MED*.

ii. By Equation (3.8), the 'vertical' deviation of $F_n(Q_p)$ from p is proportional to the 'horizontal' deviation of Q_p from $Q_{n,p}$, with f_p as the proportionality factor. Heuristically, were F_n to be differentiable with derivative matching that of F, a Taylor expansion of $F_n(Q_p)$ at $Q_{n,p}$, would yield $p + f_p \cdot (Q_p - Q_{n,p})$, as stated in Equation (3.8).

Careful justification, including the order claimed for the remainder term, requires delicate argument. A key to connecting these 'vertical' and 'horizontal' deviations is the identity "$Q_F(p) \le x$ if and only if $p \le F(x)$" (Lemma 1.1 of Chapter 1), applied with F_n instead of F; see part 3° of the proof below. To prove the theorem, we first note a *separation lemma* and then use this to prove a convergence result that will be needed later.

Lemma 3.1 *(Separation Lemma): Let v and w be two numbers with $|v - w| > 2\epsilon > 0$ Then for some integer j (positive, zero or negative) we must have $v < j\epsilon$ and $w > (j+1)\epsilon$ or vice versa, i.e. there must be some interval $A_j = [j\epsilon, (j+1)\epsilon]$ which separates v and w. See Figure 3.1, which has $\epsilon = 1$ for clarity.*

Proof By contradiction. If v and w belong to the same or adjacent intervals A_j then, clearly, $|v - w| \le 2\epsilon$. □

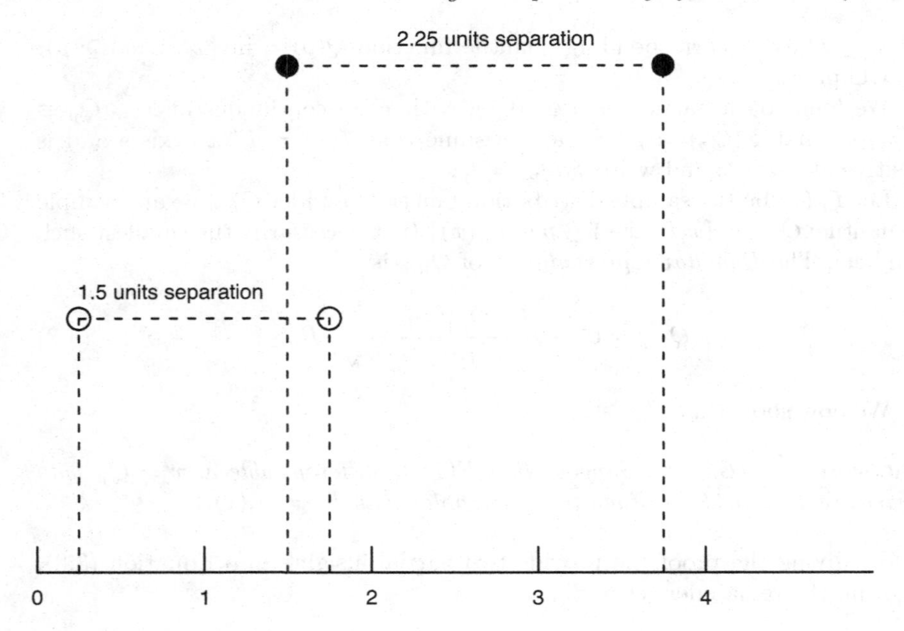

FIGURE 3.1
The separation lemma.

Corollary 3.1 *Let V_n, W_n be two sequences of random variables, and, for $\epsilon > 0$ and each real t define the events $B_n(t)$ and $C_n(t)$ as*

$$B_n(t) = \{V_n \le t, W_n \ge t + \epsilon\}, \quad C_n(t) = \{W_n \le t, V_n \ge t + \epsilon\}. \tag{3.9}$$

Then, if

$$(i)\ W_n \quad = O_p(1) \ \text{ and } \ (ii) \lim_{n\to\infty} \text{pr}\{B_n(t)\} = \lim_{n\to\infty} \text{pr}\{C_n(t)\} = 0,$$

for all t and ϵ then $V_n - W_n = o_p(1)$.

Proof Taking the union of $B_n(t)$ and $C_n(t)$ over all t yields the event $\{|V_n - W_n| > \epsilon$, with probability that we want to show tends to zero. It is sufficient to show that for each $\delta > 0$ and $\epsilon > 0$, an n can be found such that $m > n$ implies $\text{pr}(|V_m - W_m| > 2\epsilon) < 2\delta$. A careful argument is needed to replace the uncountable union with a finite union whose probability can be shown to be sufficiently small. First, using (i), choose a finite interval I and integer n_1 such that $\text{pr}(W_n \in I) > 1 - \delta$ if $n > n_1$.

Now split I into $k - 1$ subintervals (a_{j-1}, a_j) of width less than ϵ and consider the events $B_{nj} = B_n(a_j)$ and $C_{nj} = C_n(a_j)$. By (ii), for each j there exist n_{1j}, n_{2j} such that $\text{pr}(B_{mj}) < \delta/(2k)$ and $\text{pr}(C_{mj}) < \delta/(2k)$ for $m > n_{1j}$ and $m > n_{2j}$ respectively. By the separation lemma

$$\Big\{|V_n - W_n| > 2\epsilon\Big\} \subset \Big\{W_n \notin I\Big\} \cup \left[\bigcup_{j=1}^{k} (B_{nj} \cap \{W_n \in I\}) \cup (C_{nj} \cap \{W_n \in I\})\right],$$

a union of $2k+1$ events. So for $m > \max(n_1, n_{1j}, n_{2j})$ Bonferroni's inequality gives

$$\text{pr}\{|V_n - W_n| > 2\epsilon\} \leq \text{pr}(W_n \notin I) + 2k\frac{\delta}{2k} < \delta + \delta = 2\delta. \quad \square$$

We now return to the proof of the main theorem. For simplicity, drop the subscript p. First let $Q_n \equiv \inf\{x | F_n(x) \geq p\}$ be the smallest sample p-quantile. Define

$$V_n \equiv \sqrt{n}(Q_n - Q), W_n \equiv -\sqrt{n}\frac{F_n(Q) - p}{f}$$

Then $R_n = V_n - W_n$, and we need to prove that $V_n - W_n = o_p(1)$.

Extend W_n to $W_n(t)$, defined as

$$\begin{aligned} W_n(t) &\equiv -\sqrt{n}\frac{F_n\Big(Q + \frac{t}{\sqrt{n}}\Big) - p - \frac{t}{\sqrt{n}}f}{f} \\ &= -\sqrt{n}\frac{F_n\Big(Q + \frac{t}{\sqrt{n}}\Big) - p}{f} + t \end{aligned}$$

Note that $W_n = W_n(0)$. We break the proof into several parts.

1°. $F_n(Q)$ is a sample proportion, and hence an average, with expectation $F(Q) = p$ and variance $p(1-p)/n$. So

$$W_n \to_d N(0, \tau^2), \quad \text{with} \quad \tau^2 \equiv p(1-p)/f^2.$$

2°. We next show that, for each t, $W_n - W_n(t) = o_p(1)$, and hence, from 1°, $\text{pr}\{W_n(t) \leq t\} \to \Phi(t/\tau)$. To see this, let $s_n = t/\sqrt{n}$ and $P_n(s_n) = F_n(Q + s_n) - F_n(Q)$, a binomial proportion (or the negative thereof if $t < 0$). Then, writing $P(s, n) = E\{P_n(s_n)\}$ we have

$$P(s_n) = F(Q + s_n) - F(Q) = f s_n + o(s_n), \quad \text{by Taylor}$$

and

$$\text{var}\{P_n(s_n)\} = \{|P(s_n)|(1 - |P(s_n)|)/n\} = O(s_n/n) = o(1/n).$$

Then

$$\begin{aligned} f^2 E\{W_n - W_n(t)\}^2 &= nE\{P_n(s_n) - f s_n\}^2 \\ &= n\big[\text{var}\{P_n(s_n)\} + \{P(s_n) - f s_n\}^2\big] \\ &= n\{o(1/n) + o(s_n^2)\} = o(1) \end{aligned}$$

This proves that $W_n - W_n(t) \to 0$ in mean-square, and hence in probability.

3°. We now show that, for each t, $\{V_n \leq t\} = \{W_n(t) \leq t\}$. Specifically,

$$\{V_n \leq t\} = \Big\{Q_n \leq Q + \frac{t}{\sqrt{n}}\Big\} = \Big\{p \leq F_n\Big(Q + \frac{t}{\sqrt{n}}\Big)\Big\} = \{W_n(t) \leq t\} \quad (3.10)$$

(the second equality following from Lemma 1.1 in Chapter 1 applied to Q_n and F_n and the other equalities being essentially notational).

As a consequence of 2° and 3°, $\text{pr}(V_n \leq t) \to \Phi(t/\tau)$, that is, $V_n \to_d N(0, \tau^2)$. This conclusion is of major interest. Still, we have not yet proved the 'near-equality-with-high-probability' of V_n and W_n, stated in the Theorem, but only that they have the same asymptotic distributions.

4°. To complete the proof, we use 2° and 3° to verify condition (i) in the corollary proved earlier. (Condition (ii) is implied by 1°.) From Equations (3.9) and (3.10), we have

$$\text{pr}\{B_n(t)\} = \text{pr}\{W_n(t) \leq t, W_n > t + \epsilon\} \leq \text{pr}\{W_n - W_n(t) \geq \epsilon\}$$

which tends to zero by 1°. Similarly, $\text{pr}\{C_n(t)\} \to 0$. Application of the corollary then completes the proof, for this choice of definition of Q_n.

5°. If Q_n is defined as the largest inverse of F_n instead of the smallest, a similar argument holds except now $\{V_n \leq t\} = \{W_n(t) < t\}$ and minor changes are necessitated throughout. Since the theorem holds for either of these extreme choices for Q_n, it holds for any choice.

This completes the proof of the Bahadur representation. □

3.4 U-statistics:- Central Limit Theorem and Asymptotic Linearity

This section gives a detailed proof of the asymptotic normality of a wide class of statistics called *U-statistics*, illustrating the use of Hájek projection. Let $\psi(x_1, \ldots, x_k)$ be a symmetric function of k variables. Here we will assume that each x_i is scalar although the results easily extend to vector-valued x's. By "symmetric" we mean that ψ is invariant under permutation of its arguments, thus if $(\alpha_1, \ldots, \alpha_k)$ is a permutation of the integers $(1, \ldots, k)$ then

$$\psi(x_{\alpha_1}, \ldots, x_{\alpha_k}) = \psi(x_1, \ldots, x_k).$$

Suppose now that $n \geq k$, that $X_1, \ldots, X_n$ are independent and identically distributed random variables and that

$$U_n = \binom{n}{k}^{-1} \sum_{\alpha \in \mathcal{C}(n,k)} \psi(x_{\alpha_1}, \ldots, x_{\alpha_k}).$$

Here $\mathcal{C}(n,k)$ is the set of all ordered k-tuples from the integers $1,\ldots,n$, for example if $n = 3, k = 2$ then $\mathcal{C}(n,k) = \{(1,2),(1,3),(2,3)\}$. Then U_n is called a U-statistic. The divisor $\binom{n}{k}$ is simply the number of elements in $\mathcal{C}(n,k)$. The name arises because U_n is an unbiased estimate of the quantity $E\psi(X_1,\ldots,X_k)$, which we will call θ.

As a simple example suppose that $k = 2$ and $\psi(x_1,x_2) = \frac{1}{2}(x_1 - x_2)^2$. If the X_i have finite variance σ^2, it is easily seen that $E\{\psi(X_1,X_2)\} = \sigma^2$. We may suppose that the mean of each X_i is zero; if not, we may subtract the common mean μ from both X_1 and X_2 without changing $\psi(X_1,X_2)$. Then $E\{\frac{1}{2}(X_1 - X_2)^2\} = \frac{1}{2}E(X_1^2 + X_2^2) = \frac{1}{2}(\sigma^2 + \sigma^2) = \sigma^2$. The corresponding U-statistic is

$$U_n = \binom{n}{2}^{-1} \sum_{i<j} \tfrac{1}{2}(X_i - X_j)^2.$$

It turns out that $U_n = S_X^2$, the usual sample variance. This equality can be proved by a few lines of algebra (*Exercise 13*) or more quickly by noting that U_n and S_X^2 are both homogeneous, symmetric, quadratic functions of the X_i and so must be of the form $a\sum_i X_i^2 + b\sum_{i<j} X_iX_j$, $a'\sum_i X_i^2 + b'\sum_{i<j} X_iX_j$ respectively. Since both expressions are unbiased estimates of σ^2 for any value of μ it follows that $a = a'$ (set $\mu = 0$) and $b = b'$ (set $\sigma = 0$), so they must be identical.

A U-statistic is a sample average, but over quantities that are not mutually independent (unless $k = 1$) since terms corresponding to k-tuples $\alpha^{(1)}$ and $\alpha^{(2)}$ that have elements in common will be dependent. The covariance between the two such terms depends only on the number ℓ of elements they have in common. Writing, temporarily, $X_{[i,j]} = \{X_i,\ldots X_j\}$ for $i \le j$ to simplify the notation, we have, for $1 \le \ell < k$

$$\begin{aligned}
&E\{\psi(X_{[1,\ell]}, X_{[\ell+1,k]})\psi(X_{[1,\ell]}, X_{[k+1,k+l]}) - \theta^2\} \\
&= E[E\{\psi(X_{[1,\ell]}, X_{[\ell+1,k]})\psi(X_{[1,\ell]}, X_{[k+\ell],[k+2\ell]})|X_{[1,\ell]}\}] - \theta^2 \\
&= E[E\{\psi(X_{[1,\ell]}, X_{[\ell+1,k]})|X_{[1,\ell]}\}E\{\psi(X_{[1,\ell]}, X_{[k+\ell],[k+2\ell]})|X_{[1,\ell]}\}] - \theta^2 \\
&= \mathrm{var}[E\{\psi(X_{[1,k]})|X_{[1,\ell]}\}] = \sigma_\ell^2,
\end{aligned}$$

say. These terms are all finite if $\sigma_k^2 = \mathrm{var}\{\psi(X_1,\ldots,X_k)\} < \infty$ which we will now assume.

For $k < \ell$ the number of pairs of k-tuples with exactly l elements in common is

$$\binom{n}{k}\binom{k}{\ell}\binom{n-k}{k-\ell} = \frac{n!}{\ell!(k-\ell)!(k-\ell)!(n-2k+\ell)!}.$$

Pairs $(\alpha^{(1)},\alpha^{(2)})$ and $(\alpha^{(2)},\alpha^{(1)})$ are counted separately in the enumeration if $\alpha^{(1)}$ and $\alpha^{(2)}$ are not identical. This calculation gives an explicit formula for the variance of U in terms of the σ_ℓ^2, namely

$$\operatorname{var}(U_n) = \binom{n}{k}^{-2} \sum_{\ell=1}^{k} \binom{n}{k}\binom{k}{\ell}\binom{n-k}{k-\ell}\sigma_\ell^2 = \binom{n}{k}^{-1} \sum_{\ell=1}^{k} \binom{k}{\ell}\binom{n-k}{k-\ell}\sigma_\ell^2. \tag{3.11}$$

Since $n!/(n-k)! \sim n^k$ as $n \to \infty$, the ℓ'th term in the sum is $O(n^{k-\ell})$, which is decreasing in ℓ. The first term is

$$\begin{aligned} \binom{n}{k}^{-1}\binom{k}{1}\binom{n-k}{k-1}\sigma_1^2 &= \frac{k!(n-k)!}{n!}k\frac{(n-k)!}{(n-2k+1)!(k-1)!}\sigma_1^2 \\ &= \frac{k^2}{n}\sigma_1^2 + O(n^{-2}), \end{aligned}$$

and the terms in $l = 2, \ldots, k$ are of higher order in n. So

$$\operatorname{var}(U_n) = \frac{k^2}{n}\sigma_1^2 + O(n^{-2}).$$

The combinatorics here are a little intricate, but the essential point is that when n is large, only the term with $\ell = 1$ matters. The other terms are all of smaller order in n. We can use this fact to show that a U-statistic, suitably normalized, can be approximated closely by a normalized sum of independent identically distributed terms with remainder tending to zero in probability, that is, is asymptotically linear. The only condition needed is the finiteness of variance.

Write $\eta(X_1) = E\{\psi(X_1, \ldots, X_k)|X_1\}$ so that $E\{\eta(X_1)\} = \theta$, $\operatorname{var}\{\eta(X_1)\} = \sigma_1^2$, and set

$$A_n = \theta + \sum_{i=1}^{n}\{E(U_n|X_i) - \theta\}.$$

The proportion of all k-tuples from the integers $1, \ldots, n$ that contain a specific integer i is k/n, so

$$E(U_n|X_i) - \theta = \frac{k}{n}\{\eta(X_i) - \theta\}, \tag{3.12}$$

and

$$A_n = \theta + \frac{k}{n}\sum_{i=1}^{n}\{\eta(X_i) - \theta\}.$$

So A_n, although defined as a sum, is a sample average of independent and identically distributed random variables and has mean θ and variance $(k^2/n)\sigma_1^2$. These are the same to $o(n^{-1})$ as the mean and variance of U_n. We shall see that more is true, $\sqrt{n}(U_n - A_n) = o_p(1)$, so that $\sqrt{n}(U_n - \theta)$ has the same limiting distribution as $\sqrt{n}(A_n - \theta)$. Since the central limit theorem applies to the A_n, the common limiting distribution is normal with mean zero and variance $(k^2/n)\sigma_1^2$.

Theorem 3.2 *(Hoeffding): Let $X_1, X_2, \ldots$ be independent and identically distributed and let*

$$U_n = \frac{1}{\binom{n}{k}} \sum_{\alpha \in \mathcal{C}(n,k)} \psi(X_{\alpha_1}, \ldots, X_{\alpha_k})$$

where $\psi(x_1, \ldots, x_k)$ is a symmetric function of k arguments $x_1, \ldots, x_k$. Suppose that $\psi(X_1, \ldots, X_k)$ has mean θ and finite variance. Then

$$\frac{\sqrt{n}(U_n - \theta)}{k\sigma_1} \rightarrow_d N(0, 1),$$

where

$$\sigma_1^2 = \text{var}[E\{\psi(X_1, \ldots, X_k)|X_1\}],$$

(assumed to be strictly positive).

Note (i) It can happen that $\sigma_1^2 = 0$. See, for example, *Exercise 17*. In that case $\sqrt{n}(U_n - \theta) \rightarrow_p 0$ and multiplication by a higher power of n is needed to obtain a proper limiting distribution, which is typically non-normal.
Note (ii) The asymptotic variance σ_1^2 is *not* equal to $\text{var}\{\psi(X_1, \cdots, X_k)\}$ - in fact this quantity does not enter into the limiting distribution, although it does enter into the exact finite-sample variance formula (3.11).

Proof To show that $\sqrt{n}(U_n - A_n) = o_p(1)$ it is sufficient to show that $E(U_n - A_n)^2 = o(n^{-1})$ and apply Chebichef's inequality.

From the previous work, we have

$$\text{var}(A_n) = \frac{k^2}{n}\sigma_1^2,$$

$$\text{var}(U_n) = \frac{k^2}{n}\sigma_1^2 + o(n^{-1}),$$

so the required result will follow if it can be shown that

$$\text{cov}(A_n, U_n) = \frac{k^2}{n}\sigma_1^2 + o(n^{-1}).$$

Now

$$\begin{aligned}
E(A_n - \theta)(U_n - \theta) &= \frac{k}{n}\sum_{i=1}^{n} E[\{\eta(X_i) - \theta\}(U_n - \theta)] \\
&= kE[\{\eta(X_1) - \theta\}(U_n - \theta)] \\
&= kE\left(E[\{\eta(X_1) - \theta\}(U_n - \theta)|X_1]\right) \\
&= kE\left[\{\eta(X_1) - \theta\}E(U_n - \theta|X_1)\right] \\
&= \frac{k^2}{n}E\{\eta(X_1) - \theta\}^2 = \frac{k^2}{n}\sigma_1^2
\end{aligned}$$

using the iterated expectation formula and Equation (3.12). This completes the proof. □

3.5 Gâteaux Differentiation

As indicated earlier, the Gâteaux derivative of a function $\nu(F)$ of the distribution F is the function of y obtained by differentiating the mixture distribution $F_{\epsilon,x}(y) = (1-\epsilon)F(y) + \epsilon\mathbf{1}[y \geq x]$, where $0 \leq \epsilon \leq 1$, in x. Keep in mind that y denotes the argument of the distribution function $F(\cdot)$, while x is the location (direction) of the perturbation of F of interest. The function $\mathbf{1}[y \geq x]$, as a function of y for a fixed x, takes the value zero for $y < x$ and one for $y \geq x$. Explicitly,

$$\nu'(F)(x) = \lim_{\epsilon \to 0+} \frac{\partial \nu(F_{x,\epsilon})}{\partial \epsilon}.$$

As examples, we consider (a) the sample mean, (b) the sample quantile, (c) the α - trimmed mean and (d) the so-called "Winsorized mean", a variant of the trimmed mean. Of these the sample mean needs no introduction, and the sample quantile has already been studied in Section 3.3. The trimmed mean and, to a lesser extent, the Winsorized mean are important in *robust estimation.* As will be seen (and it is obvious from the way they are defined), they are less sensitive than the ordinary mean to outlying observations. For a specified proportion α $(0 \leq \alpha \leq 0.5)$ the α-trimmed mean is the mean of the middle $1-2\alpha$ observations, that is, the mean of the remaining observations after those in the upper α and lower α tails of the sample are omitted. The extreme cases $\alpha = 0$ and $\alpha = 0.5$ correspond to the sample mean and sample median respectively. If $F(\cdot) = F_0(\cdot - \theta)$, where $F_0(x) + F_0(-x) = 1$, so that $F(\cdot)$ represents a distribution that is symmetric about θ, then $\nu(F) = \theta$. The Winsorized mean is similar to the trimmed mean, except that the observations in the lower and upper tails of the distribution are not omitted but instead moved inwards to the α'th and $(1-\alpha)$'th sample quantiles, respectively.

Example 3.1 *(a): Sample Mean .*

Here $\nu(F) = \int y dF(y) = \mu$, the population mean, so that with

$$F_{\epsilon,x}(y) = (1-\epsilon)F(y) + \epsilon\mathbf{1}[y \geq x],$$

$$\nu(F_{\epsilon,x}) = (1-\epsilon)\mu + \epsilon \int y \cdot d\mathbf{1}[y \geq x] = (1-\epsilon)\mu + \epsilon x.$$

since the (Stieltjes) integral with respect to the distribution $d\mathbf{1}[y \geq x]$ simply replaces y by x in the integrand. Hence $\nu'(F) = x - \mu$. Here the representation (3.7) is exact, with R_n identically zero.

Example 3.2 *(b): Sample Quantile.*

Suppose that $\nu(F) = Q(\alpha)$, the α quantile of $F(y)$ and that $F(y)$ has density $f(y)$ at $y = Q(\alpha)$. Recall that by the inverse function theorem, $Q'(u) = 1/F'\{Q(u)\} = 1/f\{Q(u)\}$ at any point where the derivative exists.

For $x \leq Q(\alpha)$, $F_{\epsilon,x}\{Q(\alpha)\} = \epsilon + (1-\epsilon)\alpha$ and for $x > Q(\alpha)$, $F_{\epsilon,x}\{Q(\alpha)\} = (1-\epsilon)\alpha$. Dropping the subscript x for simplicity (x is fixed in what follows),

$$\frac{\partial}{\partial \epsilon} F_\epsilon\{Q(\alpha)\} = \mathbf{1}[x \leq Q(\alpha)] - \alpha.$$

Now

$$\nu'(F) = \lim_{\epsilon \to 0} \frac{\partial}{\partial \epsilon} Q_\epsilon(\alpha).$$

To calculate this partial derivative, we note that, for each ϵ, $F_\epsilon(\cdot)$ and $Q_\epsilon(\cdot)$ are inverse functions so that

$$Q_\epsilon\{F_\epsilon(y)\} \equiv y.$$

This identity can be used to express $\partial Q_\epsilon / \partial \epsilon$ in terms of F_ϵ and its derivatives. If y is fixed but ϵ is allowed to vary, the total derivative of $Q_\epsilon\{F_\epsilon(y)\}$ is zero.

Now ϵ enters the left side equation both directly as a subscripted argument of Q and indirectly as a subscripted argument of F, so the total derivative in ϵ is the sum of the corresponding two terms. We obtain

$$\frac{\partial Q_\epsilon\{F_\epsilon(y)\}}{\partial \epsilon} + Q'_\epsilon\{F_\epsilon(y)\} \frac{\partial F_\epsilon(y)}{\partial \epsilon} = 0.$$

Putting $y = Q(\alpha)$ and letting $\epsilon \to 0+$ gives

$$\lim_{\epsilon \to 0+} \frac{\partial Q_\epsilon(\alpha)}{\partial \epsilon} + \frac{\mathbf{1}[x \leq Q(\alpha)] - \alpha}{f\{Q(\alpha)\}} = 0,$$

so that

$$\nu'(F) = \lim_{\epsilon \to 0} \frac{\partial Q_\epsilon(\alpha)}{\partial \epsilon} = -\frac{\mathbf{1}[x \leq Q(\alpha)] - \alpha}{f\{Q(\alpha)\}}.$$

This is the same influence function as was derived using the Bahadur representation in Section 3.3. Note that $E_F\{\nu'(F)(X)\} = 0$.

Example 3.3 *(c): Trimmed Mean.*

Suppose that F is symmetric about its median *MED* say and continuous with density $f(\cdot)$. Write $Q(\cdot)$ for the quantile function, so that $F\{Q(u)\} = u$ for $0 < u < 1$. Then the functional $\nu(F)$ is

$$\nu(F) = \frac{1}{1-2\alpha} \int_{Q(\alpha)}^{Q(1-\alpha)} x dF(x) = \frac{1}{1-2\alpha} \int_\alpha^{1-\alpha} Q(u) du,$$

making the transformation $x = Q(u)$ in the integral. Since the sample distribution function F_n is necessarily discrete, there is some minor ambiguity over the definition of $\nu(F_n)$. A natural choice is

$$\nu(F_n) = \frac{1}{n - 2\lfloor \alpha n \rfloor} \sum_{k=\lfloor \alpha n \rfloor}^{n-\lfloor \alpha n \rfloor} X_{(k)},$$

where the $X_{(k)}$ are the ordered observations $X_{(1)} < \ldots < X_{(n)}$, and $\lfloor\cdot\rfloor$ is the floor function. However, this ambiguity does not affect the asymptotic properties of the estimate.

As discussed in Example 3.2 (*b*), the derivative of $Q_\epsilon(u)$ in ϵ at $\epsilon = 0$ is

$$-\frac{\mathbf{1}[x \leq Q(u)] - u}{f\{Q(u)\}}.$$

Assuming that differentiation in ϵ and integration over u can be interchanged,

$$\nu'(F) = -\frac{1}{1-2\alpha}\int_\alpha^{1-\alpha} \frac{\mathbf{1}[x \leq Q(u)] - u}{f\{Q(u)\}} du.$$

The change of variable $y = Q(u)$ — note that we must distinguish between this transformed variable of integration and the argument x of $\nu'(F)$ — gives

$$\nu'(F) = -\frac{1}{1-2\alpha}\int_{Q(\alpha)}^{Q(1-\alpha)} \{\mathbf{1}[x \leq y] - F(y)\} dy.$$

To evaluate the integral of the first term, we must consider the three cases (i) $x \leq Q(\alpha)$, (ii) $Q(\alpha) < x \leq Q(1-\alpha)$, (iii) $x > Q(1-\alpha)$ separately. In (i) $\mathbf{1}[x \leq y] = 1$ for all relevant values of y, so the integral is $Q(1-\alpha) - Q(\alpha)$. In (iii) $\mathbf{1}[x \leq y] = 0$ for all relevant values of y so its integral is also zero. In (ii) the indicator function is unity over the range $x < y \leq Q(1-\alpha)$ and is zero otherwise, so the integral is $Q(1-\alpha) - x$. The second term does not depend on x, and using the symmetry of the distribution F about its median MED we have that $Q(\alpha) + Q(1-\alpha) = 2MED$, and that the mean value of $F(x)$ over the range $Q(\alpha), Q(1-\alpha)$ is $\frac{1}{2}$, so putting the two terms together and simplifying gives

$$(1-2\alpha)\nu'(F)(x) = \begin{cases} Q(\alpha) - MED & \text{if } x < Q(\alpha); \\ x - MED) & \text{if } Q(\alpha) \leq x \leq Q(1-\alpha); \\ Q(1-\alpha) - MED & \text{if } x > Q(1-\alpha). \end{cases}$$

The graph of $\nu'(F)(x)$ against x consists of horizontal lines in the tail regions outside the two selected quantiles, joined by a linear segment in the region between them, see the solid curve in Figure 3.2. It is continuous.

A rigorous proof of the asserted asymptotic normality of the trimmed mean can be constructed from the representation

$$\int_{Q_n(\alpha)}^{Q_n(1-\alpha)} x dF_n(x) = \int_{Q(\alpha)}^{Q(1-\alpha)} x dF_n(x) + \int_{Q_n(\alpha)}^{Q(\alpha)} x dF_n(x) + \int_{Q_n(1-\alpha)}^{Q(1-\alpha)} x dF_n(x),$$

and showing that the sum of the last two terms may be approximated by

$$\int_{Q_n(\alpha)}^{Q(\alpha)} x dF(x) + \int_{Q_n(1-\alpha)}^{Q(1-\alpha)} x dF(x).$$

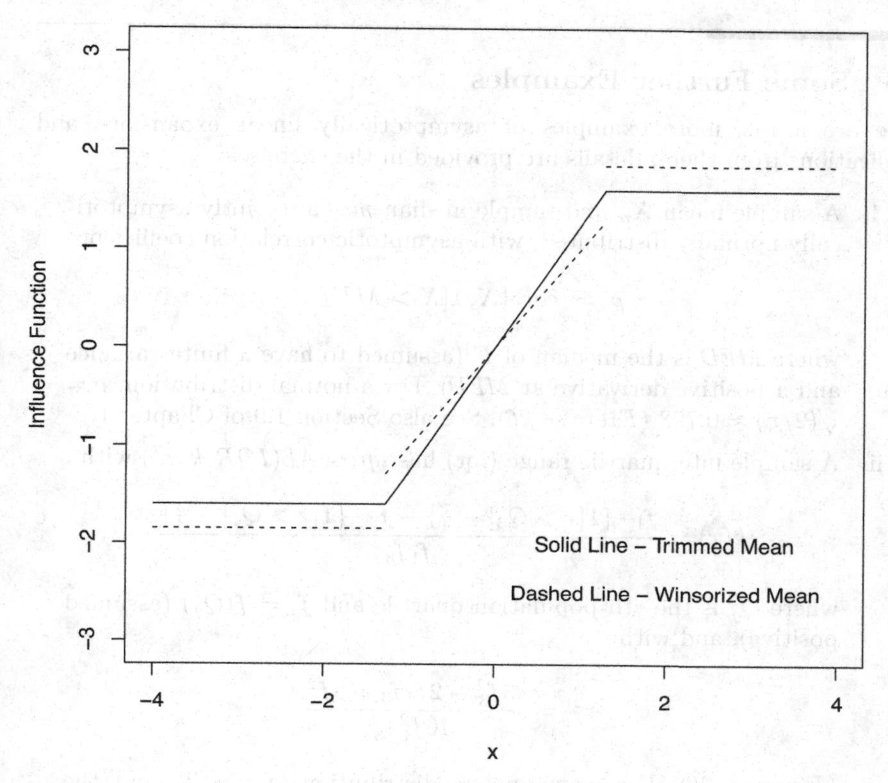

FIGURE 3.2
Influence functions, alpha = 0.1.

The first term is the average of truncated variables $X_i' = X_i \mathbf{1}[Q_\alpha \leq X_i \leq Q_{1-\alpha}]$, so the ordinary central limit theorem applies. The Bahadur representation of the sample quantile provides asymptotically linear representations of the two remaining terms. Notice that these two terms are not themselves asymptotically negligible.

Example 3.4 *(d): Winsorized Mean.*

The Winsorized mean, which shrinks the observations in the lower and upper tails to the α'th and $1-\alpha$'th quantile respectively, can be expressed as a weighted average of the trimmed mean and the two sample quantiles, with weights $1-2\alpha$, α and α, respectively. Its influence function is the corresponding average of the individual influence functions.

The two quantiles combined give an influence function which is negative for $x < Q(\alpha)$, zero for $Q(\alpha) < x < Q(1-\alpha)$ and positive for $x > Q(1-\alpha)$. So the full influence curve is linear between $Q(\alpha)$ and $Q(1-\alpha)$, has jumps at these two points, and is constant in each of the two tail regions outside these quantiles. See the dashed curve in Figure 3.2.

3.6 Some Further Examples

Here are a few more examples of asymptotically linear expansions and implications from them; details are provided in the exercises:

i. A sample mean $\bar{X}_n$ and sample median *med* are jointly asymptotically normally distributed, with asymptotic correlation coefficient

$$\rho = \text{corr}\{X, \mathbf{1}[X > MED]\},$$

where *MED* is the median of F (assumed to have a finite variance and a positive derivative at *MED*). For a normal distribution, $\rho = \sqrt{(2/\pi)} \approx 0.798$ (*Exercise 26*). See also Section 1.9 of Chapter 1.

ii. A sample interquartile range (iqr) has $iqr \sim AL(IQR, k, \tau^2)$ with

$$k(x) = \frac{f_1 \cdot \{\mathbf{1}[x > Q_3] - \frac{1}{4}\} - f_3 \cdot \{\mathbf{1}[x > Q_1] - \frac{3}{4}\}}{f_1 f_3}$$

where Q_j is the jth population quartile and $f_j = f(Q_j)$ (assumed positive), and with

$$\tau^2 = \frac{3f_3^2 - 2f_1 f_3 + 3f_1^2}{16 f_1^2 f_3^2}$$

(*Exercise 20*). For a symmetric distribution, $f_3 = f_1$ and the expression simplifies to $\tau^2 = 1/(4f_1^2)$.

iii. A sample mean and variance, and a sample mean and standard deviation, each has large-sample correlation coefficient $SK/\sqrt{KUR+2}$ where *SK* is the population *coefficient of skewness* μ_3/σ^3 and KUR is the population *coefficient of excess kurtosis* $\frac{\mu_4}{\sigma^4} - 3$. Hence, they are asymptotically independent if and only if $\mu_3 = 0$ (assuming that $\mu_4 < \infty$) (*Exercise 21*).

iv. Consider a random sample of (X, Y) values with the conditional distribution of Y given $X = x$ having density $f(y - \alpha - \beta x)$ for some density f, and suppose the marginal density of X is g. Then the *least squares* estimates of the *regression coefficients* α and β are asymptotically linear (*Exercise 24*). Note, however, that inference about the regression parameters is usually done conditionally on given values for the X's, whereas the limiting argument here requires that the X's be random as well as the Y's.

We end this section by illustrating asymptotic linear expansions of two alternative estimates of a parameter in a parametric model and the comparisons afforded thereby. Consider a random sample from an exponential distribution with rate parameter θ; that is, the density is $f_\theta(x) = \theta \exp(-\theta x)$ for $x > 0$.

The maximum likelihood estimate, the same as a moment estimate, of θ is $\hat{\theta} = 1/\bar{X}$. Using Proposition 3.1, we find that $\hat{\theta} \sim AL\{\theta, -\theta^2(x - 1/\theta), \theta^2\}$. For comparison, a quantile estimate, based on medians, is $\tilde{\theta} = \log 2/med$. (Maximum likelihood, moment, and quantile estimates will be defined in Chapter 5). Again using Proposition 3.1, we find that $\tilde{\theta} \sim AL\{\theta, -2\theta/\log 2 \cdot (\mathbf{1}[x > \log 2/\theta] - \frac{1}{2}), (\theta/\log 2)^2\}$. The ratio of large-sample variances, a measure of *relative efficiency* (see Chapter 5), is therefore $(\log 2)^2 \approx 0.48$; that is, the large-sample variance of $\tilde{\theta}$ is about twice that of $\hat{\theta}$.

The large-sample correlation coefficient ρ between these two estimates is that between their respective kernels. Equivalently, $\rho = \operatorname{corr}\big(X, \mathbf{1}[(X > MED]\big)$. We do the calculation with $\theta = 1$, but the same result holds for general θ. We find that $E(X \cdot \mathbf{1}[X > \log 2]) = \frac{1}{2}(1 + \log 2)$ and $E(X) \cdot E(\mathbf{1}[X > \log 2]) = \frac{1}{2}$ so that $\rho = \frac{1}{2}\log 2/(1 \cdot \frac{1}{2}) = \log 2$. Thus, the relative efficiency, found above to be $(\log 2)^2$, coincides with ρ^2. This coincidence of relative efficiency with a squared correlation coefficient will appear again; see Chapter 6.

3.7 Final Comments

We have stated that many statistics are approximately normally distributed; in fact, statistics T_n, T_n^2 and $\log T_n$ may all three be approximately normally distributed! These are only 'first order' approximations, and one of these normal approximations may be much better than another, for any particular sample size n. One consideration is to transform the statistic so that its range is not grossly limited; this will often lead to a better normal approximation. Thus, if T_n is strictly positive, and with small values not entirely unlikely, $\log T_n$ is a likely candidate for an improved normal approximation; examples include sample variances, sample standard deviations, and ratios of asymptotically linear positive-valued statistics. If $T_n \in (0, 1)$, a logit transform might be considered. For a sample correlation coefficient r, which must lie in the range $-1 \leq r \leq 1$, R. A. Fisher proposed the transformation $z = \frac{1}{2}\log\{(1 + r)/(1 - r)\}$, see *Exercise 34*. Characteristics of the distribution of the original T_n may then be obtained by back transformation. In particular applications, it is usually wise, when feasible, to explore the quality of any approximation, or to perform a second-order analysis. (See Kolassa (1997), for example.) Simulations are often helpful.

What happens if the observations are not independent and identically distributed? Many useful statistics may be represented as the average of independent, but not identically distributed, terms, plus an error of smaller order. Linear regression coefficients provide examples, when the explanatory variables are treated as non-random. However, additional assumptions are needed to conclude weak law and central limit consequences; see Section 2.7 of Chapter 2. We limit the terminology "asymptotically linear" to the case of independent and identically distributed observations. Extensions to two or more independent random samples are fairly straightforward, but not detailed here.

Exercises

1. Estimation of third central moments

   (a) Show that the third central moment about the population mean μ, satisfies $\sum(X_i - \mu)^3/n \sim AL\{\mu_3, (x-\mu)^3 - \mu_3, \mu_6 - \mu_3^2\}$.

   (b) Show that the third central moment about the sample mean $\bar{X}$ satisfies

   $$\sum(X_i - \bar{X}_i)^3/n \sim AL\{\mu_3, (x-\mu)^3 - \mu_3 - 3\sigma^2(x-\mu), \mu_6 - \mu_3^2 - 6\sigma^2\mu_4 + 9\sigma^6\}.$$

   Using divisor n instead of n does not affect this result. [*Hint:* Use the expression $a^3 - b^3 = (a-b)(a^2 + ab + b^2)$ with $a = X_i - \bar{X}$, $b = X_i - \mu$, sum over i, note that $a - b$ does not depend on i and show that the sample means of the terms a^2, ab and b^2 all converge in probability to σ^2].

   (c) Show that if $2\mu_4 > 3\sigma^4$, an inequality which holds when the X_i are normally distributed, substitution of the estimate $\bar{X}$ for the true value μ *reduces* the variance of the asymptotic distribution. For another example of this phenomenon, see *Exercise 33*.

2. Use Proposition 3.2 to verify the asymptotically linear representation of m_{2n}.

3. *Continuation of Exercise 1.* Show that the sample skewness coefficient $sk \equiv m_{3n}/m_{2n}^{3/2}$, where m_{rn} is the sample rth central moment, has $sk \sim AL\{SK, y^3 - SK - 3y - \frac{3}{2}SK(y^2-1), \tau^2\}$ where $SK = \mu_3/\sigma^3$, the population skewness coefficient, $y = (x-\mu)/\sigma$ and τ^2 is a function of λ_3, λ_4, λ_5 and λ_6 where $\lambda_r = \mu_r/\sigma^r = E(Y^r)$.

4. The *sample coefficient of excess kurtosis* is $kur \equiv (m_{4n}/m_{2n}^2) - 3$ where m_{rn} is the sample rth central moment. Derive its asymptotically linear expansion.

5. Consider a bivariate random sample of (X, Y) values. Let p_n be the proportion of them for which $\{X \in A\} \cap \{Y \in B\}$ and q_n the proportion for which $\{X \in A\}$. Let $r_n \equiv p_n/q_n$, the proportion of those observations with $X \in A$ which also have $Y \in B$. Derive the asymptotically linear expansion for r_n. [Note that r_n is a sample conditional probability.]

6. Prove Proposition 3.2 (Asymptotic linearity of differentiable functions of two or more asymptotically linear statistics).

7. Extend Proposition 3.2 (or equivalently, Proposition 3.1) to a vector-valued version, with $\mathbf{T}_X$ d-dimensional and $\mathbf{h}(\cdot): \mathcal{R}^d \to \mathcal{R}^r$ for some $r \leq d$.

8. (a) What is the large-sample distribution of the sample median from a *Weibull distribution*?

   (b) Repeat (a) when the distribution is modified so that any observation greater than a specified value c is set equal to c. Consider the cases $F(c) > \frac{1}{2}$, $F(c) = \frac{1}{2}$ and $F(c) < \frac{1}{2}$.

9. (a) Show that the asymptotic correlation (Acorr say) between two sample quantiles is the same as that between the sample distribution functions evaluated at the corresponding population quantiles, that is, for $p < q$

$$\begin{aligned} \mathrm{Acorr}\{Q_n(p), Q_n(q)\} &= \mathrm{Acorr}[F_n\{Q(p)\}, F_n\{Q_n(q)\}] \\ &= \sqrt{[p(1-q)/\{(1-p)q\}]}. \end{aligned}$$

   In particular, the asymptotic correlation coefficient between the first and third sample quantiles is $\frac{1}{3}$.

   (b) Derive the asymptotically linear expansion for the *mid-quartile range*, m_1 say, defined as the average of the first and third sample quartiles. Note that this is actually a measure of location.

10. Refer to *Exercise 21* in Chapter 1 for some relevant results. Consider a sequence $(X_1, Y_1), (X_2, Y_2), \ldots$ of independent observations from a bivariate normal distribution with means zero, unit standard deviations and correlation coefficient $\rho \geq 0$.

    (a) Let A_n denote the sample median of $X_1, \ldots, X_n$ and B_n the sample median of $Y_1, \ldots, Y_n$. Write down the asymptotic linear expansions for A_n and B_n and use these to find the asymptotic joint distribution of (A_n, B_n).

    (b) By considering the form of the graph of the arcsine function over $(0, 1)$ show that the asymptotic correlation coefficient between the two sample medians is always less than or equal in magnitude to that between the two sample means, with equality if and only if $\rho = 0$ or $\rho = 1$.

    (c) What is the asymptotic joint distribution of $(\bar{X}_n, B_n)$?

11. (a) Find the asymptotically linear representation of the weighted average $m_\lambda = \lambda med + (1-\lambda)m_1$ of the sample median m and the mid-quartile range m_1 (see *Exercise 9*). Show that for both the Cauchy distribution with density $f(x;\mu) = 1/[\pi\{1 + (x-\mu)^2\}]$ and the Laplace distribution with density $f(x;\mu) = \frac{1}{2}\exp\{-|x-\mu|\}$ the

optimal choice of λ (in the sense of minimizing the variance of m_λ) is $\lambda = 1$. Note the result of *Exercise 14* in Chapter 1.

(b) Show also that for the Laplace distribution, but not the Cauchy, this result is still true if we replace the mid-range by the average of any two complementary quartiles (say the αth and $1 - \alpha th$).

(c) Show that for a distribution with $f_1 = f_2 = f_3$, for example any density that is constant over an interval including the first and third quartiles, the optimal choice of λ is $\lambda = 0$.

12. Consider the *skewed Laplace distribution* with density $f(x, \theta) = f_0(x - \theta)$ where

$$f_0(x) = \begin{cases} \frac{2}{3}e^{2x} & \text{if } x \leq 0; \\ \frac{2}{3}e^{-x} & \text{if } x > 0; \end{cases}$$

(a) Sketch the function $f_0(x)$ and verify that it integrates to unity. Show that the median (M) and upper (U) and lower (L) quartiles of the distribution with this density are $M = \log(4/3)$, $U = \log(8/3)$ and $L = \frac{1}{2}\log(3/4)$, and that the corresponding quantile density functions are $f_0(M) = \frac{1}{2}$, $f_0(U) = \frac{1}{4}$, $f_0(L) = \frac{1}{2}$.

(b) Let $(\hat{M}_n, \hat{U}_n, \hat{L}_n)$ denote the corresponding sample quartiles calculated from a random sample of n independent observations from the distribution with density $f(x, \theta)$. Use the Bahadur representation to calculate the asymptotic joint distribution of $\sqrt{n}(\hat{L}_n - L, \hat{M}_n - M, \hat{U}_n - U)$, including its full covariance matrix.

(c) Consider the estimator $\tilde{\theta} = a\hat{L}_n + b\hat{M}_n + c\hat{U}_n + d$ of θ, where a, b, c, d are specified constants. What conditions on these constants are needed to ensure consistency of this estimator?

(d) Find the values of a, b, c that minimize the asymptotic variance of $\tilde{\theta}$ [*Hint:* Use the asymptotic covariance matrix from (b) to derive a quadratic form in a, b, c for the asymptotic variance of $\tilde{\theta}$. Use the condition in part (c) to reduce this to a function of a and c only and show that for any a, the optimal value of c is zero. Now find the optimal a].

13. Verify directly that

$$S_X^2 = \binom{n}{2}^{-1} \sum_{i<j} \tfrac{1}{2}(X_i - X_j)^2.$$

14. (a) Suppose that $0 < a < 1$ and U is uniformly distributed over $(0, 1)$. Find the density of $X = |U - a|$ and show that $E(X) = \frac{1}{2} - a(1 - a)$. [*Hint:* First consider the case that $a < \frac{1}{2}$].

(b) Now suppose that a is replaced by a random variable A, also uniformly distributed over $(0, 1)$. Show that $E(X) = 1/3$.

(c) Suppose now that $U_1, U_2, \ldots, U_n$ are independent and identically distributed uniform random variables over $(0,1)$ and let

$$U_n = \frac{1}{\binom{n}{2}} \sum_{i<j} |U_i - U_j|.$$

Show that $\sqrt{n}(U_n - 1/3)$ is asymptotically normal with mean zero and variance $1/45$.

15. Verify that if X_1, X_2 and X_3 are independent and identically distributed with finite third moment then $T = X_1(X_1 - X_2)(X_1 - 2X_3)$ is an unbiased estimator of the third population cumulant $\kappa_3 = E(X - \mu)^3$. For a random sample of size n use T to propose a U-statistic that is an unbiased estimate of κ_3. Note that T is not symmetric in its arguments, but may easily be made so. (R. A. Fisher derived symmetric unbiased estimators of population cumulants called k-statistics; it can be shown that each k-statistic is a U-statistic).

16. Suppose that $X_1, X_2, \ldots, X_n$ are independent and identically distributed with a Pareto distribution,

$$F(x) = 1 - \left(\frac{1}{1+x}\right)^{\beta}, \quad (x > 0, \beta > 0).$$

(a) What is the distribution of $\min(X_1, X_2)$?

(b) Consider the U-statistic

$$U_n = \binom{n}{2}^{-1} \sum_{i<j} \min(X_i, X_j).$$

Show that U_n has (i) finite mean if and only if $\beta > \frac{1}{2}$ and (ii) finite variance if and only if $\beta > 1$.

(c) Calculate $E\{\min(X_1, X_2)|X_1 = x\}$. Consider the three cases $\beta < 1$, $\beta = 1$ and $\beta > 1$ separately. [*Hint:* First show that if the random variable X has survivor function $\bar{F}(\cdot)$ then $E\{\min(X, c)\} = \int_0^c \bar{F}(x)dx$].

(d) Show that $\sigma_1^2 = \text{var}[E\{\min(X_1, X_2)|X_1\}]$ is finite if and only if $\beta > 2/3$.

17. Suppose that $X_1, X_2, \ldots, X_n$ are independent and identically distributed, each uniform over $(0, 2\pi)$. Let

$$U = \binom{n}{2}^{-1} \sum_{i<j} \cos(X_i - X_j).$$

This might be proposed as a test statistic of the hypothesis of uniformity of data distributed on the unit circle against the possibility of clustering, since values of $X_i - X_j$ close to zero (mod 2π) will give values of $\cos(X_i - X_j)$ close to its maximum possible value of unity.

(a) Show that under the asserted uniform distribution, $E(U) = 0$ and also $\sigma_1^2 = 0$.

(b) Show that nU_n has a non-degenerate limiting distribution of the form $\chi_2^2 - 1$ (i.e. $nU_n + 1 \to_d \chi_2^2$). The limiting distribution has mean zero as it should, but is very far from symmetric. This reflects the fact that the n points on the circle could all be very close to each other but cannot all be very far from each other. [*Hint:* Use the identity $\cos(A - B) = \cos(A)\cos(B) + \sin(A)\sin(B)$ and the bivariate central limit theorem applied to $\{\sum \cos(X_i), \sum \sin(X_i)\}$.]

18. This question relates to the Wilcoxon signed-rank statistic. The connection will be discussed and more detailed calculations presented in Chapter 8. Suppose that $X_1, \ldots, X_n$ are independent and identically distributed with symmetric density $f(x)$, so that $f(x) = f(-x)$. Consider the "Walsh averages" $Y_{ij} = \frac{1}{2}(X_i + X_j)$, defined here just for $i < j$.

    (a) Show that $\text{pr}(Y_{12} > 0) = \frac{1}{2}$ and find an expression in terms of $f(\cdot)$ for $\text{pr}(Y_{12} > 0, Y_{13} > 0)$.

    (b) Let U_n denote the proportion of all Walsh averages that are negative. Appeal to U-statistic theory to find the limiting distribution of U_n.

    (c) In the special case that the X_i are uniformly distributed (over $[-1, 1]$ say) show that $U_n - \bar{X}_n = o_p(1/\sqrt{n})$.

    (d) Find an exact expression for $\text{var}(U_n)$ in this case. How large must n be for this to be within 10% of the asymptotic variance?

19. (Asymptotic joint distribution of the sample mean and median). Verify (i) in Section 3.6.

20. (Asymptotic distribution of the sample interquartile range). Verify (ii) in Section 3.6.

21. (Asymptotic joint distribution of the sample mean and variance). Verify (iii) in Section 3.6.

22. Determine asymptotically linear expansions

    (a) for $\bar{X}_n$ and S_n when sampling from a two-parameter *gamma distribution*;

    (b) for *med* and *iqr* from a two-parameter *Cauchy distribution*;

    (c) for $\bar{X}_n$ and *med* from a two-parameter *Laplace distribution*.

In each case, determine the large-sample correlation coeffcicient between the two statistics

23. Under what conditions are the sample median and the sample interquartile range asymptotically independent?

24. (Linear regression) Verify (iv) in Section 3.6.

25. Show that if $T_n \sim AL\{0, k(\cdot), \tau^2\}$, then nT_n^2/τ^2 converges in distribution to chi-squared with one degree of freedom.

26. (a) For standard normal X, show that the covariance of X and $\mathbf{1}[X > 0]$ is $1/\surd(2\pi)$, and hence that the corresponding correlation coefficient is $\surd(2/\pi)$. [*Hint:* The indefinite integral of $x\varphi(x)$ is $-\varphi(x)$, with ϕ the standard normal density.]

    (b) Extend (a) to the case of $X \sim N(\mu, \sigma^2)$.

    (c) What are the implications about the relative efficiency of the sample median as an estimate of the population mean when sampling from a normal distribution?

27. Consider two estimates of the *pth* quantile Q_p of a normal distribution with unknown mean μ and unknown standard devation σ, based on a random sample $X_1, \ldots, X_n$ from the distribution:
    (i) $\hat{Q}_{p1} = \bar{X} + z_p S_X$, where $\bar{X}$ and S_X are the sample mean and sample standard deviation of the X_i and z_p is the population quantile of the standard normal distribution;
    (ii) $\tilde{Q}_{p2}$, the *pth* sample quantile of the X_i.

    Calculate the asymptotic variances σ_{p1}^2 of $\hat{Q}_{p1}$ and σ_{p2}^2 of $\hat{Q}_{p2}$. Note that the ratio $\sigma_{p1}^2/\sigma_{p2}^2$ is free of μ and σ, so depends only on p and plot it against against p in the range $0.01 \leq p \leq 0.99$.

    Even when $\hat{Q}_p$ has a smaller variance than $\tilde{Q}_p$ it may be preferable when the assumption of normality is not certain. Roughly, for what values of p would this be reasonable? Briefly explain your choice[1]

28. Consider sampling from a two-parameter *Laplace distribution*, centered at θ with known scale parameter. Compare the large-sample variances of the sample mean and the sample median as estimates of θ. (Compare with *Exercise 26c.*)

29. Many asymptotically linear expansions have been illustrated. Consider each in turn and determine under what conditions the corresponding statistic is *robust*, in the sense that its influence function is bounded (Section 3.1).

[1]I thank Anthony Almudevar for this interesting exercise—something to get one's teeth into, as it were.

30. Asymptotically linear statistics have asymptotic variances of order $1/n$. Show that, if $T_n \sim AL\{0, k(\cdot), \tau^2\}$, then T_n^2 has asymptotic variance of order $1/n^2$ and hence is not asymptotically linear [*Note:* We say: 'the asymptotic variance of T_n is of order $1/n^r$' if, for some μ_n, $n^{r/2}(T_n - \mu_n)$ converges in distribution to a distribution with finite positive variance. Note that this is not the same as saying that the variance of T_n is of order $1/n^r$.]

31. Consider a random sample of size n from a distribution F on $\mathcal{R}^+$ with $F(0) = 0$ and $dF(x)/dx\big|_{x=0} = f_0 > 0$; for example F may be exponential or uniform. Show that the first order statistic $X_{(1)n} \equiv \min\{X_1, \ldots, X_n\}$ has asymptotic variance of order $1/n^2$ (as defined in *Exercise 30*) and hence is not asymptotically linear. [*Hint:* Start by considering the case that F is uniform on $(0, 1)$].

32. Here, we compare, theoretically and empirically (by Monte Carlo), the quality of a normal approximation to the distribution of the sample standard deviation S_X to those derived from normal approximations to various transforms of S_X.

   (a) For a sample of size 20 from a standard normal distribution, determine the probability π that the sample standard deviation exceeds 1.25 (by reference to the appropriate chi-squared distribution). Then determine π approximately by reference to (i) the large-sample distribution of the sample standard deviation, to (ii) the large-sample distribution of the sample variance and to (iii) the large-sample distribution of $\log S_X$. Which approximation is best for this particular case? Repeat for samples of size 50.

   (b) Use R to generate 1000 samples of size 20 from a normal distribution. In each of the 1000 samples, find the sample standard deviation S_X, its square v and its logarithm w. Examine histograms of the 1000 values of S_X, v and w. Which appears closer to normality? Repeat for samples of size 50.

   (c) Repeat (b) but sampling from an exponential distribution with unit variance.

   [*Note:* A *skewness-reducing transformation* — see WILSON & HILFERTY (1931) — is another possibility.]

33. Consider sampling pairs (X, Y) from a bivariate normal distribution with respective means μ_X and μ_Y, variances σ_X^2 and σ_Y^2 and correlation coefficient ρ.

   (a) Derive the asymptotically linear expansion for the sample correlation coefficient r. Show in particular that

$$\sqrt{n}(r - \rho) \to_d N\{0, (1 - \rho^2)^2\}.$$

(b) Suppose that the values of the parameters $\mu_1, \mu_2, \sigma_1, \sigma_2$ are known. It is tempting to substitute these values for the corresponding estimates into r, and use $\tilde{r} = \sum(x_i - \mu_X)(y_i - \mu_Y)/(n\sigma_1\sigma_2)$ to estimate ρ. Show however that if $\rho \neq 0$ this *increases* the asymptotic variance, to $1 + \rho^2$. See STUART (1955).

34. (Continuation)

(a) Fisher's z-transform for the correlation coefficient r is defined as $z(r) \equiv \tanh^{-1}(r) = \frac{1}{2}\log\{(1+r)/(1-r)\}$. What is its asymptotically linear expansion? [*Note:* R. A. Fisher showed that the distribution of $Z \equiv z(R)$ is very well approximated by $N\{z(\rho), 1/(n-3)\}$.]

(b) For the simulated data in Figure 1.2, $n = 200$ and $r = 0.811$. Verify that the calculation $(0.758, 0.854)$ of the 95% confidence interval for r using Fisher's transformation [*Hint:* Transform to the z-scale, evaluate the confidence interval on this scale and transform back to the r-scale].

35. (a) Approximate the probability that, in a sample of size $n = 100$ from a bivariate normal distribution with $\rho = 0$, the sample correlation coefficient R exceeds 0.2 (i) when using the raw data as in *Exercise 32(a)* and (ii) the transformed values as in *Exercise 33(a)*. Also compute the exact probability, obtainable from the fact (not proved here) that $R/\sqrt{\{(1-R^2)/(n-2)\}}$ is distributed as Student's t with $n-2$ degrees of freedom.

(b) Repeat the two approximations in (a), but now for the event $\{R > 0.7\}$ and assuming $\rho = 0.5$. (The exact distribution is not readily available when $\rho \neq 0$.)

(c) Generate 10000 samples of size $n = 100$ each from the bivariate normal distribution in (c) and check empirically the proportion of times $R > 0.2$. Repeat for the event $R > 0.7$ and $\rho = 0.5$ as in (b). Without loss of generality, set both population means to zero and both variances to unity. [*Hint:* Note that if U and V are independent standard normal, then $X = U$ and $Y \equiv \rho U + \sqrt{(1-\rho^2)}V$ are bivariate normal with unit standard deviations and correlation coefficient ρ. See *Exercise 27* in Chapter 1.]

36. Suppose that $(X_1, Y_1), (X_2, Y_2), \ldots, (X_n, Y_n)$ are independent pairs of observations from a five-parameter bivariate normal distribution. Suppose that interest centers on the ratio of variances, $\theta = \sigma_Y^2/\sigma_X^2$ say.

(a) Show that the corresponding ratio of sample variances, $\hat{\theta} = s_Y^2/s_X^2$, is asymptotically linear with $\tau^2 = 4\theta^2(1-\rho^2)$. Note that $\tau^2 \to 0$ as $\rho \to \pm 1$. This is to be expected. Why?

(b) Show that $\hat{\theta}$ is asymptotically independent of the sample correlation coefficient [*Note:* The exact joint distribution of $\hat{\theta}$ and r for a sample of size n can be calculated explicitly using results derived by R. A. Fisher. See KENDALL, STEWART & ORD (1977), Exercise 16.16. The two variables are not independent, but their dependence decreases to zero as n increases].

37. (a) From the asymptotically linear representation developed in Section 3.5 show that the asymptotic variance of the α-trimmed mean for a distribution with density $f(x-\theta)$, where $f(\cdot)$ is symmetric, is τ^2/n where,

$$\tau^2 = \frac{2}{(1-2\alpha)^2}\left\{\int_0^c x^2 f(x)dx + c^2\alpha\right\},$$

and $c = Q(1-\alpha) = F^{-1}(1-\alpha)$ with Q and F denoting the usual quantile and distribution functions from $f(x)$.

(b) Calculate this as a function of α when $f(x)$ is

(i) the Laplace density $\frac{1}{2}\exp(-|x|)$,

(ii) the standard normal density $\phi(x)$ (recall that the integral of $x^2\varphi(x) = -x\{d\varphi(x)/dx\}$ can expressed in terms of the $\Phi(x)$ using integration by parts) and

(iii) the Cauchy density $f(x) = \{\pi(1+x^2)\}^{-1}$. Recall that for the Cauchy distribution, $F(x) = (1/\pi)\tan^{-1}(x) + \frac{1}{2}$, so that $Q(1-\alpha) = \tan\{\pi(\frac{1}{2}-\alpha)\}$.

(c) Plot each of these as a function of α in the range $0.1 \leq \alpha \leq 0.5$.

4

Local Analysis

Our approach to the large-sample theory of statistical inference is through *local analysis*, involving the use of *moving parameters*. We start with an overview of the meaning of this and motivations for it. We then introduce Lucien LeCam's concept of *Local Asymptotic Normality*, and the attendant notions of *scores* and *information* developed by R. A. Fisher. Finally, we present *LeCam's third lemma*, the primary tool in deriving large-sample distributions under moving parameter assumptions. This chapter expands material in Hall & Mathiason (1990).

4.1 Motivation

First, we introduce some notation: The potential data are $\mathbf{X}_n$, with n an index of the amount of data, usually sample size. Typical data values are denoted by $\mathbf{x}_n$ and the actual observed data by $\mathbf{x}_n^*$. Usually, $\mathbf{X}_n = (X_1, \ldots, X_n)$, a random sample, but not necessarily.

The statistical model for the potential data is defined by a distribution with (joint) density $p_{n,\theta}(\mathbf{x}_n)$ with parameter $\theta \in \Theta$, an open subset of $\mathcal{R}^d$. We require that the model be *identifiable*, that is, $p_{n,\theta}(\cdot) \neq p_{n,\theta'}(\cdot)$ if $\theta \neq \theta'$. The model distribution may be absolutely continuous, discrete or mixed, but any discrete component must be confined to a parameter-free set $\mathcal{X}$ of mass points. More formally, the density is with respect to a parameter-free dominating measure. The object is to use the data $\mathbf{x}_n^*$ to make some inference about the unknown value of θ either to estimate it, or some function of it, or to test a hypothesis about it. Later chapters (Chapters 7 and 9), address the question of whether or not the assumed model is correct, but for now we assume that there is a true value of $\theta \in \Theta$ which generates the data. Boundary points of Θ are not included; for example the open interval $\Theta = (0, \infty)$ is allowed but the closed interval $[0, \infty)$ is not.

So far, the index n indicates the actual amount of data; perhaps we should have used n^* since we have embedded the actual inference problem in a sequence of problems indexed by n. In this sense, the model depends on n. The distinguishing feature of *local analysis* is to allow the value of the parameter θ also to depend on n, but to ensure that θ_n is close to an anchoring

DOI: 10.1201/9780429160080-4

value θ as $n \to \infty$. We will make this statement more precise shortly. We also assume that the actual n is large, large enough so that various limiting results, developed for the sequence of inference problems, will provide useful approximations for the actual inference problem. Important aspects of the theory of hypothesis testing and estimation, popular for much of the past century, require such a *moving parameter* model. Sometimes it may be useful to let the nature of the inference vary with n; thus, we might require that the *confidence coefficient* of a confidence interval increase with increasing sample size, or that the *significance level* of a test decrease with increasing sample size. However, we do not allow such possibilities here.

The need for moving parameter values θ_n in a theory of large-sample hypothesis testing is as follows. Typically, parameter estimates are approximately normally distributed with large-sample variances of order $O(1/n)$, and therefore standard error $O(1/\sqrt{n})$. Parameter values are then known, with high probability, to within, say, three standard errors of their true values. Therefore, if we want to test whether some real parameter has one of two specific values, we can determine the answer with near certainty with a sufficiently large n, one that is large enough to ensure that the sampling distributions of the estimate under the two hypotheses barely overlap. To make the problem non-trivial for increasing n, we move the two hypothesized values closer together, at a distance $O(1/\sqrt{n})$ apart. In particular, this approach allows useful power approximations. A further technical reason, to be explored later, is that allowing moving values for nuisance parameters obviates the necessity for restrictions to tests that are unbiased with respect to nuisance parameters, an approach often used in small-sample theory (LEHMANN & CASELLA, 2006). One drawback of the approach is that to allow for a neighborhood around each possible value of θ, parameter spaces must of necessity be open, so that values on the boundary are excluded.

In the context of estimation, the use of moving parameters provides some smoothness or uniformity (in θ) of various large-sample performance properties. This enables a rich theory. It also rules out a strange anomaly known as *superefficiency* at isolated parameter values. This phenomenon may be illustrated by a famous example of J. L. Hodges, Jr.:

Hodges's estimate: Consider a random sample from the distribution $N(\theta, 1)$, with sample mean $\bar{X}_n$. Consider, as an estimate of θ,

$$\tilde{\theta}_n = \begin{cases} 0, & \text{if } |\bar{X}_n| < n^{-1/4}, \\ \bar{X}_n & \text{otherwise.} \end{cases}$$

Write A_n for the event $\{|\bar{X}_n| < n^{-1/4}\}$. Then

$$\begin{aligned} P_\theta(A_n) &= P_\theta\big(-n^{1/4} - \sqrt{n}\theta < \sqrt{n}(\bar{X}_n - \theta) < n^{1/4} - \sqrt{n}\theta\big) \\ &= \Phi\big(n^{1/4} - \sqrt{n}\theta\big) - \Phi\big(-n^{1/4} - \sqrt{n}\theta\big) \end{aligned}$$

If the true θ is zero, we conclude that $P_0(A_n) \to 1$. But if $\theta \neq 0$, $-\sqrt{n}\theta$ is the dominant term in each argument of Φ, and $P_\theta(A_n) \to 0$.

Let $Z_n = \sqrt{n}(\tilde{\theta}_n - \theta)$. First suppose that $\theta = 0$. Then $Z_n = 0$ in A_n, and since A_n has probability tending to one, $Z_n \to_p 0$ (and hence $Z_n \to_d 0$). Now suppose that $\theta \neq 0$. Then in A_n^C, $Z_n = \sqrt{n}(\bar{X}_n - \theta)$; since A_n^C has probability tending to one, $Z_n \to_d N(0,1)$ by the central limit theorem.

Hence, $\sqrt{n}(\tilde{\theta}_n - \theta)$ converges in distribution to zero if the true value of θ is zero, and converges in distribution to standard normal otherwise. The large-sample variance, that is, the variance of the limiting distribution, of $\tilde{\theta}_n$ is, respectively, zero or $1/n$. Hence, the estimate $\tilde{\theta}_n$ is *super-efficient* at the isolated point $\theta = 0$. It is more efficient there than is the estimate $\bar{X}_n$, which has variance $1/n$ whatever the value of θ. But continuity (in θ) of the large-sample distributional property is lost at the isolated point $\theta = 0$. For any finite n, the estimate $\tilde{\theta}_n$ is biased, except when $\theta = 0$, but its large-sample distribution is always centered correctly.

Such examples distort the theory, in ruling out $\bar{X}_n$ as a uniformly efficient estimate of the mean θ in large samples from a normal distribution, but have little practical significance since they make a distinction between the true value of θ being zero or $\pm 10^{-10}$, for example. We eliminate them by making certain regularity requirements, using moving parameter values. This enables an optimality theory to be developed, without the distraction of isolated superefficiency. This will be pursued in Chapter 5.

[*Note*: Hodges's example can easily be modified so that super-efficiency obtains at each of a number of isolated values of θ (see *Exercise 2*), but not for any interval of values. As a related example, it may be shown that there exist estimates of the Bernoulli parameter in a sequence of Bernoulli trials which are super-efficient at every rational value of the parameter; see Appendix III to Chapter 5 and the associated *Exercises*. However LeCam (1960) showed, quite generally, that a set where super-efficiency obtains must be sparse, in the sense of having Lebesgue measure zero.]

In a moving parameter model, we consider parameter values, whether real- or vector-valued, of the form

$$\theta_n = \theta + \frac{h}{\sqrt{n}} + o\Big(\frac{1}{\sqrt{n}}\Big), \tag{4.1}$$

moving towards some particular value θ from *direction* h at *rate* $1/\sqrt{n}$. Sometimes, the particular θ is a hypothesized value and θ_n a potential alternative value. At other times, θ simply anchors the neighborhood of possible parameter values to be considered. Alternatively, if $|h|$ denotes the length of h, or the magnitude if h is real, we might call $h/\|h\|$, the *direction* and $\|h\|/\sqrt{n}$ the *rate*, but we use the former terminology. We find that the indicated rate is just right for the construction of a useful theory. We usually omit the $o(1/\sqrt{n})$ term in Equation (4.1), but such a term may be required to assure that $\theta_n \in \Theta$ for small values of n, or when transforming from one parameterization to another.

When taking expectations, variances, stochastic limits and the like, we do so with θ being the true value of the parameter unless noted otherwise.

4.2 Local Asymptotic Normality—Random Sampling Case

We start with the special (but primary) case of independent identically distributed observations $X_1, X_2, \ldots$ and model density $f_\theta(x)$ for each X. At first reading, it is simpler to consider θ as real, but the exposition applies also to a parameter with dimensional $d > 1$.

R. A. FISHER (1925) introduced the concepts of *scores* and *information.* They are defined as follows, here the *score per observation* and *information per observation*, with dots indicating differentiation in θ:

Definition 4.1 *(Scores and Information): Suppose that the log density $l_\theta(x) \equiv \log f_\theta(x)$ is twice differentiable in θ for (almost) every x, and let*

$$s(x, \theta) = \dot{l}_\theta(x) \ \text{(a column vector) and } i(x, \theta) = -\ddot{l}_\theta(x) \quad \textit{a } d \times d \textit{ matrix).}$$

Also, let

$$B(\theta) = \mathrm{var}_\theta s(X, \theta),$$

assumed finite and positive definite. The function s is called the score per observation and i the sample information per observation; the variance, B, of the score is called the (expected) information per observation, as explained below.

We sometimes write $l(x, \theta)$ instead of $l_\theta(x)$; when treated as a function of θ at an observed $x = x^*$, it is called the *log likelihood for* x^*.

It is typically true that $E_\theta\{s(X, \theta)\} = 0$ and $E_\theta\{i(X, \theta)\} = B(\theta)$, the latter justifying the name 'expected information'; indeed, this formula is often used to define $B(\theta)$, but we choose to use the variance of the score as the definition because this turns out to be slightly more fundamental. Thus, B is both the variance (matrix) of the score and the expected value of the sample information (matrix). These two basic facts (set out in Equation (4.2) below) may be proved by interchanging the order of integration (in the expectation operation) and differentiation, when justifiable; but they hold somewhat more generally. We find it easier to take them as assumptions to be verified in each application.

Here is the argument (in the absolutely continuous case) for the first of these. Note that the interchange of integration over x and differentiation in θ requires that $\mathcal{X}$, the support of $f(x, \theta)$, be free of θ.

$$E_\theta\{s(X, \theta)\} = \int \{\partial \log f_\theta(x)/\partial\theta\} f_\theta(x)dx = \int \{\partial f_\theta(x)/\partial\theta\}dx$$
$$= (d/d\theta) \int f_\theta(x)dx,$$

all integrals being over $\mathcal{X}$. Sufficient conditions for the interchange, given by the *Lebesgue dominated convergence theorem*, are that the partial derivative

$|\partial f_\theta/\partial\theta|$ exist and be bounded by a parameter-free integrable function, or $|\partial \log f_\theta/\partial\theta|$ be bounded by a parameter-free function with finite expectation (e.g., CRAMÉR, 1946). The right side of the equation is $(d/d\theta)1 = 0$, completing the argument. The discrete case is essentially the same, and the dimension of θ is irrelevant. Verification of the second basic fact in Equation (4.2) is only slightly more complicated (*Exercise 3*). Several examples are given below.

Two more assumptions are needed, one a Lipschitz condition on the sample information and the other continuity of B in θ. These form the

Basic Assumptions, Random Sampling Case:

(A) For each θ, scores and information are defined, and

$$E_\theta\{s(X,\theta)\} = 0, \quad E_\theta\{i(X,\theta)\} = \mathrm{var}_\theta\{s(X,\theta)\} \equiv B(\theta) > 0. \quad (4.2)$$

(B) For each θ, each element $i(\cdot,\cdot)$ in the sample information matrix satisfies

$$|i(x,\theta+\epsilon h) - i(x,\theta)| < c(\epsilon,\theta)\cdot M(x,\theta) \quad \text{for (almost) all } x \quad (4.3)$$

for every h ($\|h\| < b$, say) and small positive ϵ for some $c(\epsilon,\theta)\downarrow 0$ as $\epsilon \downarrow 0$ and some $M(x,\theta)$ for which $E_\theta\{M(X,\theta)\} < \infty$.

(C) $B(\theta)$ is continuous in θ.

When θ is multidimensional, $B(\theta) > 0$ means that the matrix B is positive definite; equivalently, that the variance of every linear combination of the coordinates of the score is positive. In (B), it is convenient to say: '(4.3) holds uniformly for bounded h' since a supremum over $\|h\| < b$ could be inserted on the left in Equation (4.3). However, this bound b may well depend on θ.

These conditions (A)–(C) are essentially the classical regularity conditions for maximum likelihood theory, as in CRAMÉR (1946), SERFLING (1980), and LEHMANN & CASELLA (2006), for example. Condition (B) is stronger than continuity of $i(x,\theta)$ in θ but slightly weaker than an alternative condition based on third derivatives of the log density. Specifically, taking $c(\epsilon,\theta) = \epsilon b$ and applying the mean value theorem to the left side in Equation (4.3), we find that (B) is implied by the stronger version

(B′) As in (B) with (4.3) replaced by

$$|\dot{i}(x,\theta+\epsilon h)| < M(x,\theta) \quad \text{for each partial derivative } \dot{i} \text{ (in } \theta)$$

of each element of the sample information matrix.

We now verify these assumptions in the case of a one-parameter exponential family, in *natural* form. We assume that

$$f_\theta(x) = \exp\{t(x)\theta - d(\theta) - v(x)\}\mathbf{1}_A(x)$$

where the support A is free of θ, and that $d(\cdot)$ is twice differentiable with $\ddot{d}$ continuous and positive. Then $s(x,\theta) = t(x) - \dot{d}(\theta)$ and $i(x,\theta) = \ddot{d}(\theta)$, constant in x ($\in A$). From properties of exponential families (see BICKEL & DOKSUM (2015) for example), $\dot{d}$ and $\ddot{d}$ are the mean and variance of $t(X)$, and hence $B = \ddot{d} > 0$; and so (A) holds. But with $i(x, \theta) = \ddot{d}(\theta) = B(\theta)$, Condition (B) simplifies to just continuity of $\ddot{d}$, coinciding with (C). The argument extends to k-parameter exponential families.

For an exponential family *not* in natural form, the sample information is no longer constant, but the Basic Assumptions do still hold under minimal assumptions (*Exercise 10*). We show later that verification in the natural form is actually sufficient for many purposes.

Next, we consider a two-parameter example, namely that of sampling from $N(\mu, \sigma^2)$ with $\theta = (\mu, \sigma) \in \mathcal{R} \times \mathcal{R}^+$. Then

$$\log f_\theta(x) = \text{const} - \log \sigma - \frac{(x-\mu)^2}{2\sigma^2}.$$

The score vector has components

$$s_\mu = z/\sigma, s_\sigma = (z^2 - 1)/\sigma,$$

with $z = (x - \mu)/\sigma$. The sample information matrix (per observation) is

$$i = \begin{pmatrix} i_{\mu\mu}, & i_{\mu\sigma} \\ i_{\sigma\mu} & i_{\sigma\sigma} \end{pmatrix} = \begin{pmatrix} 1/\sigma^2 & 2z/\sigma^2 \\ 2z/\sigma^2 & (3z^2-1)/\sigma^2 \end{pmatrix}$$

and hence the expected information matrix is

$$B = \begin{pmatrix} B_{\mu\mu} & B_{\mu\sigma} \\ B_{\sigma\mu} & B_{\sigma\sigma} \end{pmatrix} = \begin{pmatrix} 1/\sigma^2 & 0 \\ 0 & 2/\sigma^2 \end{pmatrix}$$

and (A), (B′) and (C) are seen to hold.

Now let us consider a distribution that is not of exponential family form. Verification of the conditions need not be so immediate. We illustrate with a *Weibull distribution* with shape parameter θ but known scale parameter, which we take to be unity:

$$f_\theta(x) = \theta x^{\theta-1} e^{-x^\theta}, \quad F_\theta(x) = 1 - e^{-x^\theta} \quad \text{for } x > 0, \theta > 0.$$

Then

$$s(x, \theta) = \frac{1}{\theta} + (\log x) \cdot (1 - x^\theta) \quad \text{and} \quad i(x, \theta) = \frac{1}{\theta^2} + (\log x)^2 x^\theta.$$

Many calculations involving this distribution are simplified by transforming to $Y = X^\theta$ which has a standard exponential distribution. Then $\theta s(x, \theta) = 1 + (\log y)(1 - y)$ and $\theta^2 i(x, \theta) = 1 + (\log y)^2 y$. Noting that $d^k\Gamma(r)/dr^k = \int (\log y)^k y^{r-1} e^{-y} dy$, we then find that

$$\theta \cdot E(s) = 1 + \Gamma'(1) - \Gamma'(2)$$

$$\theta^2 \cdot E(s^2) = 1 + 2\Gamma'(1) - 2\Gamma'(2) + \Gamma''(1) - 2\Gamma''(2) + \Gamma''(3)$$
$$\theta^2 \cdot E(i) = 1 + \Gamma''(2).$$

From properties of the *digamma* and *trigamma functions* (see the Appendix to Chapter 1), it can be shown that

$$\Gamma'(1) = -\gamma, \quad \Gamma'(2) = -\gamma + 1, \quad \Gamma'(3) = -2\gamma + 3$$

and

$$\Gamma''(1) = \zeta + \gamma^2, \quad \Gamma''(2) = \zeta - 2\gamma + \gamma^2, \quad \Gamma''(3) = 2(\zeta + 1 - 3\gamma + \gamma^2)$$

where $\gamma \approx 0.5772$ is (*Euler's constant*) and $\zeta = \pi^2/6$. Then (A) follows with $B(\theta) = [\zeta + (1-\gamma)^2]/\theta^2 \approx 1.8237/\theta^2$, from which condition (C) follows.

We now verify the stronger version (B′) of (B), thereby completing the verification of the Basic Assumptions. Since the parameter θ is positive, we restrict the neighborhood to $\theta + \epsilon h > 0$ for $|h| < b$ and therefore to $\epsilon < \delta\theta/b$ for some $\delta \in (0,1)$. Hence, $\theta(1-\delta) < \theta + \epsilon h < \theta(1+\delta)$. We therefore find

$$|\dot{i}(x, \theta + \epsilon h)| = \left| \frac{-2}{(\theta + \epsilon h)^3} + (\log x)^3 x^{\theta + \epsilon h} \right|$$
$$< \frac{2}{\theta^3 (1-\delta)^3} + |\log x|^3 \{x^{\theta(1+\delta)} + x^{\theta(1-\delta)}\} \equiv M(x, \theta)$$

(the two powers of x taking care of cases with $0 < x < 1$ and $x > 1$). Now

$$E_\theta\{M(X,\theta)\} = \frac{1}{\theta^3}\left\{\frac{2}{(1-\delta)^3} + \int_0^\infty |\log y|^3 (y^{1+\delta} + y^{1-\delta}) e^{-y} dy\right\} < \infty$$

since $\Gamma'''(r) = \int (\log y)^3 y^{r-1} e^{-y} dy$ is known to converge for any $r > 0$.

We now revert to the general case and develop what we will call the *Local Asymptotic Normality* of the model. We start with a Taylor expansion of the log likelihood for $\mathbf{x}_n$ in a neighborhood of θ. To this end, we find the first and second derivatives of the log likelihood at $\mathbf{x}_n$:

$$\frac{\partial}{\partial \theta} \log p_{n,\theta}(\mathbf{x}_n) = \frac{\partial}{\partial \theta} \sum_{j=1}^n l_\theta(x_j) = \sum_{j=1}^n s(x_j, \theta)$$

(‘*total scores*’) and

$$-\frac{\partial^2}{\partial \theta^2} \log p_{n,\theta}(\mathbf{x}_n) = \sum_{j=1}^n i(x_j, \theta)$$

(‘*total sample information*’) when θ is real; more generally, we have a vector of partial derivatives and a matrix of second-order partials (but the reader may do well to treat θ as real at first reading). Now consider $\log p_{n,\theta+\epsilon h}(\mathbf{x})$

as a function of h and expand in a Taylor series (Lagrange's form) through second-order derivatives—or, perhaps more simply, treat it as a function of ϵ and expand—, obtaining either way

$$\log p_{n,\theta+\epsilon h}(\mathbf{x}_n) = \log p_{n,\theta}(\mathbf{x}_n)+\epsilon h^T \sum_{j=1}^{n} s(x_j,\theta)-\tfrac{1}{2}\epsilon^2 h^T\{\textstyle\sum_{j=1}^{n} i(x_j,\theta+\epsilon\xi_n h)\}h \tag{4.4}$$

for some $\xi_n = \xi_n(\mathbf{x}_n,\theta) \in (0,1)$.

Let us examine the matrix in the last term in Equation (4.4), incorporating the factor ϵ^2 with $\epsilon = 1/\sqrt{n}$. We may write the uvth element in the matrix as the sum of three terms:

$$\begin{aligned}\frac{1}{n}\sum_{j=1}^{n} i_{uv}(x_j,\theta+\frac{\xi_n}{\sqrt{n}}h) = B_{uv}(\theta) &+ \Big\{\frac{1}{n}\sum_{j=1}^{n} i_{uv}(x_j,\theta) - B_{uv}(\theta)\Big\}\\ &+\frac{1}{n}\sum_{j=1}^{n}\{i_{uv}(x_j,\theta+\frac{\xi_n}{\sqrt{n}}h) - i_{uv}(x_j,\theta)\}.\end{aligned} \tag{4.5}$$

Replace the x's in Equations (4.4) and (4.5) by X's. Then, by the weak law of large numbers and (4.2), the term in braces in Equation (4.5) tends to zero in probability. The absolute value of the last term is

$$\leq \frac{1}{n}\sum_{j=1}^{n}\left|i_{uv}(X_j,\theta+\frac{\xi_n}{\sqrt{n}}h) - i_{uv}(X_j,\theta)\right| < c(\frac{1}{\sqrt{n}},\theta)\cdot\frac{1}{n}\sum_{j=1}^{n} M_{uv}(X_j,\theta) \tag{4.6}$$

by Equation (4.3) with $\epsilon = 1/\sqrt{n}$ and using the uniformity in h. The bound in Equation (4.6) being h-free permits taking the supremum over $\{h|\|h\| < b\}$. The average on the right in Equation (4.6) converges in probability (by the weak law of large numbers) to a finite constant (condition (B)), while $c(1\sqrt{n},\theta)$ tends to zero (condition (B)), and so the right side in Equation (4.5) is $B_{uv}(\theta)+o_p(1)$, uniformly for $\|h\| < b$. As this holds for each term in the matrix, the last term in Equation (4.4), with $\epsilon = 1/\sqrt{n}$, is

$$-\tfrac{1}{2}h^T\{B+o_p(1)\}h = -\tfrac{1}{2}h^T Bh + o_p(1)$$

uniformly.

Now move the first term on the right side of (4.4) to the left side, again with $\epsilon = 1/\sqrt{n}$, and define

$$S_n(\mathbf{X}_n,\theta) = \frac{1}{\sqrt{n}}\sum_{j=1}^{n} s(X_j,\theta), \tag{4.7}$$

which converges in distribution by the central limit theorem. Collecting together these results gives, in the random sampling case, an expansion for the *local log likelihood-ratio* as a certain asymptotically normal random variable:

Theorem 4.1 *(Uniform Local Asymptotic Normality): Let*

$$L_n = L_n(\theta, h) = \log p_{n,\theta_n}/(\mathbf{X}_n)p_{n,\theta}(\mathbf{X}_n) \quad \text{for } \theta_n = \theta + \frac{h}{\sqrt{n}} + o\Big(\frac{1}{\sqrt{n}}\Big). \tag{4.8}$$

Then, under the Basic Assumptions, and when the parameter is θ*,*

$$L_n(\theta, h) = h^T S_n(\mathbf{X}_n, \theta) - \tfrac{1}{2} h^T B(\theta) h + o_p(1) \tag{4.9}$$

and

$$S_n(\mathbf{X}_n, \theta) \to_d N\{0, B(\theta)\}; \tag{4.10}$$

moreover, the convergence of the error term in Equation (4.9) is uniform for bounded h*. All probability statements are for* θ *being the true value of the parameter.*

The vector $S_n(\mathbf{X}_n, \theta)$ in Equation (4.7) is called simply the *score*—for the full data rather than 'per observation'. This is neither the 'total score' nor the 'average score', but is the most useful form for theoretical purposes, especially when extending beyond the random sampling case. Since arguments assume θ to be the true value of the parameter, the denominator density in $p_{n,\theta}$ is positive; so that $L_n \in [-\infty, \infty)$.

Note that Equations (4.9) and (4.10) imply

Corollary 4.1 *Under the conditions of Theorem 4.1,*

$$L_n \to_d N(-\tfrac{1}{2}\delta^2, \delta^2) \quad \text{with } \delta^2 = h^T B h.$$

We will return to this in Section 4.4.

Expanding the score at $\theta + \epsilon h$ in the same fashion as the expansion (4.4) above, we are led to

$$S_n(\mathbf{X}_n, \theta + \epsilon h) = S_n(\mathbf{X}_n, \theta) - \sqrt{n}\epsilon\{B(\theta) + o_p(1)\}h, \tag{4.11}$$

with the $o_p(1)$ term uniform for bounded h. Setting $\epsilon = 1/\sqrt{n}$, we obtain (*Exercise 7*)

Theorem 4.2 *(Regular Scores property): Under the Basic Assumptions, and when the true parameter value is* θ*, and* θ_n *is as in Equation (4.8),*

$$S_n(\mathbf{X}_n, \theta_n) = S_n(\mathbf{X}_n, \theta) - B(\theta)h + o_p(1), \tag{4.12}$$

moreover the error term converges uniformly for bounded h*.*

When θ is multidimensional, Equations (4.11) and (4.12) may be interpreted as holding for each coordinate of the score vector.

The uniformity of convergence in Equation (4.12), that the supremum over h for which $\|h\| < b$, of the absolute value of the error term, converges in probability to zero, allows us to apply the expansion (4.12) to *random, data-dependent*, $h = h_n(\mathbf{X}_n)$, so long as h_n is bounded in probability. This is a subtle but important statement which we now prove using an 'epsilon-delta' argument. It is the sole use of the Regular Scores property.

Proof Let $V_n(h) \equiv S_n(\theta + h/\sqrt{n}) - S_n(\theta) + Bh$, and $U_n \equiv V_n\{h_n(\mathbf{X}_n)\}$. We need to show that $U_n = o_p(1)$.

Specifically, we need to show that for any $\epsilon > 0$ and $\delta > 0$ there exists an N such that $n > N$ implies that $\text{pr}(|U_n| > \delta) < \epsilon$. Let $A_n \equiv A_n(b) \equiv \{\mathbf{x}_n | \|h_n(\mathbf{x}_n)\| < b\}$. Since $h_n(\mathbf{X}_n)$ is bounded in probability there exist b and N_1 such that $P_{n,\theta}(A_n) > 1 - \frac{1}{2}\epsilon$ for $n > N_1$. Let $V_n(h) \equiv S_n(\theta + h/\sqrt{n}) - S_n(\theta) + Bh$ and $U_n \equiv V_n(h_n)$. Then $\mathbf{1}[A_n] \cdot |U_n| \leq \sup_{\|h\|<b} |V_n|$, which is $o_p(1)$ by the uniformity in the Regular Scores theorem. Therefore, for any $\delta > 0$ there exists an N_2 such that for all $n \geq N_2$, $\text{pr}(\mathbf{1}[A_n] \cdot |U_n| > \delta) < \epsilon$. But

$$\begin{aligned}\text{pr}(|U_n| > \delta) &= \text{pr}\{(|U_n| > \delta) \cap A_n\} + \text{pr}\{(|U_n| > \delta) \cap A_n^C\} \\ &\leq \text{pr}\big\{\mathbf{1}[A_n] \cdot |U_n| > \delta)\big\} + \text{pr}(A_n^C) \ < \ 2\epsilon\end{aligned}$$

for all $n \geq \max(N_1, N_2)$. Since δ and ϵ are arbitrary, this shows that $U_n = o_p(1)$. □

We therefore have:

Corollary 4.2 *If $h_n(\mathbf{X}_n) = O_p(1)$, then (4.12) holds with $\theta_n = \theta + h_n(\mathbf{X}_n)/\sqrt{n}$.*

The expansion (4.9) is essentially a Taylor expansion, through second derivatives, of $\log p_{n,\theta_n}$ (around $\theta_n = \theta$), with nS_n being the first derivative and $-nB(\theta)$ approximating the second derivative (being its expectation in the random sampling case); Theorem 4.1 makes a claim about the error term. Likewise, (4.12) is essentially a one-term Taylor expansion of nS_n, employing the same approximation, and Theorem 4.2 makes a claim about the error term.

Sometimes it is easier to verify Equations (4.11) and (4.12) directly, by Taylor expansion, rather than verifying the Basic Assumptions. Or it may be easier to verify the Basic Assumptions in a different parameterization than the one chosen. We now show that the latter is sufficient:

Proposition 4.1 *(Parameter Transformation): Suppose that γ is an invertible transformation of the d-dimensional vector parameter θ having a continuous second derivative. Specifically, write*

$$\dot{\gamma} = \begin{pmatrix} \partial\gamma_1/\partial\theta_1 & \cdots & \partial\gamma_d/\partial\theta_1 \\ \vdots & \ddots & \vdots \\ \partial\gamma_1/\partial\theta_d & \cdots & \partial\gamma_d/\partial\theta_d \end{pmatrix}$$

for the matrix of first derivatives of γ in θ and for each k let $\ddot{\gamma}_k$ denote the matrix of second derivatives $\partial^2\gamma_k/\partial\theta_i\partial\theta_j$.

Appending the symbol ^{o} *for the* γ *parameterization, and with dots denoting differentiation in* θ, *(4.9),(4.10) and (4.11) hold in the* θ *parameterization if and only if they hold in the* γ *parameterization with* $h^o = \dot\gamma h$ *and*

$$s(x,\theta) = \dot\gamma s^o(x,\gamma), i(x,\theta) = \dot\gamma i^o(x,\gamma) - \ddot\gamma s^o(x,\gamma), \text{ and } B(\theta) = \dot\gamma B^o(\gamma)\dot\gamma^T. \tag{4.13}$$

Here $\ddot\gamma s^o(x,\gamma)$ *is interpreted as* $\sum_k \ddot\gamma_k s^o(x,\gamma_k)$.

Proof The expressions for $s(x,\theta)$ and $i(x,\theta)$ follow immediately from the chain rule for partial derivatives. The expression for $B(\theta)$ follows because $E\{s^o(x,\gamma)\} = 0$. Continuity of $B(\theta)$ implies continuity of $B(\gamma)$ by the assumed continuity of the first derivatives of γ. Continuity of the second derivatives is needed to ensure that properties (B) or (B$'$) of the observed information i are inherited under the transformation. □

An immediate consequence of this result is that the Regular Scores and Uniform Local Asymptotic Normality properties hold for an exponential family in *nonnatural* form since the Basic Assumptions have been verified for exponential families in *natural* form.

<u>An Example: Location-Scale Families</u>

This simple but important example of an information calculation will be extended in the next section and we will return to it in Chapter 5. Suppose that the observations $Y_1, \ldots, Y_n$ are independent and identically distributed with common density

$$f(y,\alpha,\sigma) = \frac{1}{\sigma} g\left(\frac{y-\alpha}{\sigma}\right),$$

where $g(\cdot)$ is a twice-differentiable density function with finite information. For notational convenience let $z = (y-\alpha)/\sigma$ denote the standardized variable and write $s(z) = -g'(z)/g(z)$. This would equal the score for α when $\sigma = 1$. For general (α,σ) the scores per observation are $s^o_\alpha = (1/\sigma)s^1(z)$, $s^o_\sigma = -(1/\sigma)\{1 - zg(z)\}$ since $\partial z/\partial\sigma = -z/\sigma$. The joint observed information matrix $i(\alpha,\sigma)$ for one observation is

$$\sigma^2 i^o = \begin{pmatrix} s'(z) & zs'(z) + s(z) \\ \cdot & 2zs(z) + z^2 s'(z) - 1 \end{pmatrix}$$

with expectation given by

$$\sigma^2 B^o = \begin{pmatrix} \int s'(z)g(z)dz & \int zs'(z) + s(z) \\ \cdot & 2zs(z) + z^2 s'(z) - 1 \end{pmatrix}$$

and the right side of this equation is free of unknown parameters.

When the density $g(\cdot)$ is symmetric about zero $s(z) = -s(z)$ and $s'(z) = s'(-z)$ for all z, so the integrand in the off-diagonal term is an odd function of z, making its integral equal to zero. This will not generally be true when $g(\cdot)$ is asymmetric. As will be seen in Chapter 5, diagonal information matrices offer certain advantages. Fortunately, in this example a simple reparameterization achieves a diagonal information matrix. Writing $\gamma = (\alpha, \sigma)$, we change to new parameters $\theta = (\nu, \tau)$ given by $\nu = \alpha - \lambda\sigma$, $\tau = \sigma$, where λ is a number to be selected. The inverse transformation is $\alpha = \nu + \lambda\tau$, $\sigma = \tau$

The derivative of the this transformation is

$$\dot{\gamma} = \begin{pmatrix} 1 & 0 \\ \lambda & 1 \end{pmatrix},$$

so that

$$\tau^2 B = \begin{pmatrix} 1 & 0 \\ \lambda & 1 \end{pmatrix} \begin{pmatrix} a & b \\ b & d \end{pmatrix} \begin{pmatrix} 1 & \lambda \\ 0 & 1 \end{pmatrix},$$

where a, b, c are the elements of $B^o(\gamma)$ calculated earlier. Multiplying out gives

$$\tau^2 B = \begin{pmatrix} a & \lambda a + b \\ \lambda a + b & \lambda^2 a + 2\lambda b + d \end{pmatrix}.$$

The choice $\lambda = -b/a$ makes the off-diagonal term zero. As we noted, this choice depends on the form of the function $g(\cdot)$, but it does not depend on the true values of α and σ. Also, the leading element (a) of this information matrix is unchanged by the transformation. This makes perfect sense, as this element represents the information about the location parameter when the scale parameter is known, and this information is the same for α as for $\alpha + \lambda\sigma$ when λ is specified and σ is known.

Extension for Smoothly Approached Threshold Parameters: When a random sampling model contains a *threshold parameter*, that is a parameter determining the support of the distribution, as for a uniform distribution on $(0, \theta)$, conditions (A)–(C) may fail; yet the properties of Uniform Local Asymptotic Normality and Regular Scores may still hold. The following extension covers all known cases: Let $\mathcal{X}(\theta)$ be the support of $f_\theta(\cdot)$. Instead of requiring (4.3) for (almost) all x, we require it only for x in a set $\mathcal{X}_n(\theta)$ having probability $1 - o(1/n)$ and in which $f_{\theta+\epsilon h}(x) > 0$ for some $\epsilon > 0$ and all $\|h\| < b$. A sufficient condition is that the density and its derivative (in x) tend to zero as the argument approaches a threshold. (The proofs are omitted). This permits the introduction of a threshold shift parameter in each of the following models: *lognormal*, *Weibull* (with shape parameter exceeding 2), and *gamma* (with shape parameter exceeding 2); and a *beta* model on an unknown interval can also be handled whenever both shape parameters exceed 2. The stated restrictions on the parameters are needed to ensure that the density approaches zero smoothly at thresholds, ruling out in particular uniform distributions. Alternatively, these cases can be handled by extending the definition of scores through use of *mean-square derivatives*, which obviates the necessity for a condition like (4.3). See IBRAGIMOV & HAS'MINSKII (1981) or BICKEL, KLAASSEN, RITOV, & WELLNER (1993). *Exercise 8* gives an illustration.

4.3 Local Asymptotic Normality—General Case

The conclusions of the two theorems from the previous section turn out to hold more generally than for random sampling, and they provide the basis for a large-sample theory of inference. Hence, we state these results as assumptions. We show below how they can be satisfied in a regression model with non-random explanatory variables and indicate that the two-sample shift problem can be handled similarly. A somewhat similar demonstration applies to *censored* data models, with non-random censoring values. In each of these cases, the data consist of a number of independent components, typically needed for local asymptotic normality. Without such independence, a modified form of the property including a so-called mixing condition may apply, but this is not considered here.

1. **Uniform Local Asymptotic Normality property:** With $L_n(\theta, h)$ defined in Equation (4.8), the expansion (4.9) and convergence in Equation (4.10) hold for some random vector $S_n(\mathbf{X}_n, \theta)$, the *score*, and some non-random positive-definite symmetric matrix $B(\theta)$, the *information*. Moreover, the convergence of the error term in Equation (4.9) is uniform for bounded h.
2. **Regular Scores property:** The score S_n satisfies (4.12), uniformly for bounded h.
3. **Continuity of Information assumption:** $B(\theta)$ is continuous in θ.

Thus, by the two theorems in the previous section, these three properties, Uniform Local Asymptotic Normality, Regular Scores and Continuity of Information are implied by the Basic Assumptions in the case of a random sampling model. We sometimes drop the requirement of uniformity from Local Asymptotic Normality, but as indicated earlier have no need for a non-uniform Regular Scores assumption.

Here, the score S_n and information B have not been directly defined, and n is some convenient index of the amount of data available. But the score may be identified by differentiating the log likelihood $\log p_{n,\theta}(\mathbf{x}_n)$ in θ and dividing by $\sqrt{n}$. The information must be the limit in probability of $-1/n$ times the second partial derivative of the log likelihood — the *average sample information* — when such derivatives exist, but may be more easily identified as the large-sample variance of the score. Either way, the validity of Equations (4.9), (4.10) and (4.12), and the continuity of $B(\theta)$, must be verified.

Since $\log p_{n,\theta}$ is additive across independent parts of the data, it is usually simpler to work with each independent factor in $p_{n,\theta}$ and relate scores and information for the full model to those for the various factors. See for example *Exercises 15–16*.

We now indicate assumptions that will imply these three properties in a regression model outside the scope of random sampling.

A regression model: We extend the location scale model discussed in the previous section to include dependence on an explanatory variable, here denoted by x, keeping Y for the outcome variable as in the earlier discussion. Let $x_1, x_2, \ldots$ be a sequence of given constants (real-valued, for simplicity here), and suppose that $Y_1, Y_2, \ldots$ are independent with Y_j having density

$$f_j(y; \alpha, \beta, \sigma, x_j) = \frac{1}{\sigma} g(z_j), \; z_j = \frac{1}{\sigma}(y - \alpha - \beta x_j),$$

where $g(z)$ is a density (for example the standard normal, $N(0,1)$) with finite expected information I_g, $s(z) \equiv -g'(z)/g(z)$ and $I_g \equiv \int s(z)^2 g(z) dz$. Here, $\theta = (\alpha, \beta, \sigma)$, with dimension $d = 3$.

The scores per observation can again be defined, but now depend on x_j. By differentiation, using the chain rule, we find that

$$\sigma s_\alpha(z_j) = s(z_j), \; \sigma s_\beta(z_j) = x_j s(z_j), \; \sigma s_\sigma(z_j) = z_j s(z_j) - 1.$$

The elements of i, the 3×3 sample information matrix (for each observation) are found by differentiating the scores and changing the signs. Dropping the subscript j we find that with

$$i = \begin{pmatrix} i_{\alpha\alpha}(z) & i_{\alpha\beta}(z) & i_{\alpha\sigma}(z) \\ \cdot & i_{\beta\beta}(z) & i_{\beta\sigma}(z) \\ \cdot & \cdot & i_{\sigma\sigma}(z) \end{pmatrix},$$

then

$$\sigma^2 i = \begin{pmatrix} s'(z) & x s'(z) & z s'(z) + s(z) \\ \cdot & x^2 s'(z) & x[z s'(z) + s(z)] \\ \cdot & \cdot & 2 z s(z) + z^2 s'(z) - 1 \end{pmatrix}.$$

It may now be shown that the log likelihood for $\mathbf{y}_n$ can be expanded, as before, and put into the form Equation (4.8) as long as $\sum_{j=1}^n x_j^4/n$ converges to a finite positive limit and the score $s(z)$ is sufficiently smooth and regular. For example, $i_{\beta\beta}$, the $\beta\beta$ term of the average information is $(\sigma^2 n)^{-1} \sum_{j=1}^n x_j^2 s'(z_j)$, which converges in probability to a corresponding term in the expected information of $\sigma^{-2} a_2 \cdot E\{s'(Z)\}$ by Chebyshev's inequality if $\text{var}\{s'(Z)\} < \infty$ and $a_2 = \lim \sum_{j=1}^n x_j^2/n$. Assuring uniformity in h in Equations (4.11) and (4.12) may be more troublesome. Some of the calculations are simplified by reparameterizing with $\gamma = 1/\sigma$ before applying the Transformation Proposition. In particular, if $g(\cdot)$ is the standard normal density and $a_{rn} = \sum_{j=1}^n x_j^r/n \to a_r$ for positive integers $r \leq 4$, then (4.9) holds with

$$S_n = \frac{1}{\sigma n} \sum_{j=1}^{n} (Z_j, x_j Z_j, Z_j^2 - 1)^T$$

and $B(\alpha, \beta, \sigma) = \sigma^{-2} A$ where

$$A = \begin{pmatrix} 1 & a_1 & 0 \\ a_1 & a_2 & 0 \\ 0 & 0 & 2 \end{pmatrix}.$$

Here, $Z_j = \sigma^{-1}(Y_j - \alpha - \beta x_j)$. The details of the derivation are omitted here.

The technical difficulties involved in handling regularity conditions for non-identically distributed observations may be avoided by using an unconditional regression model in which the x_j are also viewed as random variables, with known density $k(x)$, say. See *Exercise 16.* With f above now defined as the conditional density for Y given $X = x$, this formulation implies a random sampling model for (X, Y) with joint density $f_\theta(x, y) = f(y; \theta, x)k(x)$, and the Basic Assumptions may be applied directly. Since the function $k(x)$ disappears when the log likelihood is differentiated in the parameters, the scores and observed information are the same in the unconditional as in the conditional model. However, the calculation of the Fisher information matrix B requires taking the expectation of the previous expression over the density $k(x)$ of X.

4.4 LeCam's Third Lemma

LeCam's third lemma provides a device for determining the large-sample distribution of a statistic when the parameter θ_n varies slightly with n—specifically, as specified in Equation (4.1). There are alternative methods, utilizing *uniform convergence in distribution* of statistics; see for example the treatment of *Pitman efficiency* in SERFLING (1980) or LEHMANN (1975). However, LeCam's third lemma is a natural companion to the our approach using local asymptotic normality.

We start with a preliminary lemma whose proof is straightforward algebra. A useful introductory exercise is to prove the same lemma with the role of the first argument u omitted; that is, to show that $g(v) = e^v f(v)$ whenever f and g are normal densities with means equal to $\mp$ one-half the common variance. However, we go directly to the general case.

Lemma 4.1 *(Preliminary): Suppose that f and g are bivariate normal densities with means $(0, -\frac{1}{2}\delta^2)$ and $(\beta, +\frac{1}{2}\delta^2)$, common nonsingular variance matrices with variances σ^2 and δ^2 and covariance β. Then $g(u, v) = e^v f(u, v)$ for all (u, v).*

That is, the logarithm of the density ratio $g : f$ is simply the second argument of the densities.

Proof The bivariate normal density with mean vector μ and covariance matrix Σ is

$$f(x; \mu, \Sigma) = \frac{1}{2\pi \; \{\det(\Sigma)\}^{\frac{1}{2}}} \exp\{-\tfrac{1}{2}(x - \mu)^T \Sigma^{-1} (x - \mu)\}.$$

See Equation (21) of Chapter 1 or *Exercise 1.22.* The lemma now reduces to a straightforward, though tedious, verification that the two quadratic forms in (u,v) in the exponents of the left and right side of the asserted equation are identical (*Exercise 17*).

Inversion of the covariance matrix can be avoided by showing instead that the bivariate moment generating functions

$$\tilde{M}_f(t_1,t_2)=\int_u\int_v \exp(ut_1+vt_2)e^v f(u,v)dudv,$$

and

$$M_g(t_1,t_2)=\int_u\int_v \exp(ut_1+vt_2)g(u,v)dudv,$$

or their logarithms, are identical and appealing to the relevant uniqueness theorem (Chapter 1). For a bivariate normal distribution with mean vector μ and covariance matrix Σ, $M(t_1,t_2)=\exp(\mu^T t+\frac{1}{2}t^T\Sigma t)$. Since

$$\tilde{M}_f(t_1,t_2)=\int_u\int_v f(u,v)\exp\{t_1u+(t_2+1)v\}dudv=M_f(t_1,t_2+1),$$

where $M_f(t_1,t_2)$ is the moment generating function of $f(u,v)$, we must show that the expressions $M_f(t_1,t_2+1)$ and $M_g(t_1,t_2)$ are identical.

The negatives of their logarithms are respectively.

$$(0,-\tfrac{1}{2}\delta^2)\begin{pmatrix}t_1\\ t_2+1\end{pmatrix}+\tfrac{1}{2}(t_1,t_2+1)\begin{pmatrix}\sigma^2 & \beta\\ \beta & \delta^2\end{pmatrix}\begin{pmatrix}t_1\\ t_2+1\end{pmatrix}$$

and

$$(\beta,\ \tfrac{1}{2}\delta^2)\begin{pmatrix}t_1\\ t_2\end{pmatrix}+\tfrac{1}{2}(t_1,t_2)\begin{pmatrix}\sigma^2 & \beta\\ \beta & \delta^2\end{pmatrix}\begin{pmatrix}t_1\\ t_2\end{pmatrix}.$$

These expressions simplify to

$$\tfrac{1}{2}\delta^2\{(t_2+1)^2-(t_2+1)\}+\tfrac{1}{2}t_1\sigma^2t_1+\beta t_1(t_2+1),$$

and

$$\beta t_1+\tfrac{1}{2}\delta^2t_2+\tfrac{1}{2}t_1\sigma^2t_1+\beta t_1t_2+\tfrac{1}{2}\delta^2t_2^2,$$

respectively, which are indeed identical. □

Note This proof applies when Σ is singular, and easily extends to vector u.

Here is a sketch of the basic idea of LeCam's third lemma. Suppose that U_n is a statistic which converges in distribution to $N(0,\sigma^2)$ when the true parameter value is θ. We seek its large-sample distribution when θ is replaced by a local value θ_n. We already know, when the true parameter is θ, that the log-likelihood ratio statistic $L_n\equiv V_n$ converges in distribution to $N(-\frac{1}{2}\delta^2,\delta^2)$ for δ^2 defined in the Corollary to Theorem 4.1. We need to suppose slightly more: that U_n and V_n jointly converge in distribution, not just marginally, and that the limit distribution is bivariate normal; let the covariance parameter be β. Let A_n represent a set of possible values for (U_n, V_n). Then, by the *change-of-measure* theorem of Chapter 1, we expect that

$$P_{n,\theta_n}(A_n) = E_{n,\theta}(e^{V_n}\mathbf{1}[A_n]). \tag{4.14}$$

The right side is, approximately, an integration with respect to a bivariate normal density f (in the notation of the Preliminary Lemma). By this lemma, we should be able to drop the exponential factor in Equation (4.14) if the means of U_n and V_n are shifted, by β and δ^2, as indicated in the lemma but with bivariate normality preserved; the right side is then just the probability of A_n under this mean-shifted bivariate normal distribution. We thus conclude that (U_n, V_n) has, approximately, this bivariate normal distribution when the value of the parameter is θ_n, and, in particular, that U_n is $AN(\beta, \sigma^2)$. *The moving parameter has simply shifted the large-sample mean of U_n by the amount β, with β being the large-sample covariance between U_n and the log likelihood-ratio.* The joint convergence of U_n and V_n can conveniently be established by use of the *Cramér-Wold device*—consideration of the marginal distribution of the combination $aU_n + bV_n$ for every (a, b).

We now develop this argument. In more general notation, let P_n and Q_n be two alternative models for potential data $\mathbf{X}_n$ with (joint) densities p_n and q_n, and let L_n be the log likelihood-ratio q_n/p_n (in $[-\infty, \infty)$ when P_n is the model and in $(-\infty, \infty]$ when Q_n is the model). In our applications, p_n is $p_{n,\theta}$ and q_n is p_{n,θ_n}, but this structure is not essential here.

Suppose that u_n and v_n are real-valued functions of $\mathbf{x}_n$, and write $U_n = u_n(\mathbf{X}_n)$ and $V_n \equiv L_n(\mathbf{X}_n)$, defined above. Let F_n and G_n represent the respective joint distribution functions of (U_n, V_n) under p_n and q_n respectively, and let F and G represent bivariate normal distribution functions corresponding to f and g in the Preliminary Lemma above.

Proposition 4.2 *(LeCam's third lemma): If $F_n \to F$ (everywhere), then $G_n \to G$ (everywhere).*

A corollary fact is that $L_n \sim AN(\frac{1}{2}\delta^2, \delta^2)$ under Q_n—just as under P_n except for a sign change in the mean. This corollary is consistent with what we would expect from symmetry: interchanging the roles of P_n and Q_n changes the sign of L_n. We illustrate the lemma below. But first:

Proof Fix (u, v), the argument of the distribution function G_n. For simplicity, assume P_n and Q_n have the same support. (To extend beyond this, use *LeCam's first lemma*; see Appendix A in Hall & Mathiason (1990).) We find that

$$\begin{aligned} G_n(u, v) &= \int \mathbf{1}[u' \le u, v' \le v]dG_n(u', v') \\ &= \int \mathbf{1}[u_n(\mathbf{x}_n) \le u, L_n(\mathbf{x}_n) \le v]dQ_n(\mathbf{x}_n) \end{aligned}$$

$$= \int \mathbf{1}[u_n(\mathbf{x}_n) \le u, L_n(\mathbf{x}_n) \le v] e^{L_n(\mathbf{x}_n)} dP_n(\mathbf{x}_n)$$

$$= \int \mathbf{1}[u' \le u, v' \le v] e^{v'} dF_n(u', v').$$

A two-dimensional version of the *Helly-Bray Theorem*, extended to allow discontinuities in the integrand (see Chapter 2), then implies convergence to the corresponding integral with dF replacing dF_n. But $e^{v'} dF(u', v') = dG(u', v')$, by the Preliminary Lemma. The resulting integral of an indicator $\mathbf{1}[u' \le u, v' \le v]$ with respect to $dG(u, v)$ is precisely $G(u, v)$, completing the proof. □

The lemma and its proof are adapted from material in HÁJEK & ŠIDÁK (1967), which in turn was adapted from the original paper of LECAM (1960).

Extensions:

(1) The matrix Σ can be singular. The only difference here is that we need to verify that $dG(u, v) = e^v dF(u, v)$ without using a bivariate normal density. The proof using moment generating functions applies without change. Alternatively, with $\rho = \pm 1$, say $U = a + bV$ (with probability one), we may apply the Preliminary Lemma with the role of u missing, and use the corollary stated after Proposition 4.2 to complete the proof.

(2) The statistic U_n may be d-dimensional. This case is easily done by applying the one-dimensional case to $a^T U_n$ for every d-vector a, as in the *Cramér-Wold device*.

Now let us apply LeCam's third lemma upon assuming Local Asymptotic Normality. Then L_n in Equation (4.9) has $L_n \sim AN(-\frac{1}{2}\delta^2, \delta^2)$ with $\delta^2 = h^T B h$ under $P_{n,\theta}$, as required by the hypothesis of the lemma. The corollary stated after Proposition 4.2 asserts that L_n has the same limiting distribution, except for a sign change in the mean, under P_{n,θ_n}. Since L_n is linear in the score vector S_n (asymptotically—see Equation (4.9)), it is slightly simpler to replace the role of L_n by S_n, which has $L_n \sim AN(0, B)$ under $P_{n,\theta}$, and $\text{Acov}(U_n, L_n) = h \cdot \text{Acov}(U_n, S_n)$. This leads to:

Proposition 4.3 *(Score Version of LeCam's third lemma): Suppose that Local Asymptotic Normality holds with scores S_n. If (U_n, S_n) (of dimensions d' and d) converges in distribution under $P_{n,\theta}$ to bivariate normality with all means zero and positive-definite variance matrix in block form with A the asymptotic variance for U_n, B for S_n, with C their covariance matrix ($d' \times d$), then the same limiting distribution holds under P_{n,θ_n} with $\theta_n = \theta + h/\sqrt{n}$, except the asymptotic means are Ch for U_n and Bh for S_n.*

Another corollary fact comes from a degenerate case with $\sigma = 0$. Then, Local Asymptotic Normality and $U_n = o_p(1)$ under $P_{n,\theta}$ together imply $U_n = o_p(1)$ under P_{n,θ_n}. Convergence in probability does not distinguish between θ and θ_n. It is also true that under Local Asymptotic Normality any sequence of events with probabilities tending to zero under $P_{n,\theta}$ has probabilities tending to zero under P_{n,θ_n}. (This is a consequence of *LeCam's first lemma*). Sequences P_{n,θ_n} with this latter property are said to be *contiguous* to $P_{n,\theta}$. This concept of LeCam's is discussed further in HÁJEK & ŠIDÁK (1967), but not formally used here.

Here is an application of LeCam's third lemma: Let M_n be the sample median in a random sample of size n from an exponential distribution with rate parameter θ (according to model $P_{n,\theta}$) or with rate parameter $\theta_n = \theta + h/\sqrt{n}$. Let $U_n = \sqrt{n}(M_n - MED)$ where $MED = MED(\theta)$ is the population median $(\log 2)/\theta$. We know that M_n is approximately normal for fixed θ; how about for a 'moving' θ_n'? and with what parameters?

We proceed as follows. Details are left to the reader, see *Exercise 22*:

1. For fixed θ, $M_n \sim AL\{MED, k(\cdot), \tau^2\}$ with kernel $k(x) = -\{\mathbf{1}[x \leq MED(\theta)] - \frac{1}{2}\}/\frac{1}{2}\theta$; $\tau^2 = \text{var } k(X) = 1/\theta^2$ by the Bahadur Representation. Theorem; see Chapter 3. Hence, $U_n \sim AN(0, 1/\theta^2)$.
2. The score per observation (derivative of the log density in θ) is $s(x) = \frac{1}{\theta} - x$, and hence the average score $(S_n/\sqrt{n})$ is $\bar{s}_n = \sum_{j=1}^{n}(1/\theta - x_j)/n$; this has $\bar{s}_n \sim AL\{0, s(\cdot), B = 1/\theta^2\}$.
3. The asymptotically linearity of M_n and of $\bar{s}_n$ assures asymptotic bivariate normality (see Chapter 3) with asymptotic covariance $C = E_\theta\{k(X)s(X)\}$—that is, $(U_n, S_n) \sim AN\{(0,0), \Sigma\}$ when the parameter is θ, where Σ has variance terms τ^2 and B (each $1/\theta^2$) and covariance term C.
4. The score version of LeCam's third lemma then implies that, when the value of the parameter is θ_n, $U_n \sim AN(Ch, 1/\theta^2)$, so that only the covariance C is yet to be determined explicitly.
5. We find that

$$C = \frac{2}{\theta} E_\theta\{(\mathbf{1}[X \leq \frac{\log 2}{\theta}] - \tfrac{1}{2}) \cdot (X - \tfrac{1}{\theta})\}$$

which equals

$$\frac{2}{\theta}\int_0^{\frac{\log 2}{\theta}} x\theta e^{-x\theta} dx - \frac{1}{\theta^2}$$

yielding, after integration, $C = -(\log 2)/\theta^2$.

This result may look strange at first. However, we might expect the distribution of M_n to be centered at

$$MED(\theta_n) = (\log 2)/\theta_n = MED(\theta) - \frac{\log 2}{\theta^2}\frac{h}{\sqrt{n}} + o(\frac{1}{\sqrt{n}})$$

(by Taylor's theorem); that is, U_n should be centered at $-(1/\theta^2)(\log 2)h + o(1)$, as was just proved! Thus, it turns out that $C = (d/d\theta)MED(\theta)$. Also, we might expect the large-sample variance of M_n to be $1/(n\theta_n^2)$; but this is $1/(n\theta^2) + o(1/n)$, and so is essentially unchanged.

The above is an application of a general result, proved (*Exercise 21*) in the same way:

Proposition 4.4 *(Asymptotic Linearity Under Local Models): Consider a model for independent identically distributed observations from a density with parameter θ (of dimension d) which satisfies the Basic Assumptions, and suppose that $Y_n = y_n(\mathbf{X}_n) \sim AL(\nu_\theta, k_\theta(\cdot), \tau_\theta^2)$. Then*

$$\sqrt{n}(Y_n - \nu_\theta) \sim AN\left(Ch, \tau_\theta^2\right) \quad \text{under} \quad \theta_n = \theta + \frac{h}{\sqrt{n}} \tag{4.15}$$

where $C \equiv C(\theta) = \mathrm{cov}[k_\theta(X), s(X,\theta)]$ and $s(\cdot,\theta)$ is the score per observation.

Proposition 4.4 may be extended to d'-dimensional Y_n in an obvious way. If $h = 0$, (4.15) is essentially a consequence of the central limit theorem and Slutsky's theorem (see Chapter 3). Hence, Proposition 4.4 may be viewed as an extended central limit theorem, extended to moving parameters (*Exercise 23*).

As in the above example, it typically turns out that $C = \dot{\nu}_\theta$, but a direct proof is elusive except in special cases (*Exercise 23*). This identification is reassuring since we might expect $Y_n \sim AN(\nu_{\theta_n}, \tau_{\theta_n}^2/n)$ under θ_n and expect $\nu_{\theta_n} \approx \nu_\theta + \dot{\nu}_\theta h/n$ (by Taylor's theorem) and $\tau_{\theta_n}^2 \approx \tau_\theta^2$.

Exercises

1. Consider a test about the success probability θ in Bernoulli sampling, perhaps evaluating possible bias when flipping a coin. As background, DIACONIS, HOLMES, & MONTGOMERY (2007) have claimed that flipped coins caught in the hand have probability 0.51 of coming up as started. How many flips do we need to conduct to reject the hypothesis that $\theta = 0.50$ when the true value is $\theta = 0.51$? The test is of the null hypothesis that $\theta = 0.50$, rejecting if the proportion of successes is sufficiently different from 0.50, with significance level $\alpha = 0.01$ (based on the normal approximation to the binomial distribution of the success count). Calculate the sample size n, that would give power exceeding 0.99 of rejecting the null hypothesis when $\theta = 0.51$. How about for power 0.999? For more discussion of hypothesis testing, see Chapter 7.

2. Consider a random sample from $N(\theta, 1)$, and modify $\bar{X}_n$ as an estimate of θ as follows: Round $\bar{X}_n$ to 3 decimal places whenever the resulting rounding error is $< n^{-1/4}$ in magnitude, but otherwise do not round $\bar{X}_n$. Show that this estimate of θ is super-efficient at every θ for which $\theta \times 10^3$ is an integer. [Distinguishing between statistics that are rounded to a finite number of decimal places and those that are not can hardly have practical relevance!]

3. Verify, in the random sampling case, that the expected sample information equals the variance of the score ((2) of the Basic Assumptions). You may confine attention to real-valued θ. Assume that the orders of differentiation and integration can be interchanged, as needed.
4. Consider a Poisson random sampling model with

$$f_\lambda(x) \;=\; e^{-\lambda}\lambda^x/x! \quad \text{for} \quad x = 0, 1, 2, \ldots \quad \text{and} \quad \lambda > 0.$$

(a) Determine the score and sample information per observation and the expected information.

(b) Re-parameterize with $\gamma \;=\; \log\lambda$, repeat (a) in the new parameterization and check Equation (4.13).

Check (2) in both parameterizations.

5. Verify the Basic Assumptions for a *Cauchy location model*, with

$$f_\theta(x) = \frac{1}{\pi} \cdot \frac{1}{1 + (x - \theta)^2}.$$

[*Hint:* Show that the derivative of the sample information is bounded.]

6. Verify the Basic Assumptions for a *Cauchy scale model*, with

$$f_\tau(x) = \frac{1}{\pi} \cdot \frac{\tau}{\tau^2 + x^2}.$$

7. Prove the Regular Scores Theorem.
8. Consider a random sample from the *Laplace (double exponential)* density

$$f_\theta(x) \;=\; \tfrac{1}{2}\exp(-|x - \theta|), \theta \in \mathcal{R}.$$

(a) Show that the score (per observation) is $s(x, \theta) = \text{sign}(x - \theta)$. What is the information B?

(b) Show that

$$L_n(\theta, h) - \frac{h}{\sqrt{n}}\sum_{j=1}^{n} \text{sign}(X_j - \theta) \;+ \tfrac{1}{2}h^2$$

converges in mean-square (and hence in probability) to 0, and hence that Local Asymptotic Normality holds. [Note, however, that condition (A) fails, as does the Regular Scores property].

The following formulas may be useful in (b): When $\theta = 0$,

$$E|X - a| = 1 + \tfrac{1}{2}a^2 + O(a^3), E|X - a|^2 = 2 + a^2,$$

$E\{|X| \cdot |X-a|\} = 2 + O(a^3), E\{\text{sign}(X) \cdot |X-a|\} = -a + O(a^2).$

Verify these results. For some, you will need to partition the domain of integration into regions over which the integrand is differentiable.

9. Show that the Uniform Local Asymptotic Normality and Regular Scores properties hold for the two-parameter normal random sampling model.

10. Consider random sampling from a one-parameter exponential family in nonnatural form

$$f_\theta(x) = \exp\{c(\theta)t(x) - d(\theta) - s(x)\}\mathbf{1}_A(x)$$

for θ in an open interval Θ where $f_\theta(\cdot)$ is a density (of a discrete or absolutely continuous distribution).

(a) Show that the Basic Assumptions hold. Make any necessary assumptions about $c(\cdot)$, $t(\cdot)$, $d(\cdot)$, etc.

(b) Verify directly that the Uniform Local Asymptotic Normality and the Regular Scores properties hold (under modest conditions).

11. Consider a random sampling model with density $f_\theta(x)$ and real parameter θ. The *entropy* (per observation) is defined as $H(\theta) = E_\theta[-\log f_\theta(X)]$ (when existent). Argue (without rigor) that

$$\frac{1}{n}\log p_{n,\theta}(\mathbf{X}_n) = -H(\theta_o) - \tfrac{1}{2}B(\theta_o)\cdot(\theta-\theta_o)^2 + o_p(1)$$

when θ_o is the true parameter and θ is nearby. Hence, in a neighborhood of θ_o, the log likelihood function is, approximately, symmetric around θ_o, where its maximum occurs, and with a parabolic shape. This provides heuristic support for maximum likelihood estimation.

12. Prove the Parameter Transformation Proposition of Section 4.2.

13. Consider the Weibull distribution with shape parameter θ, as given in Section 4.2. Show that a logarithmic parameter transformation is *information stabilizing* in that the information for $\gamma = \log\theta$ is parameter-free.

14. Consider a two-sample model for a sample of size n_1 from a population with density $f_\theta(x)$ with $\theta = \theta_1$ and a sample of size $n_2 = [np_2]$ from the same population but with $\theta = \theta_2$. Each sample satisfies the Basic Assumptions of Section 4.2 with score (per observation) $s(x, \theta)$ and information $B(\theta)$. Now consider a joint model for the two samples, with n the total number of observations and write $n_1 = [np_1]$ and $n_2 = [np_2]$ with $p_1 + p_2 = 1$. Let ξ be the parameter vector (θ_1, θ_2).

(a) Identify the score vector and information matrix for the combined, two-sample, model, and show that Uniform Local Asymptotic Normality and Regular Scores properties hold. [*Hint:* Identify the component scores as $\sqrt{p_j}$ times the jth-sample score. You may assume θ to be real, for simplicity, but clearly this is not required.]

(b) Be specific in the case of sampling from a Poisson distribution (see *Exercise 4*) or from the exponential family (16).

Note: Each of the two components needs not be random samples as long as they are independent and each satisfies the Uniform Local Asymptotic Normality, Regular Scores and Continuity of Information properties.

15. Formulate a two-sample shift model, with sample sizes $n_j(n) \sim p_j n$ ($j = 1, 2$) and $p_1 + p_2 = 1$ ($p_j > 0$ and known), and *location parameters* μ and ν and common *scale parameter* σ — that is, sampling from $(1/\sigma)g\{(x-\mu)/\sigma\}$ and $(1/\sigma)g\{(x-\nu)/\sigma\}$. Find some sufficient conditions on the density $g(\cdot)$ for Uniform Local Asymptotic Normality, the Regular Scores Property and Continuity of Information to hold. [This is the regression problem of Section 4.3 but with a binary explanatory variable].

16. Consider the unconditional regression model (last paragraph of Section 4.3), and derive the scores and information for $\theta = (\alpha, \beta, \sigma)$.

17. Prove the Preliminary Lemma directly, without using moment generating functions.

18. Extend the Preliminary Lemma to k-dimensional $\mathbf{u}$; v remains real-valued.

19. What is the large-sample distribution of the score

    (a) in *Exercise 10* above when the parameter is $\theta + h/\sqrt{n}$ rather than θ?

    (b) in *Exercise 14* above with θ_j replaced by $\theta_j + h_j/\sqrt{n}$ for $j = 1, 2$?

20. Continuation of *Exercise 4*.

    (a) Find the 'moving parameter' large-sample distribution of the maximum likelihood estimate of λ.

    (b) Compare the maximum likelihood estimate with the *analog estimate* based on the proportion of zero values in the sample. What is the ratio of large-sample variances?

    (c) Determine the large-sample squared-correlation coefficient between these two estimates. [*Note:* It turns out that this coincides with the *asymptotic relative efficiency* of the analog estimate relative to the efficient estimate.]

21. Prove Proposition 4.4.

22. Find the 'moving parameter' large-sample distribution of the maximum likelihood estimate of θ when sampling from the exponential

density $f_\theta(x) = \theta \exp(-\theta x)\mathbf{1}[x > 0]$. Compare it with that for the quantile estimate (based on medians). [Note that, in both cases, the normalized error distribution is h-free—that is, the limit distribution of $\sqrt{n}(\hat{\theta}_n - \theta_n)$ under $\theta_n = \theta + h/\sqrt{n}$. This is a typical fact—as we shall see later.]

23. Suppose that $\bar{X}_n$ is the mean of a random sample of size n from a distribution with parameter θ (real), mean μ_θ and finite positive variance σ_θ^2. When the parameter is $\theta_n = \theta + h/\sqrt{n}$, show that $\bar{X}_n \sim AN(\mu_\theta + \dot{\mu}_\theta h/\sqrt{n}, \sigma^2/n)$. [You may assume differentiation may be passed under the integral sign.]

24. Consider a random sampling model with parameter θ and a statistic T_n which is $AL\{\theta, k(\cdot,\theta), \tau(\theta)^2\}$. Assuming that both $k(\cdot,\cdot)$ and $k(\cdot,\cdot)^2$ satisfy conditions like (B) or (B′) in the Basic Assumptions, show that
$$\frac{1}{n}\sum_{j=1}^{n} k(X_j, T_n)^2 = \tau^2 + o_p(1)$$
and that
$$\frac{1}{n-1}\sum_{j=1}^{n}\{k(X_j, T_n) - \bar{k}_n(\mathbf{X}_n, T_n)\}^2 = \tau^2 + o_p(1).$$
You may confine attention to the case of real-valued θ and T_n. [*Note:* Choosing T_n to be an estimate of θ, this provides a basis for estimating the standard error of T_n, alternative to $\tau(T_n)/n$.]

25. As in *Exercise 24*, but with an assumption only about $k(\cdot,\cdot)$, and assuming $E_\theta M(X,\theta)^2 < \infty$. [*Hint:* With $k_n = k(x, T_n)$ and $k = k(x,\theta)$, write $k_n^2 - k^2 = (k_n - k)^2 + 2k(k_n - k)$, and use the Cauchy-Schwarz inequality.]

5

Large Sample Estimation

Consider a model for $\mathbf{X}_n$, often a random sample, with d-dimensional parameter θ, specified by a density $p_{n,\theta}(\mathbf{x}_n)$, where θ belongs to some open set $\Theta \subset \mathcal{R}^d$. We first consider estimation of θ, that is 'fitting the model' by using the observed data $\mathbf{x}_n^*$ to choose an appropriate value of the parameter. Let $\tilde{\theta}_n$, a function of $\mathbf{x}_n$, to be evaluated at $\mathbf{x}_n^*$, represent a typical estimate of θ.

There are many ways of generating estimates. Popular methods include the *method of moments* (and other *analog methods*), *maximum likelihood*, *least squares* and *minimum chi-squared.* These are described in textbooks on mathematical statistics; for example, BICKEL & DOKSUM (2015), LARSEN & MARX (2018) and CASELLA & BERGER (2002). We first review the first of these methods. We next introduce and discuss large-sample properties of estimates, especially *consistency* and *asymptotic normality.* We then develop a theory of *efficient estimation*, characterizing and constructing such estimates. For this, we confine attention to a local analysis, assuming local asymptotic normality (see Chapter 4) and limiting attention to so-called *regular* estimates. We begin by presenting a closely parallel development, without asymptotics, of the *Cramér-Rao bound* on the variance of estimators that are exactly unbiased in finite samples.

It turns out that maximum likelihood estimates, or more precisely estimates closely related to maximum likelihood, are asymptotically efficient. Still, maximum likelihood itself can go wrong; see the counterexamples in *Exercises 8* and *22* and the more elaborate example in the Appendix. In this chapter, we assume that the full parameter vector θ is of interest. Section 5.1 presents some common approaches to deriving estimates. Section 5.2 outlines some exact theories of unbiased estimation in finite samples. The large-sample theory, with examples, is developed in Sections 5.3 to 5.5. Section 5.6 introduces M-estimation and *estimating equations.* The Appendix presents an unusual counter-example.

Extensions of the theory to address inference about only certain components of θ, functions of θ and under constraints on θ are addressed in Chapter 6. Limited attention to some other estimation problems appears in later chapters.

DOI: 10.1201/9780429160080-5

5.1 Moment and Quantile Estimation

Consider a random sampling model for a real-valued random variable X, with a d-dimensional parameter θ. Select d population moments, that is, moments of the distribution of X. Ordinary moments, central moments or cumulants may be used. The only condition is that these d moments, as functions of θ, be jointly invertible, at least in a neighborhood of the true θ. *Moment estimates* $\tilde{\theta}_n$ are obtained by setting the population moments equal their sample analogs and solving the resulting equations.

Alternatively, we may choose d population quantiles, that is quantiles of the distribution of X, or functions of these, and equate them to the corresponding sample quantile functions, and again solve for θ. Again we must assume an invertible relationship. The resulting solutions are called *quantile estimates.*

For a two-parameter model, it is often convenient to equate the sample mean and sample variance to the population mean and population variance or to equate the sample median and sample interquartile range to the population median and population interquartile range.

An additional modification is to use moments or quantiles of some real-valued function $T = t(X)$ rather than of X itself or to use the first moment of each of d such functions. This approach also allows extension of the quantile method to multivariate X; the moment method also extends readily to multivariate X. In particular, we might take $t(\cdot)$ to be the indicator variable of a set in the sample space of X and equate the probability of that set, the mean of the indicator variable, as a function of θ, to the sample proportion in that set. For example, to estimate the Poisson parameter λ, we may equate the Poisson probability $\text{pr}(X = 0) = e^{-\lambda}$ to the proportion p_n of values in the sample that are zero, solving to obtain $\tilde{\lambda}_n = -\log p_n$ as the resulting moment estimate. For several parameters, we may use indicator variables for several different sets. Specifically, we may equate the empirical distribution function F_n to the model distribution function F_θ at each of d specific arguments and solve for the d coordinates of θ. Again we equate sample proportions to probabilities. Although it is a special case of moment estimation, we sometimes distinguish this approach as the *distribution function method.* It is closely related to quantile estimation, where the proportions are fixed instead of the arguments of the distribution function.

As another example of moment estimation based on functions of X, the parameters of the Weibull distribution may be estimated by equating moments of $\log X$ to sample moments thereof. This turns out to be more convenient than using moments of X itself. However, whenever there is a simple closed-form expression for the distribution function of X, as in the Weibull case, estimates based on quantiles or estimates or distribution functions are usually more convenient than estimates based on moments. For bivariate normal X, the easiest method is to equate population and sample means, variances and covariance (or correlation).

Recall from Chapter 3 that sample moments, quantiles and proportions are all asymptotically linear under minor conditions. If the corresponding population values are smooth functions of θ, then the resulting estimates of θ will also be asymptotically linear. See Propositions 1 and 2 in Chapter 3. Hence, their large-sample behavior is readily available.

Specifically, consider a moment estimate $\tilde{\theta}_n$ based on $t(X)$; the dimension is unimportant, but the real-valued case is simpler. Suppose that $E_\theta\{t(X)\} = \nu(\theta)$ and $\text{var}_\theta\{t(X)\} = \Sigma(\theta)$, and that $\nu(\cdot)$ is invertible; write $\theta = \theta(\nu)$ for the inverse function, with derivative $\dot{\theta}\{\nu(\theta)\} = \{\dot{\nu}(\theta)\}^{-1}$. Then $\tilde{\theta}_n$ solves $\bar{t}_n \equiv (1/n)\sum_{j=1}^n t(x_j) = \nu(\tilde{\theta}_n)$; i.e., $\tilde{\theta}_n = \theta(\bar{t}_n)$. Hence, $\tilde{\theta}_n \sim AL\big[\theta, \dot{\theta}\cdot\{t(\cdot) - \nu(\theta)\}, \dot{\theta}\Sigma\dot{\theta}^T\big]$.

From the asymptotically linear representation of $\tilde{\theta}_n$ it follows that the asymptotic covariance matrix C of $\sqrt{n}(\tilde{\theta}_n - \theta)$ and the score statistic is the covariance of the kernel $\dot{\theta}\cdot t(X)$ and $s(X, \theta)$. This gives

$$C = \text{cov}_\theta\{\dot{\theta}\cdot t(X), s(X,\theta)\} = \dot{\theta}\cdot\text{cov}_\theta\{t(X), s(X,\theta)\}.$$

Assuming we can pass differentiation of $E_\theta\{t(X)\}$ under the integral sign, $\text{cov}_\theta\{t(X), s(X,\theta)\} = \int ts f_\theta = \int t\dot{f}_\theta = \dot{\nu}$. Therefore, $C = (\dot{\nu})^{-1}\dot{\nu} = I$, the identity matrix. If the model is discrete, the integral is replaced by a sum. This seemingly strange result will be important later. It may be helpful on first reading to go through the details of these two paragraphs when θ is real. The Weibull model with a single shape parameter discussed in Section 4.2 of Chapter 4, with $t(x) = \log x, \nu = 1/\theta$ gives an example.

A similar demonstration works for quantile estimates, also resulting in $C = I$; see *Exercise 11*. The distribution function estimate can be viewed as a special type of moment estimate as it equates the sample and population proportions of observations less than or equal to the specified x. We return to the apparently strange result that $C = I$ in the next section.

The large-sample variance of a moment, quantile or distribution function estimate may be estimated by substituting estimates of θ into the formula for the large-sample variance. Alternative estimates are also available; see *Exercises 24* and *25* in Chapter 4.

All of these estimation methods may be considered to be special cases of a more general *analog method* of estimation, whereby we equate a sufficient number of population and sample characteristics to uniquely determine the population parameters. The name arises because we estimate population characteristics by the analogous sample characteristics and then invert to determine corresponding parameter values. We must emphasize that much of the preceding development assumes that the support of X is free of θ. This assumption would be violated, for example, for the family of uniform distributions for which $f_\theta(x) = 1/\theta$ for $0 \le x \le \theta$. Also, asymptotic linearity may not be preserved when population threshold parameters are equated to extreme order statistics, such as the sample maxima and minima. We confine attention here to the moment and quantile versions and their variations, which are assured of asymptotically linear representations under minor assumptions, as was shown in Chapter 3.

5.2 Exact Unbiased Estimation from Finite Samples, the Cramér-Rao Bound

We start by presenting some classical theory on unbiased estimation in finite samples. This theory motivates the development in Section 5.3 of the concept of *asymptotic efficiency* within the class of *regular* estimates. Here we present a parallel approach to *efficiency* within the class of *unbiased* estimates in finite samples. Specifically, we present two characterizations of estimates that are exactly unbiased, and a lower bound on the variance of such estimates. This theory is not so successful in this nonasymptotic case, however, since the lower bound can rarely be achieved.

We assume that the potential data are $\mathbf{x}$ with model density $p_\theta(\mathbf{x})$ and unknown parameter θ. There is no need here for an index n. The dimension of θ is unimportant, but it may be advisable at first reading to assume that θ is real. We define the score to be the derivative of $\log p_\theta(\mathbf{x})$ with respect to (the coordinates of) θ; this is the 'total score', rather than the definition given in Chapter 4, which contains a divisor $\sqrt{n}$. We denote the 'total score' by $S'_\theta(\mathbf{x})$, with the prime used to distinguish it from other definitions of score. The information is here defined as the variance of this total score and denoted by B'_θ ('total information'). We assume that the score is well-defined and has expectation zero, and that B'_θ is finite and positive definite. Moreover, we assume, somewhat loosely, that the order of differentiation in θ and integration over $\mathbf{x}$ may be interchanged, as needed.

We start with a preliminary result.

Lemma 5.1 *For any statistic* $t(\mathbf{x})$, $\text{cov}_\theta\{t(\mathbf{X}), S'_\theta(\mathbf{X})\} = (d/d\theta)E_\theta\{t(\mathbf{X})\}$.

Proof In briefer notation, since $E(S') = 0$ and orders of operations can be interchanged, we find that that

$$\begin{aligned}\text{cov}(T, S') = E(TS') &= \int t(\mathbf{x})\{(\partial/\partial\theta)\log p_\theta(\mathbf{x})\}p_\theta(\mathbf{x})d\mathbf{x} \\ &= \int t(\mathbf{x})(\partial/\partial\theta)p_\theta(\mathbf{x})d\mathbf{x} = (d/d\theta)\int t(\mathbf{x})p_\theta(\mathbf{x})d\mathbf{x}. \quad \square\end{aligned}$$

Now $\tilde{\theta}(\mathbf{X})$ is *unbiased for* θ if $E_\theta\{\tilde{\theta}(\mathbf{X})\} = \theta$ for each value of θ; for later reference, we say it is *finite-variance unbiased* for θ when also $\text{var}_\theta\{\tilde{\theta}(\mathbf{X})\} < \infty$ for each θ. With $\tilde{\theta}(\mathbf{X}) = t(\mathbf{X})$, an estimate of θ, Lemma 1 leads immediately to

Proposition 5.1 $\tilde{\theta} = \tilde{\theta}(\mathbf{X})$ *is unbiased for* θ *if and only if* $\text{cov}_\theta(\tilde{\theta}, S'_\theta) = I$.

Furthermore,

Theorem 5.1 *(Orthogonal Decomposition of Unbiased Estimates): An estimate* $\tilde{\theta}$ *is finite-variance unbiased for* θ *if and only if*

$$\tilde{\theta}(\mathbf{X}) = \theta + B_\theta'^{-1} S'_\theta(\mathbf{X}) + R_\theta(\mathbf{X}) \tag{5.1}$$

for some function $R_\theta(\mathbf{X})$ *with, for each* θ, $E_\theta(R_\theta) = 0$, $\text{var}_\theta(R_\theta) < \infty$, *and* $\text{cov}_\theta(R_\theta, S'_\theta) = 0$.

That is, the error of estimation, $\tilde{\theta} - \theta$, is the sum of two uncorrelated terms, the first of which is common to all finite-variance unbiased estimates.

Proof First suppose that (5.1) holds with R satisfying the indicated properties. Taking expectations verifies unbiasedness, and the variance is clearly finite.

Now suppose that $\tilde{\theta}$ is finite-variance unbiased, and define R by $R_\theta = \tilde{\theta} - B_\theta'^{-1} S'_\theta$. Then (5.1) holds trivially. Now $E(R) = 0$ (since $\tilde{\theta}$ is unbiased and $E(S') = 0$), $\text{var}(R) < \infty$ (since $\tilde{\theta}$ and S' have finite variances and hence finite covariance), and $\text{cov}(R, S') = I - B'^{-1}\text{var}(S')$ (using Proposition 5.1); the right side is zero, completing the proof. □

Corollary 5.1 *Cramér-Rao Inequality. If* $\tilde{\theta}$ *is unbiased for* θ, *then*

$$\text{var}_\theta(\tilde{\theta}) \geq B_\theta'^{-1} \text{ for all } \theta \tag{5.2}$$

with equality holding for all θ *if and only if* $\tilde{\theta}(\mathbf{x}) = \theta + B_\theta'^{-1} S'_\theta(\mathbf{x})$ *with the right side being* θ*-free.*

When the dimension of θ is $d > 1$, the meaning of (5.2) is that the left side minus the right side is positive semi-definite. Equivalently, $\text{var}_\theta(c^T\tilde{\theta}) \geq c^T B_\theta'^{-1} c$ for every d-vector c.

Proof If $\tilde{\theta}$ has infinite variance, (5.2) is trivially true. Otherwise, from (5.1), $\text{var}(\tilde{\theta}) = (B')^{-1}B'(B')^{-1} + \text{var}(R) = (B')^{-1} + \text{var}(R) \geq (B')^{-1}$, with equality if and only if $R = 0$ (with probability one). □

Note The Corollary may be proved directly, without Theorem 1, using Proposition 1 and the fact that a squared correlation coefficient is at most unity. With a little extra effort, Theorem 1 may be derived as a corollary.

The *Cramér-Rao Inequality* (5.2) provides a lower bound on the variance of all unbiased estimates of θ. The limitation of this inequality is that the bound can be attained only in special cases, and hence it does not provide a useful basis for defining the efficiency of unbiased estimates. The problem is that $\theta + (B'_\theta)^{-1} S'_\theta$ is rarely free of θ, and hence is not a candidate for being an estimate of θ. Therefore, although $(B'_\theta)^{-1}$ is a legitimate lower bound on the variance of unbiased estimates, a *uniformly-minimum-variance unbiased estimate* may well have a larger variance. For an example, see *Exercise 4 part(b).* The value of the Cramér-Rao inequality in the theory of unbiased estimation was diminished once the Lehmann-Scheffé theory was developed, see for example LEHMANN & CASELLA (2006), as this provides an alternative method for identifying estimators with minimum variance among unbiased estimators. However, this approach requires an additional condition, completeness, that does not always hold.

The large-sample theory of regular and asymptotically efficient estimation developed in Section 4 parallels the theory presented here, with regularity playing the role of unbiasedness, each being a requirement that the distribution of the estimator be correctly centered around the true value of the parameter. There are only slight differences, due to the need for a role for n and slightly different definitions of scores and information. But in large-sample theory, the Cramér-Rao bound (in Section 4) can essentially always be attained and so provides a basis for defining *asymptotic efficiency* of *regular* estimates.

Note As a final comment, the theory presented here can be extended to unbiased estimation of a parametric function $\theta(\gamma)$ of the model parameter γ, as will be discussed in Chapter 6. Thus, $\tilde{\theta}$ is unbiased for $\theta(\gamma)$ if and only if $\text{cov}_\gamma(\tilde{\theta}, S'_\gamma) = \dot{\theta}(\gamma)$, $\tilde{\theta}$ is finite variance unbiased for $\theta(\gamma)$ if and only if $\tilde{\theta} = \theta + \dot{\theta}(B'_\gamma)^{-1}S'_\gamma + R_\gamma$ where R_γ has he same properties as R, and $\text{var}(\tilde{\theta}) \geq \dot{\theta}(B'_\gamma)^{-1}\dot{\theta}^T$, with equality if and only if $R \equiv 0$. See *Exercise 5.*

Extensions to two-sample, multi-sample and regression problems are straightforward.

5.3 Consistency, Root-n Consistency and Asymptotic Normality

We now revert to the asymptotic case. A sequence of estimates $\tilde{\theta}_n$ is *consistent for θ* if it converges in probability to θ:

$$P_{n,\theta}(|\tilde{\theta}_n - \theta| < \epsilon) \to 1 \text{ for each } \epsilon > 0.$$

Here, $|\cdot|$ represents any specified *norm*, such as the Euclidean norm. But, as noted in Chapter 2, marginal (or coordinatewise) convergence in probability is equivalent to joint convergence in probability, so it is sufficient to consider the coordinates of $\tilde{\theta}_n$ one at a time. Equivalently,

$$\tilde{\theta}_n = \theta + o_p(1) \text{ under } P_{n,\theta}. \tag{5.3}$$

Consistency is a very weak property, assuring only that the estimate is in the neighborhood of 'truth' with high probability whenever n is sufficiently large. It is therefore an essential requirement for a large-sample estimate.

In a moving-parameter model, with $\theta_n = \theta + h/\sqrt{n}$ we can extend the definition of consistency by replacing θ by θ_n (twice) in (5.3) — that is, requiring $\tilde{\theta}_n = \theta_n + o_p(1)$ under P_{n,θ_n}. But $\theta_n = \theta + O(1/\sqrt{n})$, and so the statement that $\tilde{\theta}_n = \theta + o_p(1)$ under P_{n,θ_n} is equivalent to the statement that $\tilde{\theta}_n = \theta_n + o_p(1)$ under P_{n,θ_n}. Under local asymptotic normality, the first of these statements follows from consistency under $P_{n,\theta}$. Hence, we conclude (under local asymptotic normality) that consistency under P_{n,θ_n} follows from consistency under $P_{n,\theta}$.

A stronger version of consistency is *root-n consistency*:

Given $\epsilon > 0$, there exists b for which $P_{n,\theta}(\sqrt{n}|\tilde{\theta}_n - \theta| < b) > 1 - \epsilon$ for all large n.

That is, the error $\tilde{\theta}_n - \theta$, when magnified by a $\sqrt{n}$ factor, is *bounded in probability*, or

$$\tilde{\theta}_n = \theta + O_p\Big(\frac{1}{\sqrt{n}}\Big) \quad \text{under } P_{n,\theta}. \tag{5.4}$$

This may be compared with the weaker (5.3). Again, marginal root-n consistency is equivalent. The usual method of verifying root-n consistency is through the slightly stronger condition that $\sqrt{n}(\tilde{\theta}_n - \theta) \sim AN(0, \Sigma)$ for some $\Sigma = \Sigma(\theta)$. Marginal asymptotic normality is enough. And each of these implies root-n consistency. Moreover, any asymptotically linear estimate, with the correct target value, is root-n consistent since it is asymptotically normal. To verify root-n consistency under a moving θ_n, LeCam's third lemma is the usual tool; here we need joint asymptotic normality of the estimator with the scores.

Since so many estimators are asymptotically normal, we focus on that case throughout this book; moreover, we sometimes assume or require asymptotic normality (correctly centered) when root-n consistency is sufficient. And we reserve the phrase 'asymptotic normality' to mean $\tilde{\theta}_n \sim AN\{\theta, (1/n)\Sigma\}$ for some Σ, implying correct centering and with large-sample variance of order $1/n$. We refer to properties of a $N\{\theta, (1/n)\Sigma\}$ distribution as *large-sample properties* of $\tilde{\theta}_n$. In particular, we write $\text{Avar}(\tilde{\theta}_n) = (1/n)\Sigma$, read 'asymptotic variance of $\tilde{\theta}_n$'—or 'large-sample variance'—as well as $\text{Avar}\{\sqrt{n}(\tilde{\theta}_n - \theta)\} = \Sigma$.

Once we have an estimate $\tilde{\theta}_n$ such that $\tilde{\theta}_n \sim AN\{\theta, (1/n)\Sigma\}$, the practical use of this estimate usually necessitates having an estimate of Σ since this is seldom known. Typically, $\Sigma = \Sigma(\theta)$ depends (continuously) on the value of the unknown parameter θ. Evaluation of $\Sigma(\theta)$ at $\theta = \tilde{\theta}_n$—say $\tilde{\Sigma}_n$—then provides a consistent estimate of Σ; sometimes other consistent estimates may be preferred (see later sections). For example, if $\tilde{\theta}_n \sim AL\{\theta, k(\cdot, \theta), \Sigma\}$, then the sample variance of the $k(X_j, \tilde{\theta}_n)$'s may provide a more convenient estimate (if k is sufficiently smooth in θ)—see *Exercises 25–26* in Chapter 4.

The asymptotic normality of $\tilde{\theta}_n$, along with an estimated variance, provides a gauge of the quality of the estimate. The square roots of the diagonal elements of $\tilde{\Sigma}_n$ provide estimated *standard errors* for the components of $\tilde{\theta}_n$, and the off-diagonal elements provide a basis for evaluating the joint behavior (dependencies) among the components—in particular, large-sample correlation coefficients.

One useful specific procedure is to construct a set of *confidence points* for each component of θ, namely, confidence bounds for a specified set of confidence coefficients—say, 0.1%, 0.5%, 1%, 2.5%, 5%, 10%, 25%, 50%, 75%, 90%, 95%, 97.5%, 99%, 99.5%, and 99.9%. Each confidence bound has the form $\tilde{\theta}_{jn} + z_\beta \tilde{\sigma}_{jn}$ where $\tilde{\sigma}_{jn}$ is the estimated standard error and $z_\beta = \Phi^{-1}(\beta)$ with β the confidence coefficient. The confidence coefficients are, of course, only based on large-sample approximations. And for any pair of coordinates of θ, a set of elliptical confidence contours can be constructed, based on the approximate bivariate normality of the corresponding estimates, again using a selection of confidence coefficients. Higher dimensional ellipsoidal confidence sets are available when $d > 2$, but are rarely so useful as the one- or two-dimensional versions. See Chapter 7 or CRAMÉR (1946) for the elliptical or ellipsoidal confidence sets. Still, common practice is to just present a single *confidence interval* for each coordinate of θ — e.g., 5% and 95% confidence points giving a 90% equal-tail confidence interval.

We now develop a large-sample theory of efficient estimation.

5.4 Joint Asymptotic Normality, Regular Estimators and Efficiency

We have seen in Section 1 that many estimators are asymptotically normally distributed, centered correctly at θ and with variance (or variance matrix) of order $1/n$. To simplify the development here, we shall confine attention to such estimators; in fact, we shall limit attention to estimators which are *jointly* asymptotically normal with the scores — that is, for which

$$\{\sqrt{n}(\tilde{\theta}_n - \theta), S_n\} \to_\theta N\left(\begin{pmatrix}0\\0\end{pmatrix}, \begin{pmatrix}A & C\\ C^T & B\end{pmatrix}\right) \tag{5.5}$$

for some $d \times d$ matrices $A = A(\theta)$ and $C = C(\theta)$, where '$\to_\theta$' stands for 'converges in distribution under the model $P_{n,\theta}$ with parameter-value θ'. This joint asymptotic normality is little more than the marginal asymptotic normality of $\tilde{\theta}_n$ together with the basic assumptions listed in Chapter 4, which define S_n and $B = B(\theta)$. In particular, joint asymptotic normality will hold whenever the basic assumptions hold and $\hat{\theta}_n$ is asymptotically linear. LeCam's third lemma, then allows us to deduce that $\sqrt{n}(\tilde{\theta}_n - \theta)$ is also asymptotically normal under $\theta_n = \theta + h/\sqrt{n}$; specifically,

$$\sqrt{n}(\tilde{\theta}_n - \theta) \to_{\theta_n} N(Ch, A). \tag{5.6}$$

More precisely, we limit attention to models in which the basic assumptions of Chapter 4 hold and to the *final* choice of estimates to those that satisfy the requirement of joint asymptotic normality with the scores (5.5); other estimates may play useful preliminary or comparative roles. The requirement of joint asymptotic normality can be avoided using more sophisticated methods, see Section 7 for discussion, but a detailed treatment is outside the scope of this book.

It was also found in Section 1 that, up to an appropriate $\sqrt{n}$ factor, the asymptotic covariance of an estimate with the scores is often the identity matrix. Such a fact about covariances also arises in exact estimation theory as a consequence of an unbiasedness requirement. In broad terms, unbiasedness is a requirement that the distribution of the error-of-estimation be centered at zero, with centering defined by expectation. In large-sample theory, we introduce a concept termed *regularity* which plays a similar role: It stipulates that the asymptotic distribution of the error of estimation be centered at zero, for every sequence θ_n (not just for every θ). Specifically:

Definition 5.1 *$\tilde{\theta}_n$ is <u>regular</u> if, for each θ, $\sqrt{n}(\tilde{\theta}_n - \theta_n)$ has, under $\theta_n = \theta + h/\sqrt{n}$, a limit distribution which is free of h.*

When $h = 0$, this limit distribution is $N\{0, A(\theta)\}$, see (5.5), and hence regularity requires it to be $N\{0, A(\theta)\}$ whatever the *direction* h. This is a kind of asymptotic local unbiasedness, the asymptotic distribution must be centered correctly, in all local neighborhoods of θ. Such regularity plays a role in large-sample theory analogous to the role of *unbiasedness* in small-sample theory discussed in the previous section, but it is much less restrictive.

Virtually all common estimators are regular, as we shall see. HÁJEK (1970) introduced the concept of regularity as a way of avoiding superefficient estimators such as *Hodges's estimate* (see Chapter 4), thereby providing the basis for a rigorous theory. Here is a verification of the irregularity of Hodges's estimate, building on the arguments given in Section 1 of Chapter 4. Re-define Z_n as $\sqrt{n}(\tilde{\theta}_n - \theta_n)$. Then, with $A_n = \{|\bar{X}_n| < n^{-1/4}\}$ as before and when $\theta = 0$, $\theta_n = h/\sqrt{n}$ and $P_{\theta_n}(A_n) \to 1$ as before. But $Z_n = -h$ in A_n and hence is neither asymptotically normal nor asymptotically h-free, since $Z_n \to_{\theta_n} -h$. (When $\theta \neq 0$, $Z_n \sim AN(0,1)$ as before, so irregularity occurs only at $\theta = 0$.) Another irregular estimate is *Stein's estimate* of a multinormal mean; see *Exercise 13.*

A simple characterization of regularity (compare with Proposition 1 in Section 5.2) may now be given, followed by an immediate corollary:

Proposition 5.2 *(Regular Estimates): The estimate $\tilde{\theta}_n$ is regular if and only if $C(\theta) \equiv I$.*

Here, $C \equiv \text{Acov}_\theta\{\sqrt{n}(\tilde{\theta}_n - \theta), S_n\}$, the large-sample covariance of $\tilde{\theta}_n$ with the scores.

Proof $\sqrt{n}(\tilde{\theta}_n - \theta_n) = \sqrt{n}(\tilde{\theta}_n - \theta) - h \to_{\theta_n} N(Ch, A) - h$ by (5.5). This limiting distribution is $N\{(C - I)h, A\}$, which is h-free if and only if $C = I$. □

Corollary 5.2
In a random sampling model, an estimate $\tilde{\theta}_n \sim AL\{\theta, k(\cdot,\theta), \Sigma\}$ *is regular if and only if*

$$\text{cov}_\theta\{k(X,\theta), s(X,\theta)\} = I \text{ for every } \theta. \tag{5.7}$$

This requirement looks peculiar; in particular, not only is the ith component of k uncorrelated with the jth component of s $(i \neq j)$, but the covariance between corresponding components must be unity. Regularity appears to be a demanding property! Yet, it commonly holds. It was already noted in Section 1 that (5.7) holds for all moment and quantile estimates (under minor conditions). A condition alternative to (5.7) is $-E_\theta\{\dot{k}(X,\theta)\} = I$; see *Exercise 12*.

Proposition 1 leads directly to a further characterization:

Theorem 5.2 *(Orthogonal Decomposition):* $\tilde{\theta}_n$ *is regular if and only if*

$$\sqrt{n}(\tilde{\theta}_n - \theta) = B^{-1}S_n + R_n \quad \text{where} \quad (R_n, S_n) \to_\theta N\left\{\begin{pmatrix}0\\0\end{pmatrix}, \begin{pmatrix}D & 0\\0 & B\end{pmatrix}\right\} \tag{5.8}$$

for some $D = D(\theta)$ *(possibly zero).*

Since the large-sample covariance of R_n and S_n is 0, we say they are *orthogonal*, and write '$R_n \perp S_n$'. Thus, in (5.8) the estimation error is decomposed into two orthogonal components, the first of which is common to *all* regular estimates. Compare with Theorem 1 in Section 5.2.

Proof Expression (5.8) implies (5.5) with $C \equiv \text{Acov}\{\sqrt{n}(\tilde{\theta}_n - \theta), S_n\} = B^{-1}B + 0 = I$; regularity then follows from Proposition 9.1. Conversely, if $\tilde{\theta}_n$ is regular, define R_n by (5.8): $R_n \equiv \sqrt{n}(\tilde{\theta}_n - \theta) - B^{-1}S_n$. By (5.5), R_n is asymptotically normal, centered at 0. The asymptotic covariance with S_n, using Proposition 1, is $I - B^{-1}B = 0$. □

Comparing (5.5) and (5.8), we have $A = B^{-1} + D$, and since D is a variance matrix we conclude that $A \geq B^{-1}$ with equality if and only if $D = 0$, i.e., if and only if $R_n = o_p(1)$. We therefore conclude:

Corollary (Large-Sample Cramér-Rao Inequality): *For regular* $\tilde{\theta}_n$*,*

$$\text{Avar}\{\sqrt{n}(\tilde{\theta}_n - \theta)\} \geq B^{-1} \tag{5.9}$$

with equality if and only if

$$\sqrt{n}(\tilde{\theta}_n - \theta) = B^{-1}S_n + o_p(1). \tag{5.10}$$

When $d > 1$, the inequality (5.9) between matrices means that the difference is positive semidefinite. Equivalently, the large-sample variance of $c^T\{\sqrt{n}(\tilde{\theta}_n - \theta)\}$ is at least $c^TB^{-1}c$ for every d-vector c, i.e., the large-sample variance of $c^T\tilde{\theta}_n$ is at least $c^TB^{-1}c/n$.

This corollary is the large-sample version of the *Cramér-Rao Inequality* of Section 5.2: We have replaced *unbiasedness* with *regularity* and asserted

(5.9) only for the *asymptotic* variance. In the unbiasedness case, the bound is rarely attainable, but, as we shall see in the next section, the bound for the large-sample variance can always be attained under modest regularity conditions.

Since B^{-1} is the smallest possible large-sample variance, we make the following

Definition 5.2 *An estimate $\hat{\theta}_n$ is asymptotically efficient if it is regular and*

$$\sqrt{n}(\hat{\theta}_n - \theta) \to_\theta N(0, B^{-1}). \tag{5.11}$$

Regularity would be implicit if (5.11) were replaced with $\sqrt{n}(\hat{\theta}_n - \theta_n) \to_{\theta_n} N(0, B^{-1})$ for every h, not just for $h = 0$.

An example of verifying asymptotic efficiency directly from the definition is as follows. Consider random sampling from an exponential family model in natural form:

$$f_\theta(x) = \exp\{t(x)\theta - d(\theta) - s(x)\}\mathbf{1}_A(x).$$

The dimension of θ is unimportant, but the case that θ is real is conceptually simpler. Moment estimation based on $t(X)$ is here identical to maximum likelihood, leading to the unique root $\hat{\theta}_n$ of the equation

$$\bar{t}_n(\mathbf{x}_n) \equiv \frac{1}{n}\sum_{j=1}^{n} t(x_j) = \dot{d}(\theta)$$

(unique because $\ddot{d}(\theta) = \mathrm{var}_\theta\{t(X)\} > 0$). The root may be found by Newton-Raphson iteration, starting from any initial guess at θ. Since $\hat{\theta}_n = g(\bar{t}_n)$ with $g(\theta) = \dot{d}^{-1}(\theta)$, Proposition 1 of Chapter 3 implies that

$$\hat{\theta}_n \sim AL\big[\theta, \ddot{d}(\theta)^{-1} \cdot \{t(x) - \dot{d}(\theta)\}, \ddot{d}(\theta)^{-1}\big].$$

Since $B = \ddot{d}$, (5.11) holds. And regularity of moment estimates was noted above (see remarks after (5.7). Thus, this moment estimate is asymptotically efficient.

From the definition and the corollary, we have

Theorem 5.3 *(Characterization of Asymptotic Efficiency): An estimate $\tilde{\theta}_n$ is asymptotically efficient if and only if (5.10) holds.*

Proof An estimate satisfying (5.10) is regular by Theorem 1, and (5.10) implies (5.11). The converse follows from the conditions for equality in the corollary. □

It is noteworthy, from (5.10), that all asymptotically efficient estimates are identical, up to $o_p(1/\sqrt{n})$:

$$\hat{\theta}_n = \theta + \frac{1}{\sqrt{n}} B^{-1} S_n + o_p\Big(\frac{1}{\sqrt{n}}\Big). \tag{5.12}$$

Again, note the analogies with the finite sample result in Section 5.2.

In the random sampling case, the characterization theorem may be re-expressed, using (5.12), as:

Corollary 5.3 *In a random sampling model, with score $s(x, \theta)$ per observation and expected information $B(\theta)$, an estimate $\tilde{\theta}_n$ is asymptotically efficient if and only if $\tilde{\theta}_n \sim AL\{\theta, B^{-1}s(\cdot, \theta), B^{-1}\}$.*

Hence, $B^{-1}s(\cdot, \theta)$ is sometimes called the *efficient influence function*, being the influence function (or kernel) of every efficient estimate $\tilde{\theta}_n$.

These characterizations are typically not directly useful for constructing asymptotically efficient estimators; after all, θ, B and S_n each depend on the unknown θ. However, it turns out that a version of (5.10), with θ replaced throughout by a suitable preliminary estimator, is the *primary* basis for constructing asymptotically efficient estimators (outside exponential families); see equation (5.14).

Now that asymptotically efficient estimators have been identified, it is useful to have a measure of relative efficiency of possibly inefficient estimators. For now, we confine attention to the single-parameter case, with $d_\theta = 1$. To compare any two regular estimators, the ratio of their large-sample variances provides a measure of relative efficiency. In particular, the large-sample variance of an asymptotically efficient estimator, divided by that of an estimator under consideration, is called the *asymptotic relative efficiency* of the latter estimator. Since large-sample variances are inversely proportional to sample sizes, the asymptotic relative efficiency can be interpreted as the ratio n/n', with n' being the sample size needed with the inefficient estimator to achieve the same asymptotic variance as an asymptotically efficient estimator using sample size n. We pursue this topic further in Chapter 6.

Recapitulation: In this section, we have developed the fundamental results of large-sample theory—except for practical methods of constructing asymptotically efficient estimates. Let us review what has been achieved, and its parallelism with minimum-variance unbiased estimation. The relationship between maximum likelihood and asymptotically efficient estimation will be developed in the next section.

In the formal exact theory of unbiased estimation, the focus is on the exact distribution of the *error of estimation*, namely $\tilde{\theta} - \theta$. *Unbiasedness* requires that this distribution be centered at zero (have mean zero), whatever the value of the unknown true parameter θ. The minimum-variance criterion judges such estimates by the variance of this error distribution, the smaller the better.

In large-sample theory, we give up attention to exact distributions and focus instead on large-sample approximations. In large samples, we can determine the true parameter within a shrinking neighborhood and hence only consider parameters of the form $\theta_n = \theta + h/\sqrt{n}$ for a fixed θ and varying h. The error of estimation is now approximately zero in that it converges in probability to zero. Hence, we need to deal with the *magnified error of estimation*, $\sqrt{n}(\tilde{\theta}_n - \theta_n)$. *Regularity* requires that this magnified error of estimation have a limiting distribution that is centered at 0, whatever the direction h—that is, for all θ_n in some fixed but shrinking neighborhood. This may be considered a special type of asymptotic unbiasedness. Since most estimators that we encounter are asymptotically normal, this limiting distribution is specified (without loss of generality) to be normal with mean zero.

In the exact small-sample theory, the *Cramér-Rao Inequality* provides a lower bound, related to inverse information, for the exact variance of any unbiased estimate. However, the bound is not very useful since it is attainable only in very special cases. Large-sample theory places a corresponding lower bound on the variance of the limiting distribution of the magnified error of estimation. We will find that this bound typically can be attained, enabling a useful definition of efficiency: *An estimator is asymptotically efficient if its magnified error of estimation, in a shrinking neighborhood, is asymptotically normal, centered at zero with variance equal to inverse information.*

We have seen in some examples that such efficiency is attainable. We now go on to develop general methods of constructing estimators that attain this bound—that are asymptotically efficient. We make use of various characterizations of regularity and of asymptotic efficiency developed above.

One last point: Unbiased estimates are not always available but their large-sample counterparts, regular estimates, typically are.

5.5 Constructing Efficient Estimates

In addition to the assumption of joint asymptotic normality with the scores, introduced in Section 4, we now require the properties of *Regular Scores* and *Continuity of Information, $B(\theta)$*. In random sampling models, these properties were shown to follow from the basic assumptions in Chapter 4; more generally, they must be taken as assumptions to be verified.

The Regular Scores property asserts that with $\theta_n = \theta + h/\sqrt{n}$, $S_n(\mathbf{x}, \theta_n) = S_n(\mathbf{x}, \theta) - Bh + o_p(1)$, uniformly for bounded h. As noted after Theorem 2 in Chapter 4, this uniformity allows the use of a data-dependent h so long as it is bounded in probability. We will be interested in a function $h_n(\mathbf{X})$ which moves θ to a value $\tilde{\theta}_n$ that is closer to a solution of the *score equation* $S_n(\mathbf{x}, \theta) = 0$. Consider $h_n = \sqrt{n}(\tilde{\theta}_n - \theta)$ so that $\tilde{\theta}_n = \theta + h_n/\sqrt{n}$, and suppose that h_n is bounded in probability (e.g., asymptotically normal). Then the Regular Scores expansion becomes (omitting $\mathbf{x}$ from the scores notation)

$$S_n(\tilde{\theta}_n) = S_n(\theta) - B \cdot \sqrt{n}(\tilde{\theta}_n - \theta) + o_p(1). \tag{5.13}$$

This expression (5.13) may be rewritten as

$$\sqrt{n}(\tilde{\theta}_n - \theta) = B^{-1}S_n(\theta) - B^{-1}S_n(\tilde{\theta}_n) + o_p(1). \tag{5.14}$$

Comparing with equation (5.8) in the statement of Theorem 2, we see that $\tilde{\theta}_n$ is regular if and only if $R_n = -B^{-1}S_n(\tilde{\theta}_n)+o_p(1)$. Now by Theorem 3, for such a $\tilde{\theta}_n$ to be asymptotically efficient, $R_n = o_p(1)$ and hence $S_n(\tilde{\theta}_n) = o_p(1)$ — and conversely. Note that $\tilde{\theta}_n$ is not required to be asymptotically normal jointly with the scores — marginal asymptotic normality, or just root-n consistency, is sufficient. We name this characterizing property:

Definition 5.3 *$\tilde{\theta}_n$ is a near root of the score equation if $S_n(\tilde{\theta}_n) = o_p(1)$.*

In the random sampling case, this becomes $\sum_{j=1}^{n} s(X_j, \tilde{\theta}_n) = o_p(\sqrt{n})$.

Now we state this characterization as a theorem:

Theorem 5.4 *Under the Regular Scores property, a root-n consistent estimate $\hat{\theta}_n$ is asymptotically efficient if and only if it is a near root of the score equation.*

This is reminiscent of maximum likelihood; Typically, maximum likelihood estimates solve the score equation $S_n(\theta) = 0$ [why?]. To use Theorem 5.4 for verifying the efficiency of any such root, it is necessary to verify that the root is asymptotically normal (or root-n consistent). But it is not often easy to verify that a root, or a near root, is asymptotically normal. This is especially true when the score equation must be solved by iterative numerical methods. The exponential family example above is an exception, not to the need for iteration, but to verification of asymptotic normality. This leads to consideration of *one-step* or *iterative* estimates, starting from a preliminary asymptotically normal estimate $\tilde{\theta}_n$, made explicit in the Theorem and Corollary below:

Theorem 5.5 *Suppose that the Regular Scores property holds. Then $\hat{\theta}_n$ is asymptotically efficient if and only if*

$$\sqrt{n}(\hat{\theta}_n - \tilde{\theta}_n) = B^{-1}S_n(\tilde{\theta}_n) + o_p(1) \tag{5.15}$$

for any (and hence every) root-n consistent $\tilde{\theta}_n$.

Proof The Regular Scores property applied to $\tilde{\theta}_n$ is

$$S_n(\tilde{\theta}_n) = S_n(\theta) - B\sqrt{n}(\tilde{\theta}_n - \theta) + o_p(1).$$

This may be rewritten as

$$\sqrt{n}(\tilde{\theta}_n - \theta) = B^{-1}S_n(\theta) - B^{-1}S_n(\tilde{\theta}_n) + o_p(1). \tag{5.16}$$

Applying the same procedure to $\hat{\theta}_n$ (now with $h_n(\mathbf{X}) = \sqrt{n}(\hat{\theta}_n - \theta)$) gives

$$\sqrt{n}(\hat{\theta}_n - \theta) = B^{-1}S_n(\theta) - B^{-1}S_n(\hat{\theta}_n) + o_p(1).$$

Subtracting these equations eliminates the terms involving θ, giving

$$\sqrt{n}(\hat{\theta} - \tilde{\theta}) = B^{-1}S_n(\tilde{\theta}) - B^{-1}S_n(\hat{\theta}_n) + o_p(1)$$

Since $\hat{\theta}_n$ is asymptotically efficient if and only if $S_n(\hat{\theta}_n) = o_p(1)$ by Theorem 5.4 the result follows. The "if and only if" in the previous sentence also shows that if (5.15) holds for any root-n consistent estimator $\tilde{\theta}_n$ it holds for all root-n consistent estimators.

□

Corollary 5.4 *(One-Stepping): If $\tilde{\theta}_n$ is root-n consistent, then the one-step estimate*

$$\hat{\theta}_n \equiv \tilde{\theta}_n + \tilde{B}_n^{-1} \frac{1}{\sqrt{n}} S_n(\tilde{\theta}_n), \tag{5.17}$$

where $\tilde{B}_n = \tilde{B}_n(\mathbf{x})$ is nonsingular and consistent for B, is asymptotically efficient.

In the random sampling case, (5.17) is

$$\hat{\theta}_n = \tilde{\theta}_n + \tilde{B}_n^{-1} \bar{s}_n(\tilde{\theta}_n) \quad \text{where} \quad \bar{s}_n(\theta) = \frac{1}{n} \sum_{j=1}^{n} s(X_j, \theta), \tag{5.18}$$

the 'average score'. Of course, the 'one-step' (5.17) or (5.18) may be repeated — until $S_n(\tilde{\theta}_n)$ is satisfactorily close to 0. But a single step is enough to assure asymptotic efficiency. Further steps may help to remove the dependence of the final estimator on choice of the initial estimator $\tilde{\theta}_n$, however.

Popular choices for estimates $\tilde{B}_n$ of the information B are:

i. $B(\tilde{\theta}_n)$ (assuming $B(\theta)$ is continuous),

and, in the random sampling case (see *Exercises 3–4*),

ii. 'average empirical information' $\bar{i}_n(\tilde{\theta}_n) = (1/n) \sum i(X_j, \tilde{\theta}_n)$, or

iiii. 'empirical variance of the scores'

$$\widehat{\text{var}}\{s(\tilde{\theta}_n)\} = \frac{1}{n-1} \sum \{s(X_j, \tilde{\theta}_n) - \bar{s}_n(\tilde{\theta}_n)\} \cdot \{s(X_j, \tilde{\theta}_n) - \bar{s}_n(\tilde{\theta}_n)\}^T.$$

Before further discussion, we consider an example. Consider the Weibull shape parameter example from Section 4.2 of Chapter 4. There we found

$$s(x, \theta) = \frac{1}{\theta} + (\log x)(1 - x^{\theta}), \quad i(x, \theta) = \frac{1}{\theta^2} + (\log x)^2 x^{\theta}$$

and $B = 1.8237/\theta^2$.

For a preliminary estimate, let us consider using the distribution function, which has a simple analytic form. The Weibull survival function is $\bar{F}_{\theta}(x) = \exp(-x^{\theta})$. Let p_c be the proportion of sample observations that exceed c, say; equate p_c to $\exp(-c^{\theta})$ and solve for θ, obtaining $\tilde{\theta} = \{\log(-\log p_c)\}/\log c$. Substitute this for θ in each score $s(x_j, \theta)$ and average them over the $x_1, \ldots, x_n$ to obtain $\bar{s}$. Formula (5.18) now only requires choice of $\tilde{B}$. The simple choice in this example is the substitution estimate $\tilde{B} = 1.8237/\tilde{\theta}_n^2$, choice (i) above. But the other choices are also straightforward.

The assumption that the preliminary estimate $\tilde{\theta}_n$ in (5.17) is root-n consistent is critical. Simple consistency is not enough. Root-n consistency puts $\tilde{\theta}_n$ in a neighborhood of the true value of θ with radius proportional to $1/\sqrt{n}$ (with high probability). Evidently, with high probability, the score is essentially linear in this neighborhood, so the score equation has only a single root and one-stepping gets close enough to it to ensure efficiency; iteration would locate it exactly. If there were more such roots, they must virtually coincide since asymptotically efficient estimates are unique up to $o_p(1/\sqrt{n})$. If $\tilde{\theta}_n$ were known only to be consistent, there is no assurance that it will fall in a small enough neighborhood of θ to locate the 'right' root of the score equation. Implicitly, it must be that 'wrong' roots are separated from the 'right' root by a distance of at least $O_p(1/\sqrt{n})$.

But suppose a root which locates a minimum of the log likelihood is found? Then the average empirical information, at this root, will be negative definite, and hence does not provide a reasonable estimate of B. So minimizing roots are not desirable. Since the theory guarantees that the score function is approximately linear, and its derivative approximately constant, in an $O_p(1/\sqrt{n})$ neighborhood of θ there can be no minimizing root in that neighborhood, so this is of no concern, asymptotically. If a minimizing root occurs in practice, the implication is that n is not sufficiently large for adequate large-sample approximations to be valid; or the assumed model may be incorrect. Roots that minimize the log likelihood are usually easily identifiable, but saddle points of a high-dimensional log likelihood may be hard to detect, a celebrated example being the 'errors in variable' regression model, see SOLARI (1969).

An exceptional situation occurs when the score equation has a unique root, locating a unique maximum of the likelihood. Then iteration from an arbitrary starting point is sufficient. For, assuming that an asymptotically normal estimate *does* exist (and moment or quantile methods will produce such an estimate), iteration, or repeated one-stepping, from it, would also find the unique root and confirm its efficiency. By this back door reasoning, we conclude that a maximum likelihood estimate which is a unique (and maximizing) root of the score equation is asymptotically efficient. (*Note*: Classical large-sample theory arrives at this conclusion more directly, but not the efficiency of one-stepping so easily, nor characterizations like those of Sections 5.4 and 5.5; see CRAMÉR (1946), RAO (1973), or SERFLING (1980).)

In conclusion, implementation of these theorems leads us to recommend any of the following variations on maximum likelihood:

1. Find a (near) root of the score equation and show it to be asymptotically normal (or vice versa: find an asymptotically normal estimate and show it to be a near root).

2. Find all roots (approximately) and, if there are multiple roots, choose one that is close to an asymptotically normal estimate.
3. Start with a preliminary asymptotically normal estimate, and 'one-step' or iterate towards a root.

Method (2) is especially useful in those special cases when there is a unique root. Method (3) is perhaps the most broadly useful. But however an estimate $\hat{\theta}_n$ is obtained, a corresponding information estimate $\hat{B}_n$ should be evaluated and determined to be positive definite (i.e., positive determinant), implying that $\hat{\theta}_n$ is a maximizing (near) root of the likelihood, and providing some confidence that the resulting estimate may be reasonable.

In random sampling models, we start with a moment, quantile or distribution function estimate $\tilde{\theta}_n$ and iterate (5.18) until satisfactory convergence obtains. If the score equation $\sum_{j=1}^{n} s(x_j, \theta) = 0$ can be shown to have a unique maximizing solution, iteration of (5.18) may proceed from an arbitrary initial value of θ. The estimate $\tilde{B}_n$ may be updated regularly, or perhaps only occasionally, along the way; call the final value, which should be checked to be positive definite, $\hat{B}_n$. We may then conclude that the resulting $\hat{\theta}_n$ is asymptotically efficient, and in particular, asymptotically normal around θ with large-sample variance estimated by $(1/n)\hat{B}_n^{-1}$. If the iteration appears to be unstable, a stabilizing factor of $\frac{1}{2}$ may be inserted in front of $\tilde{B}_n^{-1}$ in (5.18); but instability may also indicate that n is not sufficiently large for the asymptotic theory to apply. And although *any* asymptotically normal preliminary estimate will do, in practice it may be advisable to choose a fairly efficient one. Better yet, it is advisable to explore the log-likelihood function within a region of radius equal to several standard errors of the estimate, to identify any additional maxima, and gain confidence that the correct one has been selected.

Note that, contrary to common practice, there is no known justification for choosing the *absolute* maximum of the likelihood—that is, for choosing the root of the score equation that yields the absolute maximum of the likelihood; Methods (1) or (2) above are instead correct. LEHMANN (1980, 1999) has called such estimates *efficient likelihood estimates*—to contrast them with maximum likelihood estimates. Software packages typically produce maximum likelihood estimates not efficient likelihood estimates. Fortunately, they almost always agree.

Examples where maximum likelihood fails to be asymptotically efficient are hard to come by, at least if one restricts the definition of the maximum likelihood estimator to be an *internal* maximum (as here, since we have assumed that the parameter space is open). Likelihood functions that are unbounded near certain boundary points of the parameter space are easy to construct — in settings where an internal maximum is asymptotically efficient; see *Exercise 22*. An artificial example where an internal relative maximum of the likelihood function, but not the absolute internal maximum, is asymptotically efficient is given in the Appendix; it uses method (1) above. More commonly, maximum likelihood can also go wrong when the dimension of θ grows with n (see *Exercise 8*), but there is no general large-sample theory for this case.

We conclude this section with another example: random sampling from a two-parameter Weibull model, with density and survival functions

$$f_{\alpha,\beta}(x) = \beta\alpha^{\beta}x^{\beta-1}e^{-(\alpha x)^{\beta}}1_{\mathcal{R}^+}(x), \quad \bar{F}_{\alpha,\beta}(x) = e^{-(\alpha x)^{\beta}}\,(x > 0) \tag{5.19}$$

with parameters α and β, both positive. This model is popular in reliability and survival analysis applications; and not being in the exponential family, it provides a non-trivial example of the methodology.

Following the arguments used for a one-parameter Weibull model in Section 4.2 of Chapter 4, the scores and information per observation are found to be

$$s_{\alpha} = \frac{\beta}{\alpha}\{1 - (\alpha x)^{\beta}\}, \quad s_{\beta} = \frac{1}{\beta}\Big[1 + \{1 - (\alpha x)^{\beta}\}\beta\log(\alpha x)\Big]$$

$$i_{\alpha\alpha} = \frac{\beta}{\alpha^2}\{1 + (\beta - 1)(\alpha x)^{\beta}\}, \quad i_{\beta\beta} = \frac{1}{\beta^2}\Big[1 + \{\beta\log(\alpha x)\}^2(\alpha x)^{\beta}\Big],$$

$$i_{\alpha\beta} = \frac{1}{\alpha}\Big[(\alpha x)^{\beta}\{1 + \beta\log(\alpha x)\} - 1\Big],$$

$$B_{\alpha\alpha} = \frac{\beta^2}{\alpha^2}, B_{\alpha\beta} = \frac{\Gamma'(2)}{\alpha} \approx \frac{0.4228}{\alpha}, B_{\beta\beta} = \frac{1 + \Gamma''(2)}{\beta^2} \approx \frac{1.8237}{\beta^2}.$$

We describe two alternative approaches. In the first approach, we start with two analog estimates based on quartiles. Let q_1, m and q_3 be the sample quartiles, and Q_1, M and Q_3 the population quartiles. Then, from the survival function in (16), we find

$$Q_1 = \frac{1}{\alpha}\Big(\log\frac{4}{3}\Big)^{1/\beta}, \quad M = \frac{1}{\alpha}\big(\log 2\big)^{1/\beta}, \quad Q_3 = \frac{1}{\alpha}\big(\log 4\big)^{1/\beta}.$$

Hence,

$$\log(Q_3/Q_1) = \frac{1}{\beta}\log\left(\frac{\log 4}{\log\frac{4}{3}}\right) = \frac{1}{\beta}c(\text{say}).$$

Let $r = \log(q_3/q_1)$, and equate it to c/β yielding $\tilde{\beta} = c/r$. Also equate M to m yielding $\tilde{\alpha} = (\log 2)^{r/c}/m$. These are equivalent to quantile estimates based on the median and interquartile range, but for $\log x$ rather than x. Even simpler preliminary estimates are the distribution function estimates obtained by equating the proportion of the sample values exceeding c_j to $\exp\{-(\alpha c_j)^{\beta}\}$ for two numbers c_1 and c_2 and solving for α and β.

We can now iterate from (5.18), using $\tilde{B}_n = B(\tilde{\alpha}, \tilde{\beta})$. Alternatively, we could replace α and β by $\tilde{\alpha}$ and $\tilde{\beta}$ in the elements of the i-matrix and average over the observed x^*'s; or we could let $u_j = s_{\alpha}(x_j^*, \tilde{\alpha}, \tilde{\beta})$ and $v_j = s_{\beta}(x_j^*, \tilde{\alpha}, \tilde{\beta})$

and find the sample variances and covariance of the u_j's and v_j's; but $B(\tilde{\alpha}, \tilde{\beta})$ is the simplest in this example.

An alternative approach to finding asymptotically efficient estimators is possible here. Note that the score equation for α can be solved explicitly for α as a function of β:

$$\alpha = \left\{ \frac{1}{n} \sum_{j=1}^{n} x_j^{\beta} \right\}^{-1/\beta}. \tag{5.20}$$

Substituting this in the score equation for β yields a single equation in β which turns out to be monotone in β, so a unique root $\hat{\beta}$ — obtainable by iteration from any initial guess — is guaranteed. Then $\hat{\alpha}$ can be determined from (17), as a function of $\hat{\beta}$. This pair, $(\hat{\alpha}, \hat{\beta})$, being the unique root of the score equations, must agree with the root found by repeated one-stepping, and hence is asymptotically efficient. Computationally, the alternative approach is simpler in this example. It works, however, only when there is a unique root of the score equations. (If we had first determined that the score equations had a unique solution, we could have started the iteration in (5.18) from arbitrary initial values.) This example will be revisited in Chapter 6.

An example with multiple roots of the score equation is provided by sampling from a Cauchy distribution with known scale parameter but unknown location parameter. Here, it is advisable to one-step (or iterate) from an asymptotically linear preliminary estimate, based on quantile or sample distribution function (in this case not, of course, a sample mean).

Remark: Small-order adjustments can be made to an asymptotically efficient estimate without disturbing its efficiency. By expanding $S_n(\mathbf{x}_n, \hat{\theta}_n)$ around $\hat{\theta}_n = \theta$, for example, an order $O(1/n)$ bias of a root $\hat{\theta}_n$ of the score equation can be identified (at least formally), and $\hat{\theta}_n$ adjusted to eliminate it. See the discussion in COX & HINKLEY (1974), Section 9.2(vii). However, any transformation of θ and of a bias-reduced estimate would introduce bias once again; so if the object is 'model-fitting', little has been accomplished. However, such expansions—of both the expectation and variance of $\hat{\theta}_n$—may provide some improvement in the quality of the normal approximation.

5.6 *M*-estimation

M-estimates are analogous to maximum likelihood estimates, or more precisely, to roots of score equations, which give asymptotically efficient estimates), but with scores replaced by some 'pseudo-scores,' often called ψ-*functions.* The original motivation was to gain robustness: unbounded score functions lead to estimates that are asymptotically linear with unbounded kernels, and hence *non-robust.* The use of a bounded modification of the scores can yield robustness, possibly with only little sacrifice of efficiency. See SERFLING (1980), e.g. But there are other perspectives, which we comment on below.

M-estimates may be incorporated into a 'local' large-sample theory. Here we only summarize the theory for observations that are independent and identically distributed with model parameter θ. Although the dimension of θ may be arbitrary, it is easiest to focus on real θ.

Consider a function $\psi(x, \theta)$, of the same dimension as θ, for which

$$E_\theta\{\psi(X, \theta)\} = 0, \quad \text{var}_\theta\{\psi(X, \theta)\} \equiv A(\theta)$$

and

$$\text{cov}_\theta\{\psi(X, \theta), s(X, \theta)\} \equiv C(\theta) = -E_\theta\{\dot{\psi}(X, \theta)\}$$

(the last equality obtained by differentiating $E_\theta\{\psi(X, \theta)\}$ under the integral sign).

Definition 5.4 : *$\tilde{\theta}_n$ is an* <u>*M-estimate*</u> <u>*relative*</u> <u>*to*</u> *$\psi(x, \theta)$ if*

$$\frac{1}{\sqrt{n}} \sum_{j=1}^{n} \psi(X_j, \tilde{\theta}_n) = o_p(1). \tag{5.21}$$

That is, an M-estimate is a 'near root of the ψ-equation'. To solve the ψ-equation (5.21), we iterate from a root-n consistent estimate; or, we choose ψ to assure a unique root, and then solve iteratively from any starting point. The one-stepping formula is (5.18) with an estimate of $C(\theta)$ replacing $\tilde{B}$ and the pseudo-score ψ replacing the score s.

The large-sample behavior of an M-estimate may be developed in parallel to some of the developments for asymptotically efficient estimation. But here we need to be careful that, in contrast to working with the score equation, the covariance matrix involves both ψ and its derivative in θ. Recall the Regular Scores theorem (Theorem 2 and its Corollary) of Chapter 4. The same development would yield, upon writing the left side of (5.21) as $\Psi_n(\mathbf{x}, \tilde{\theta}_n)$ and $h_n = \sqrt{n}(\tilde{\theta}_n - \theta)$,

$$o_p(1) = \Psi_n(\tilde{\theta}_n) = \Psi_n(\theta) - C \cdot h_n + o_p(1).$$

This implies that $\sqrt{n}(\tilde{\theta}_n - \theta) = C^{-1}\Psi_n(\theta) + o_p(1)$, and hence is asymptotically normal with mean vector zero and variance matrix V given by the 'sandwich' formula $V = C^{-1}A(C^{-1})^T$ that is, with A 'sandwiched' between 'two slices of' C^{-1}. Hence, the large-sample performance of the M-estimate is judged by $V(\theta)$ or an estimate thereof. For θ real, V reduces to $1/(B\rho^2) \geq 1/B$ where $\rho = \text{corr}_\theta\{\psi(X, \theta), s(X, \theta)\}$.

Since B^{-1} is the best possible large-sample variance, we must have $V \geq B^{-1}$, or equivalently $V^{-1} \leq B$. A formal proof of this by matrix algebra appears in Appendix I of Chapter 6. Applying a Taylor expansion to (5.21), we find (under regularity conditions similar to (B) and (C) in Chapter 4) that the M-estimate $\hat{\theta}_n \sim AL\{\theta, C^{-1}\psi(\cdot,\theta), V\}$.

From another perspective, M-estimates are 'implicit moment estimates' in the following sense: A moment estimate based on a function $t(x)$ is defined by equating the empirical average of the $t(x)$'s to the expectation of $t(X)$ and solving for θ. Suppose instead we choose a function $t(x, \theta)$ and proceed similarly; or, since this t already depends on θ, subtract its expectation and call the resulting function $\psi(x,\theta)$, now with expectation zero. Equate the empirical average of the ψ's to their expectation zero, and solve for θ, thereby defining an M-estimate. Hence, the M might stand for 'moment-like' as well as 'maximum-likelihood like'.

Here is an example. We noted earlier when treating the Weibull shape-parameter example in Chapter 4 that $Y \equiv X^\theta$ has a standard exponential distribution, and hence unit expectation and unit variance. Hence, we may take $\psi(x,\theta) = x^\theta - 1$ and then $A = 1$ and $\dot{\psi} = (\log x)x^\theta = (\log y)y/\theta$. Using facts about derivatives of the gamma function given in Chapter 4, we find $C = -E_\theta\{\dot{\psi}\} = -(1/\theta)\Gamma'(2) \approx -0.4228/\theta$. Hence, $V = A/C^2 \approx \theta^2/0.4228^2 \approx 5.594\theta^2$. For comparison, recall that $B^{-1} \approx \theta^2/1.8237 \approx 0.5483\theta^2$, approximately one-tenth of V. Hence, this M-estimate is *very* inefficient!

As another example, suppose that θ is real and consider the *Huber ψ-function*

$$\psi(x,\theta) = med\{-c, s(x,\theta), c\} - d(\theta) \tag{5.22}$$

for given $c > 0$—that is, the score *censored* to lie between $-c$ and c, and then shifted by an amount $d(\theta)$ chosen to make the expectation of ψ equal to zero. When the underlying distribution is normal with unit standard deviation, then $d(\theta) = 0$ and $s(x,\theta) = x - \theta$, so that the influence function is proportional to (in fact the same as) that of the α-trimmed mean with $\alpha = \Psi(-c)$ plotted in Figure 3.2. Hence the two estimates must agree to $O_p(1/\sqrt{n})$. However, they are not identical. Huber's estimate includes outlying observations in the calculation, with nominal values $\pm c$, whereas the trimmed mean excludes them, but uses them in the calculation of the quantiles, which define the regions of exclusion.

There is also a body of literature about estimation based on *unbiased estimating equations* (Godambe, 1991); the average of the ψ's, when equated to zero, may be termed an 'unbiased estimating equation' in the sense that its expectation at the true θ is correctly zero. A root of the estimating equation is therefore an M-estimate.

5.7 Bibliographic Notes

Much of this chapter is based on the discussion of estimation in Hall & Mathiason (1990). The key Orthogonal Decomposition Theorem is a simpler version of Hájek (1970). The classical large-sample theory of estimation is presented, e.g., in Cramér (1946), in Rao (1973), in Serfling (1980), and in Lehmann (1999); a limitation is the necessity of identifying a root of the score equation that is consistent. Lehmann & Casella (2006) provided a bridge to the local theory presented here. A more sophisticated theory appears in Ibragimov & Has'minskii (1981) and in Chapter 2 of Bickel, Klaassen, Ritov, & Wellner (1993). In the latter, there is no requirement of confining attention to asymptotically normal estimates, and the resulting Orthogonal Decomposition Theorem (or 'Convolution Theorem') decomposes the limit distribution of the 'magnified' error of estimation of any regular estimate into the convolution of a quantity, the limiting version of $B^{-1}S_n$, that is distributed as $N(0, B^{-1})$ with a quantity, the limiting version of our R_n, that in their formulation may not be normally distributed. The large-sample Cramér-Rao Inequality still follows—now more generally—and so the resulting concept of optimality of estimates (asymptotic efficiency), and its characterization, is still appropriate; that is, the asymptotic variance of any regular estimate that is *not* asymptotically normal exceeds that of an asymptotically efficient (and hence asymptotically normal) estimate, and so nothing is lost by limiting attention to asymptotically normal estimates. Hampel (1974) gave an excellent introduction to influence functions and M-estimation for location-scale models.

Appendix: A Counterexample to the Efficiency of Maximum Likelihood

We present an example with the following characteristics: It is a discrete random sampling model with a real parameter θ belonging to an open subset Θ of the real line. The basic assumptions are satisfied, and there is a unique maximum likelihood estimator $\hat{\theta}_n$. But $\hat{\theta}_n$ is not even consistent, much less asymptotically efficient. The maximum likelihood estimator is a root of the score equation, but the wrong root; another root, giving a relative, but not absolute, maximum, of the likelihood function is asymptotically efficient. This example, a modification of one due to Kraft & LeCam (1956), requires knowledge of the distinctions between rational and irrational numbers, especially that the rational numbers form a countable set that can be rearranged into an ordered list.

In subsection A1, we consider a deceptively simple sampling problem with an unexpected outcome, estimation of a parameter known to be rational. This problem forms a component of the main example presented in A2, but is of some interest in itself. It might seem that the restriction to rational values would not dramatically change the estimation problem. However, this is not so. For simplicity, we express the problem in the classical language of sampling of balls from urns.

A.1 Estimating the Proportion Red When Sampling Colored Balls from an Urn

Consider sampling *with replacement* from an urn containing balls of two colors, red and black (at least one of each), but the numbers of each are unknown — as is M, the total number of balls in the urn. A prize will be awarded if the proportion p of red balls is guessed exactly; no prize is given for an incorrect guess, no matter how close.

A contestant is allowed to draw randomly from the urn, with replacement, repeatedly. The contestant is not allowed to see the urn; no information about the total number of balls in it is available, indeed, it may be extremely large.

Let $\mathbf{X}_n = (X_1, X_2, \ldots, X_n)$ represent the results of the first n draws (Bernoulli trials), and let $\hat{p}_n(\mathbf{X}_n)$ be the maximum likelihood estimate of p, namely the proportion S_n/n of red balls in the first n draws. Clearly $\hat{p}_n$ must always be rational, as is p, and eventually will be in (0, 1). As stated, this is really a two-parameter problem, say $(R, B) = (\#red, \#black)$. But only the ratio R/B or $R/(R + B) = R/M \equiv p$ is identifiable, since the model distribution depends only on p.

Conventionally, we would treat this as a Bernoulli sampling problem with p in the continuum (0, 1), even though p is known to be rational. Since p *is* rational, and hence has a countable number of possible values, we perhaps can achieve a positive probability of estimating it exactly, and thereby of winning the prize. In contrast to conventional estimation, there is no reward here for a near miss; the loss function is simply zero or one according as the guess is incorrect or correct.

We show that a contestant relying on maximum likelihood has probability tending to zero (as n increases) of winning the prize, whereas there exist strategies, one is made explicit below, which have probability of winning tending to one: We hasten to add that the result is of no practical value unless a strong prior ordering can be placed on the likely values.

Proposition A - 1 *(i) For each rational* $p \in (0,1)$, $P_p\{\hat{p}_n(\mathbf{x}_n) = p\} \to 0$.

(ii) There exists a sequence of estimates $\tilde{p}_n(\mathbf{x}_n)$ *for which, for each rational* $p \in (0,1)$, $P_p\{\tilde{p}_n(\mathbf{x}_n) = p\} \to 1$.

The limits are as $n \to \infty$; the number M is fixed but unknown throughout. Of course, the maximum likelihood estimator becomes *close* to p with high probability as n increases, but we claim that it will not exactly equal p, also with high probability.

An intuitive drawback to the maximum likelihood estimator is that the rational number guessed is among those with denominator n, the sample size, rather than with denominator M, the total, but unknown, number of balls in the urn. Whatever the true value of M, eventually n will become greater than M and the only possibility of a correct guess is when the count of times a red ball is drawn has some common factors with n. But a proof of the poor performance of the maximum likelihood estimator is even simpler.

To prove (i), we note that the probability of the maximum likelihood estimator being correct is the probability that $S_n = np$, and this is bounded above by the maximum term in the binomial (n, p) distribution. From a normal approximation, this is $\{np(1-p)\}^{-1/2}\varphi(0)+o(1) = O(n^{-1/2})$, with $\varphi(\cdot)$ is the standard normal density. (See FELLER (1968), Section VII.2, for a detailed argument.) Hence, the probability of being correct tends to zero as n increases; indeed, it is $O(n^{-1/2})$.

A Bayesian, who, with high probability, limits the total number of balls in the urn through a prior distribution, may do better. We proceed in an *ad hoc* manner instead.

Order the rationals in (0,1) into a_1, a_2, ..., using the ordering 1/2, 1/3, 2/3, 1/4, 3/4, 1/5, 2/5, 3/5, 4/5, 1/6 .... Let R_N be the set of all rationals of the form r/m with $0 < r < m \leq N$, Then R_N contains all the possible values of p when $M \leq N$. Let $\#(N)$ be the cardinality of R_N. Define the *minimum spacing* d_N *in* R_N as the minimum of $|a_i - a_j|$ over $i < j \leq \#N$. Now (for $N > 2$)

$$\begin{aligned} d_N &= \min\left\{\left|\frac{r}{l} - \frac{s}{m}\right| : rm \neq sl, 0 < r < l \leq N, 0 < s < m \leq N\right\} \\ &= \min\left\{\left|\frac{rm - sl}{lm}\right|\right\}, \end{aligned}$$

which exceeds $1/\{N(N-1)\}$, since the numerator is greater than or equal to one and the denominator is less than or equal to $N(N-1)$ unless $l = m = N$; but in the latter case, the minimum is $1/N$ which is clearly too big. Now $1/N$ and $1/(N-1)$ are both in R_N, with spacing $1/\{N(N-1)\}$, and so $d_N = 1/\{N(N-1)\}$. Note that $d_N < 1/N$, the minimal spacing from zero or one to a member of R_N, so the definition of d_N can be expanded to allow spacings from zero as well as within the a_i's (and hence also for $N = 2$).

Now let $N = [n^{1/5}]$, where $[\cdot]$ denotes the integer part, so that $d_N \geq n^{-2/5}$. Then choose as $\tilde{p}_n$ that $p \in R_N$ which is closest to the maximum likelihood estimator $\hat{p}_n$ (and the smaller p if there is a tie).

This $\tilde{p}_n$ concentrates on performing well only when p is in the finite set R_N, in particular, when the total number of balls in the urn is at most N. Since the members of R_N are separated, it turns out that we can pick out the correct member of R_N with high probability by making n large enough. We give up all hope of estimating p exactly when p is *not* in R_N, and, with certainty, estimate it incorrectly. The set R_N is then allowed to grow, slowly, with increasing n. Whatever the value of p, eventually n will be large enough so that $p \in R_N$, and by then n will also be large enough so that p may be estimated exactly with high probability. We now formalize this argument.

Write $\tilde{P}_{n,p}$ for the probability of winning when using $\tilde{p}_n$, namely $P_p(\tilde{p}_n = p)$. If $p \notin R_N$, then $\tilde{P}_{n,p} = 0$. But for n (and hence N) sufficiently large, p will be in R_N. And if $p \in R_N$, $\tilde{p}_n$ will be correct whenever $\hat{p}_n$ is sufficiently close to p, in particular, when $|\hat{p}_n - p| < \frac{1}{2}d_N$. We find for such p

$$\tilde{P}_{n,p} \geq P_p\Big(|\hat{p}_n - p| < \tfrac{1}{2}d_N\Big) = P_p\Big(|S_n^*| < \tfrac{n^{1/2}}{2\sqrt{pq}}d_N\Big) \geq P_p(|S_n^*| < n^{1/2}d_N) \tag{A-1}$$

where S_n^* is the standardized binomial variable $(S_n - np)/\sqrt{(npq)}$ with $q = 1-p$. Since $d_N \geq n^{-2/5}$, the last expression in (A-1) exceeds $P_p(|S_n^*| < n^{1/10})$. Given a small positive ϵ, and for $n \geq \{\Phi^{-1}(\frac{1}{4}\epsilon)\}^{10}$ (where Φ the standard normal survivor function), $P_p(|S_n^*| \geq n^{1/10}) \leq \frac{1}{2}\epsilon + o(1)$ and hence less than ϵ for large enough n. Hence, $\tilde{P}_{n,p} > 1 - \epsilon$ for all large enough n, proving (ii). □

Here are some numbers. Let $\delta_n = 2\bar{\Phi}(\sqrt{n}d_N)$, a normal approximation to an upper bound on the probability of error of $\tilde{p}_n$ (namely, $1 - \tilde{P}_{n,p}$) when $p \in R_N$ (from (A-1)). We find that

n	N	$\#(N)$	d_N	δ_n
250	3	3	0.167	.008
1024	4	5	0.083	.008
10^4	6	11	0.033	.0009
10^5	10	31	0.011	.0004
10^6	15	71	0.0048	.000002
10^8	39	475	0.0007	10^{-10}
10^{10}	100	3059	0.0001	10^{-22}

For example, suppose that we sample $n = 1024$ times. Then $N = [1024^{1/5}] = 4$, $R_N = \{1/2, 1/3, 2/3, 1/4, 3/4\}$, $\#(N) = 5$ and $d_N = 1/12 \approx 0.083$. The approximate bound on the error probability is $\delta_n = 2\bar{\Phi}(1024^{1/2}/12) = 2\bar{\Phi}(2.667) \approx 0.0077$. Hence, the probability of winning exceeds 99% if $p \in R_4$. This includes all cases when the number of balls is either 2, 3 or 4. However, the probability of winning is zero for all other values of p. If the urn contains at most 15 balls, and we sample one million times, we will win with probability exceeding 99.999%, but if p is not equal to one of the 71 numbers in R_{15}, we will lose with certainty. An extraordinarily large n is needed to deal effectively with the possibility that the number of balls in the urn is as large as 100.

We can modify the argument to make N increase slightly faster, and δ_n decrease more slowly, by choosing $N \approx cn^{(1/4)-\delta}$ for δ very small, instead of $N \approx n^{1/5}$. But still N grows very slowly with n. And $\delta_n \approx 2\bar{\Phi}(c^{-2}n^{2\delta})$. But for our purposes in the next section, we only need a solution with the asymptotic properties established here.

This example demonstrates that theory developed for parameters in a continuum may fall apart in a countable parameter space, even if dense! (Theory of estimation in finite parameter spaces even goes by another name, namely 'multiple decision theory', and resembles hypothesis testing more than estimation.)

A.2 A Regular Estimation Problem with an Inconsistent Maximum Likelihood Estimator

We now present a modification of an example of KRAFT & LECAM (1956). Consider random sampling from two independent Bernoulli sequences simultaneously: (X_1, Y_1), (X_2, Y_2), ..., with all variables independent, X_i Bernoulli(p) and Y_i Bernoulli(π). Both p and π are functions of θ, a real parameter belonging to one of the open intervals $A_k = (2k, 2k+1)$, $k = 1, 2, \ldots$; specifically, $p(\theta) = a_k$ if $\theta \in A_k$, and $a_1, a_2, \ldots$, is the ordering of the rationals in (0,1) defined in Section A1, and $\pi(\theta) = \theta[\text{mod}1] = \theta - [\theta]$.

Thus, the parameter of the X-sequence of X's identifies which interval A_k contains θ, and the parameter of the sequence of Y's identifies the position of θ is within the interval A_k. The pair of Bernoulli parameters, one rational and one real, have been put in one-to-one correspondence with a single real parameter θ. The set Θ of possible values of θ is a union of open intervals of real numbers and so is itself open. It is not a connected set, but this does not affect the theory.

Estimation of θ proceeds in two steps. First, the sequence of $\{X_i\}$ must be used to determine which interval A_k contains θ. In Section A1, we showed that the maximum likelihood estimator performs poorly in identifying the correct interval, and presented an alternative estimate that identifies it with high probability if n is sufficiently large. Using the latter method to identify the correct interval A_k, we then use the sequence of $\{Y_i\}$ in the usual way to estimate efficiently where θ lies within A_k. Here is a sketch of the details:

A plot of $p(\theta)$ versus θ would show p constant within each interval A_k with value a_k in A_k) and that of $\pi(\theta)$ versus θ a line segment rising from 0 to 1 in each A_k. Hence, the derivatives of $p(\theta)$ and $\pi(\theta)$ are zero and one, respectively. Now the log likelihood for a single observation (x, y) is

$$l_\theta = x \log p + (1-x)\log(1-p) + y \log \pi + (1-y)\log(1-\pi)$$

with p and π depending on θ, so that the score and information (per observation), and the expected information, are found to be

$$s(x,y) = \frac{y}{\pi} - \frac{1-y}{1-\pi}, i(x,y) = \frac{y}{\pi^2} + \frac{1-y}{(1-\pi)^2}, B(\theta) = \frac{1}{\pi(1-\pi)}$$

with $\pi = \pi(\theta)$, the same as if we sampled only the Y's. The basic assumptions all hold, and the score equation is found to be $\pi(\theta) = \bar{Y}_n$.

There is a root of this equation in every interval A_k, namely $\tilde{\theta}_{kn} = 2k + \bar{Y}_n$ in A_k. Note also that each is a maximizing root (since the score is concave in each A_k). Now the log likelihood after n pairs of Bernoulli trials is

$$l_n(\theta) = S_n \log p(\theta) + (n - S_n)\log\{1 - p(\theta)\} + T_n \log \pi(\theta) + (n - T_n)\log\{1 - \pi(\theta)\}$$

where S_n and T_n are the numbers of successes in the first n X- and Y-trials. Hence,

$$l_n(\tilde{\theta}_{kn}) = S_n \log a_k + (n - S_n)\log(1 - a_k) + g(T_n, n),$$

say, which is uniquely maximized over k by choosing the k for which $a_k = S_n/n$. This identifies the maximizing root—the unique maximum likelihood estimator $\hat{\theta}_n = 2k + T_n/n$ with k chosen as just indicated.

An alternative estimate $\tilde{\theta}_n$ chooses a different one of the roots $\tilde{\theta}_{kn}$, using the scheme described in Section A1 to choose a rational number a_k and thence k. Now suppose that $\theta \in A_k$. Then, for n sufficiently large, $\tilde{\theta}_n$ will identify k correctly with probability close to 1, whereas $\hat{\theta}_n$ will do so with probability close to 0. The large-sample behavior of $\tilde{\theta}_n$ will therefore be asymptotically identical to the behavior of $2k + T_n/n$, and hence $\tilde{\theta}_n$ will be consistent with $\theta_n \sim AN(\theta, \pi(1-\pi)/n)$, and therefore asymptotically efficient. By contrast, the maximum likelihood estimator will not even be consistent; in fact, $P_\theta(|\hat{\theta}_n - \theta| < 1) = P_\theta(\hat{\theta}_n \in A_k) = O(n^{-1/2})$.

Formally, with $E_n \equiv \{\sqrt{n}(\tilde{\theta}_n - \theta) \leq z\}$ and $F_n \equiv [\sqrt{n}\{\bar{Y}_n - \pi(\theta)\} \leq z]$, we have

$$\begin{aligned}
P_\theta(E_n) &= P_\theta\big(E_n \cap \{\tilde{\theta}_n \in A_k\}\big) + P_\theta\big(E_n \cap \{\tilde{\theta}_n \notin A_k\}\big) \\
&= P_\theta\big(F_n \cap \{\tilde{\theta}_n \in A_k\}\big) + o(1) \\
&= P_\theta(F_n) - P_\theta\big(F_n \cap \{\tilde{\theta}_n \notin A_k\}\big) + o(1) \\
&= P_\theta(F_n) + o(1) = \Phi\Big\{\frac{z}{\sqrt{\pi(1-\pi)}}\Big\} + o(1),
\end{aligned}$$

and hence $\tilde{\theta}_n$ is asymptotically normal as claimed. Regularity of $\tilde{\theta}_n$ may also be proved, since, for large n, $\theta_n \in A_k$ whenever $\theta \in A_k$ and asymptotic normality under θ_n established just as under θ. Or, alternatively, use Proposition 1.

We summarize in

Proposition A - 2 *(i) The unique maximum likelihood estimate $\hat{\theta}_n$ is a root of the score equation but is not consistent and hence not asymptotically efficient (ii) There is a root, $\tilde{\theta}_n$, of the score equation with $\tilde{\theta}_n \sim AN\{\theta, (1/n)B(\theta)^{-1}\}$ and regular, and hence asymptotically efficient.*

The trick was to put together two models, one with a countable parameter space (the first Bernoulli sequence), and the other a very simple one which is regular and with parameter space the unit interval (the second Bernoulli sequence). This was achieved by pasting together a sequence of unit intervals, without destroying the regularity. The maximum likelihood estimator in the first sequence estimates its parameter $p(\theta)$ 'closely', but not exactly, and closeness on the p-scale does not yield closeness on the θ-scale.

For some other counter-examples regarding maximum likelihood, see Chapter 6 in LEHMANN & CASELLA (2006). For a general discussion of issues related to multiplicity of roots, see SMALL, ET AL. (2000). The Cauchy location problem is a well-studied case; see REEDS (1985). He showed that there can be as many as n relative maxima, but the actual number converges in distribution to a random variable $1 + M$, where M has a Poisson distribution with mean $1/\pi$; hence about 73% of the time there is a single maximum, close to

$\exp(-1/\pi) \approx 0.73$, roughly 23% of the time there are two relative maxima, and the remaining 4% of the time additional maxima occur, possibly, but rarely, a great many.

Exercises

1. The parameter λ of a Poisson distribution can be estimated by equating the proportion of zero's in the sample to $e^{-\lambda}$ and solving for λ. Calculate the asymptotic variance of this estimate and compare it to that of the sample mean.

2. Consider estimation of the mean of a normal distribution with known variance, based on a random sample. Show that the Cramér-Rao bound on the variance of unbiased estimates is attainable.

3. Consider estimation of the Poisson parameter λ, based on a random sample. Show that $\lambda + B_\lambda^{-1} S_\lambda(\lambda)$ is free of λ, and hence is a uniform minimum variance unbiased estimator of λ.

4. (a) Consider estimation of the parameter θ in an exponential distribution parameterized as $f_\theta(x) = (1/\theta)\exp(-x/\theta)$ for $x > 0$, based on a random sample ($n > 2$). Determine the Cramér-Rao bound and see if it can be attained. Find the unique uniform minimum variance unbiased estimator of θ and its variance, using the Cramér-Rao inequality, if usable, or some other method if not.

   (b) Repeat, parametrizing as $f_\gamma(x) = \gamma \exp(-x\gamma)$ for $x > 0$ and now estimating γ. [*Hint:* The notion of completeness, not discussed here, allows attention to be restricted to estimates of the form $c_n/\bar{X}$, for suitably chosen c_n].

5. (a) Extend the Proposition, Theorem and Corollary of Section 2 to the case of estimating a parametric function, as summarized in the last paragraph of that section.

   (b) Consider the model in *Exercise 4(b)*, but now estimating $\theta = 1/\gamma$. Apply the results in *Exercise 5(a)* and show that they agree with those in *Exercise 4(a)*.

6. Consider the estimation of the correlation coefficient ρ of a bivariate normal distribution with means zero and unit variances. The joint density function is

$$f(x, y, \rho) = \frac{1}{2\pi\sqrt{(1-\rho^2)}} \exp\left\{-\frac{x^2 + y^2 - 2\rho xy}{2(1-\rho^2)}\right\}.$$

(a) Show that the score function for ρ from a single observation is

$$s_\rho(x,y) = \frac{\rho}{1-\rho^2} - \frac{\rho(x^2+y^2-2\rho xy)}{(1-\rho^2)^2} + \frac{xy}{1-\rho^2}.$$

Extend this equation to the case of estimation of ρ from a random sample $(X_1,Y_1),\ldots,(X_n,Y_n)$.

(b) Show that setting the score (or total score) equal to zero yields a cubic equation in ρ which always has at least one real root in the interval $[-1,1]$. Show that this root has the same sign as $\sum xy$. [*Hint:* $2xy \le x^2+y^2$ for all (x,y)].

(c) Verify from (a) that $E\{s_\rho(X,Y)\} = 0$.

(d) Calculate the observed information from a single observation and show that the expected information is $(1+\rho^2)/(1-\rho^2)^2$. Compare with Exercise (32) in Chapter 3.

7. A *geometric distribution* with parameter p is a distribution on the non-negative integers with

$$f_p(x) = pq^x, F_p(x) = 1 - q^{x+1} (0 < p < 1, q = 1-p).$$

Propose two alternative analog estimates of p, based on a random sample. Determine the asymptotic relative efficiency of each.

8. Suppose $X_1, Y_1, X_2, Y_2, \ldots, X_n, Y_n$ are independent random variables with X_i and Y_i both $N(\mu_i, \sigma^2)$. All parameters $\mu_1,\ldots,\mu_n$ and σ^2 are unknown. For example, X_i and Y_i may be repeated measurements on a laboratory specimen from the ith individual, with μ_i representing the amount of some antigen in the specimen; the measuring instrument is inaccurate, with normally distributed errors with common variability. Let $Z_i = (Y_i - X_i)/\sqrt{2}$.

(a) Consider the estimate $\tilde{\sigma}^2 = (1/n)\sum_{i=1}^n Z_i^2$ of σ^2. Show that it is unbiased, consistent, and asymptotically normal. [This estimate can be derived from *invariance* considerations; see for example LEHMANN & CASELLA (2006) for such methods.]

(b) Show that the maximum likelihood estimate of σ^2 is $\frac{1}{2}\tilde{\sigma}^2$. [*Hint:* First maximize the log likelihood in the μ_i's and then in σ^2.]

(c) Show that the maximum likelihood estimator of σ^2 is biased and not even consistent. Such poor behavior of maximum likelihood estimators often happens when the number of parameters increases with sample size.

9. Show that the basic assumptions in the random sampling model in Chapter 4 are sufficient to imply the consistency of $\bar{i}_n(\tilde{\theta}_n)$, for a regular estimate $\tilde{\theta}_n$, as an estimate of $B(\theta)$. Confine attention to the case that θ is real.

10. State sufficient conditions, in the random sampling case, for the empirical variance of the estimated scores (see Section 5.4) to be consistent for $B(\theta)$. [*Hint:* See *Exercises 25–26* in Chapter 4.]

11. Consider a quantile estimate $\tilde{\theta}_n$ of a real parameter θ, based on medians. Show (under mild conditions) that it is asymptotically linear and regular. [*Hint:* For the latter, differentiate the identity $1/2 = F_\theta(m(\theta))$, where $m(\theta)$ is the median of the absolutely continuous distribution F_θ, by the composite differentiation rule.]

12. Consider an estimate $\tilde{\theta}_n$, based on random sampling, which is asymptotically linear, $\tilde{\theta}_n \sim AL\{\theta, k(\cdot, \theta), \Sigma\}$. Assuming that the order of differentiation in θ and integration (or summation) over x can be interchanged, show that $\tilde{\theta}_n$ is *regular* if and only if $-E_\theta\{\dot{k}(X, \theta)\} = I$, the dot indicating differentiation in θ. [*Hint:* Differentiate the identity $E_\theta\{k(X, \theta)\} = 0$.]

13. Consider a random sample from a d-dimensional ($d > 2$) normal distribution $N(\theta, I)$, and let $\bar{\mathbf{X}}_n$ be the d-dimensional sample mean. Consider *Stein's estimate* of θ:

$$\tilde{\theta}_n = \bar{\mathbf{X}}_n \Big(1 - \frac{d-2}{n|\bar{\mathbf{X}}_n|^2}\Big).$$

Show that $\tilde{\theta}_n$ is not regular, and that it is super-efficient at the zero vector $\theta = \mathbf{0}$, in the sense that $\text{Avar}(\tilde{\theta}_n) - \text{Avar}(\bar{\mathbf{X}}_n)$ is negative definite.

14. Show that a moment estimate based on $t(X)$, with expectation $\nu(\theta)$, is asymptotically efficient if and only if $\text{var}_\theta\{t(X)\} = \dot{\nu}B^{-1}\dot{\nu}^T$, where $\dot{\nu}$ denotes the row vector of partial derivatives of $\nu(\theta)$.

15. Consider a random sample from a gamma distribution with scale parameter α and known shape parameter β (see Section 1.5, Chapter 1).

    (a) Find a moment estimate of α based on means.

    (b) Is the estimate in (a) asymptotically efficient? [*Hint:* See Exercise 14.]

16. Let $\tilde{\theta}_n$ be the quantile estimate using medians in a random sample from an exponential distribution.

    (a) Display the orthogonal decomposition (5.18); in particular, write

$$\tilde{\theta}_n = \theta + B^{-1} \cdot \frac{1}{n}\sum_{j=1}^{n} s_\theta(X_j) + R_n$$

    with $R_n = \frac{1}{n}\sum_{j=1}^{n} r_\theta(X_j) + o_p(1/\sqrt{n})$ and $E_\theta\{r_\theta(X)\} = 0$.

    (b) Show directly that $r_\theta(X)$ and $s_\theta(X)$ are uncorrelated.

    (c) Find the asymptotic relative efficiency of $\tilde{\theta}_n$.

17. Continued:

    (a) Verify that the moment estimate of θ using means is asymptotically efficient.

    (b) Determine the asymptotic relative efficiency of the quantile estimate.

18. In the Cauchy location model, with $f_\nu(x) = (1/\pi)[1/\{1+(x-\nu)^2\}]$, show that the covariance between the score per observation and the kernel of the sample median is unity, and hence that the large-sample squared correlation coefficient agrees with the asymptotic relative efficiency of the median ($8/\pi^2 \approx 0.81$). Note that this also establishes the regularity of the quantile estimate based on the median.

19. Consider the sample of size 25 from a Cauchy distribution centered at 0 but with unknown scale parameter τ say:

−9.00	1.55	1.29	3.41	−0.10	0.32	−2.57	−2.23
−1.97	0.24	0.10	−2.25	−4.64	−9.31	−2.57	−1.96
−22.28	2.31	1.14	−4.71	−2.66	−0.10	4.39	−1.59
1.04							

The density and distribution function are

$$f_\tau(x) = \frac{1}{\pi} \cdot \frac{\tau}{\tau^2 + x^2}, F_\tau(x) = \tfrac{1}{2} + \tfrac{1}{\pi}\tan^{-1}(x/\tau)(\tau > 0).$$

[For a check on whether you have downloaded or copied the data correctly, $\bar{x} = -2.0860$ and the sample standard deviation is $s = 5.3325$.]

(a) Find the seventh and nineteenth largest observations in the sample and equate them to the first and third quantile of the distribution to estimate τ. Calculate the asymptotic approximation to the variance of this quantile estimate (see Chapter 3).

(b) Now obtain an asymptotically efficient estimate by one-stepping this estimate.

(c) Find three alternative estimates of the standard error of the estimate in (a).

(d) By repeating this procedure, or using a standard software maximization routine in R, find the exact maximum likelihood estimator of τ.

20. In the Laplace location problem (see *Exercise 8* in Chapter 4), show that the sample median is asymptotically efficient by calculating its asymptotic variance. Also show that the sample median solves the score equation; hence, the conclusion of Theorem 3 holds even though the Regular Scores condition fails.

TABLE 5.1
Data from Right-Censored Weibull Distribution

25	27	32	34	48	70	73	76	78	82
87	90	100	100	100	101	104	105	108	116
118	120	120	127	130	136	143	146	146	151
158	159	159	160	163	167	171	175	180	184
186	187	187	188	190	199	200	200	207	207
217	218	220	222	231	231	235	237	238	241
243	245	246	248	254	259	261	262	263	264
267	270	273	292	293	295	301	301	311	313
323	324	328	343	348	360	366*	366*	366*	366*
366*	366*	366*	366*	366*	366*	366*	366*	366*	366*

21. Assuming (12) in Chapter 4, show that $S_n(\mathbf{X}_n, \tilde{\theta}_n) = S_n(\mathbf{X}_n, \theta) + o_p(\sqrt{n})$ for a consistent $\tilde{\theta}_n$, and hence that one-stepping a consistent estimate as in equation (5.17) preserves consistency. (However, it need not create asymptotic normality; therefore, it is necessary to start one-stepping with a root-n consistent estimate in order to achieve efficiency.)

22. Consider the following two-parameter *normal mixture model*:

$$f(x, \theta) = \tfrac{1}{2}\phi(x) + \tfrac{1}{2}\tfrac{1}{\sigma}\phi\left(\tfrac{x-\mu}{\sigma}\right)$$

with $\theta = (\mu, \sigma) \in \Theta = (-\infty, +\infty) \times (0, +\infty)$ and $\phi(x)$ the standard normal density.

(a) Show that $p_{n,\theta}(\mathbf{x}_n)$ is unbounded for θ in Θ (for $n > 1$), and hence that θ has no maximum likelihood estimate in Θ. [*Hint:* Consider $(\mu, \sigma) = (x_j, \epsilon)$ for small positive ϵ and any j.]

(b) Describe an asymptotically efficient estimate of $\theta = (\mu, \sigma)$. You may omit detailed computation of scores and information; just describe how they are to be used.

23. Given in Table 5.1 is a sample of size 100 from a two-parameter Weibull distribution, right-censored at x = 366:

That is, the density is of mixed-type, with a discrete component equal to the survivor function $\bar{F}(366)$ at x = 366. Think of the values as days until death or end of follow up of some experimental animals exposed to risk, with the experiment ending at the end of one year, so that values of 366* indicate that the day of death was at least 366. [To check that you have downloaded or copied the data correctly, $\bar{x} = 214.210$ and the sample standard deviation is $s = 99.435$.]

(a) Write down the score equations for the Weibull parameters (α, β), verify that equation (5.20), with a slight modification, and subsequent statements leading to proof of uniqueness of the solution of the score equations, still hold.

(b) Find (numerically) asymptotically efficient estimates of the two parameters. Describe how you got them. Note that since fewer than 25% of the observations are right-censored, the upper quartile of the distribution can still be estimated in the usual way.

(c) Provide a consistent estimate of the variance matrix of the estimates in (a).

24. Note that the maximum likelihood estimate $\hat{p}_n$ in Section A1 of Appendix II is asymptotically normal and hence root-n consistent. By contrast, show that $\tilde{p}_n$ (defined there) is super-efficient at every rational p.

6

Estimating Parameters of Interest

We now represent the model parameters by γ (d-dimensional), with $\gamma = (\theta, \eta)$, where θ is the *parameter of interest* to be estimated and η is a *nuisance parameter*. The dimensions of θ and η are d' and $d'' = d - d'$, respectively. The nuisance parameter is an essential part of the model, but of no intrinsic interest. We continue to treat γ as a column vector in mathematical expressions, so that the previous expression could be written more precisely as $\gamma = (\theta^T, \eta^T)^T$. From another perspective, we have added the nuisance parameter η to the model considered previously. We also consider two closely related problems, that of estimating a specified d'-dimensional function $\theta(\gamma)$ of the full parameter vector γ, and the dual problem of estimating the full parameter vector γ subject to a constraint, say $\theta(\gamma) = 0$. Block matrix formulas, summarized in Appendix I, are an important tool in all this work. Appendix II describes a very useful result on the *profile likelihood function*, defined as $g(\theta) = L\{\theta, \hat{\eta}(\theta)\}$, where $\hat{\eta}(\theta)$ is chosen to maximize $L(\theta, \eta)$ in η for the given θ, which underlies some of the ideas in this chapter.

6.1 Effective Scores

In general, there are no useful shortcuts to constructing estimates only of θ, at least efficient ones, it is necessary to estimate both θ and η simultaneously. An exception, see *Exercise 7*, is when $B_{\theta\eta} = 0$ so that S_θ and S_η are uncorrelated. We sometimes write this condition as $S_\theta \perp S_\eta$. However, characterizations of asymptotically efficient estimates of only the parameter of interest, θ, have some value, as do formulas for the resulting asymptotic variance. The associated concepts are needed in hypothesis testing, to be discussed in Chapter 7.

We write $(S_{\theta n}, S_{\eta n})$ for the components of S_n, the *scores for* θ and *for* η, from a sample of size n. Again, these are treated as column vectors in mathematical expressions. We partition the full information matrix B, assumed to be nonsingular, into component blocks,

$$B = \begin{pmatrix} B_{\theta\theta} & B_{\theta\eta} \\ B_{\eta\theta} & B_{\eta\eta} \end{pmatrix}$$

DOI: 10.1201/9780429160080-6

and note that $B_{\eta\theta} = B_{\theta\eta}^T$. We use superscripts for the corresponding blocks in B^{-1}, so that

$$B^{-1} = \begin{pmatrix} B^{\theta\theta} & B^{\theta\eta} \\ B^{\eta\theta} & B^{\eta\eta} \end{pmatrix}$$

The block matrix formulas presented in Appendix I show that

$$B^{\theta\theta} = \left(B_{\theta\theta} - B_{\theta\eta}B_{\eta\eta}^{-1}B_{\eta\theta}\right)^{-1} \text{ and } B^{\theta\eta} = -B^{\theta\theta}B_{\theta\eta}B_{\eta\eta}^{-1}. \qquad (6.1)$$

From Equation (5.10) of Chapter 5, we know that the estimate $\tilde{\theta}_n$ is asymptotically efficient, or, more precisely, is part of an asymptotically efficient estimate of γ, if and only if

$$\sqrt{n}(\tilde{\theta}_n - \theta) = B^{\theta\theta}S_{\theta n} + B^{\theta\eta}S_{\eta n} + o_p(1). \qquad (6.2)$$

Using Equation (6.1), the right side of Equation (6.2) may be written as $B^{\theta\theta}(S_{\theta n} - B_{\theta\eta}B_{\eta\eta}^{-1}S_{\eta n}) + o_p(1)$. Also, from the definition of asymptotic efficiency, the left side in Equation (6.2) is asymptotically $N(0, B^{\theta\theta})$. We introduce the following concepts and notation to re-express these facts:

Definition 6.1 *S_n^*, defined as $S_{\theta n} - B_{\theta\eta}B_{\eta\eta}^{-1}S_{\eta n}$, is the effective score for θ; B^*, defined as $B^* = (B^{\theta\theta})^{-1} = B_{\theta\theta} - B_{\theta\eta}B_{\eta\eta}^{-1}B_{\eta\theta}$, is the effective information for θ.*

The effective score is the residual from regression of $S_{\theta n}$ on $S_{\eta n}$, using large-sample distributions (see *Exercise 2*). Geometrically, it is the *orthogonal complement* of the *projection* of $S_{\theta n}$ on the space spanned by $S_{\eta n}$. In the random sampling case, $S_n^* = (1/\sqrt{n})\sum_{j=1}^n s^*(X_j)$ with $s^*(x) \equiv s_\theta(x) - B_{\theta\eta}B_{\eta\eta}^{-1}s_\eta(x)$, the *effective score per observation*, and B^* is readily checked to be its variance. We have

Theorem 6.1 *$S_n^* \to_\gamma N(0, B^*)$ and $\to_{\gamma_n} N(B^*h_\theta, B^*)$ where $\gamma_n = \gamma + h/\sqrt{n}$ and $h = (h_\theta, h_\eta)$; $S_n^* \perp S_{\eta n}$.*

Proof The first assertion is proved above. We use LeCam's third lemma to derive the distribution of S_n^* under moving alternatives. We must take into account a possible effect on the limiting distribution of the change in η as well as that of the change in θ. We find that $\text{cov}(S_n, S^*) = E\{S_n \cdot (S^*)^T\}$ equals

$$E\begin{pmatrix} S_{\theta n} \\ S_{\eta n} \end{pmatrix}(S_{\theta n}^T - S_{\eta n}^T B_{\eta\eta}^{-1}B_{\eta\theta}) = \begin{pmatrix} B_{\theta\theta} - B_{\theta\eta}B_{\eta\eta}^{-1}B_{\eta\theta} \\ B_{\eta\theta} - B_{\eta\eta}B_{\eta\eta}^{-1}B_{\eta\theta} \end{pmatrix} = \begin{pmatrix} B^* \\ 0 \end{pmatrix}$$

Replacing η by η_n does not affect the the asymptotic distribution of S_n^*. This calculation also shows that $\text{cov}(S_n^*, S_{\eta n}) = 0$ so that S_n^* and $S_{\eta n}$ areorthogonal. □

Hence, S_n^* and B^* behave very much like $S_{\theta n}$ and $B_{\theta\theta}$, the score and information if η were known. And S_n^* may be described as that part of $S_{\theta n}$, the score for θ, which is *orthogonal* to the score for η. Still, care is required in that expressions have an argument η as well as θ.

All statements in Section 5.3 of Chapter 5 have analogs stated in terms of S_n^* and B^*; see HALL & MATHIASON (1990). In particular, $\tilde{\theta}_n$ is *regular for* θ if $\sqrt{n}(\tilde{\theta}_n - \theta_n)$ converges in distribution under γ_n to an h-free limit distribution, holding if and only if $\text{Acov}_\gamma\{\sqrt{n}(\tilde{\theta}_n - \theta), (S_n^*, S_{\eta n})\} = (I, 0)$; an orthogonal decomposition again holds, and $(nB^*)^{-1}$ is a lower bound on the asymptotic variance of regular estimates of θ; such an estimate is, therefore, defined to be asymptotically efficient if the asymptotic distribution of the error is $N\{0, (nB^*)^{-1}\}$. Finally, the Characterization Theorem becomes

Theorem 6.2 *(Characterization of Asymptotically Efficient Estimates): An estimate $\hat{\theta}_n$ is asymptotically efficient if and only if*

$$\sqrt{n}(\hat{\theta}_n - \theta) \;=\; (B^*)^{-1}S_n^* + o_p(1). \tag{6.3}$$

Proof: As before, if $\hat{\theta}_n$ is a regular estimator of θ, write

$$\sqrt{n}(\tilde{\theta}_n - \theta) \;=\; (B^*)^{-1}S_n^* + R_n + o_p(1),$$

so that

$$R_n = \sqrt{n}(\tilde{\theta}_n - \theta) \; - (B^*)^{-1}S_n^*.$$

This is precisely (6.2), in different notation.

However, we still need to show that the fact that $\hat{\theta}$ is part of an asymptotically efficient estimator for γ, as stated earlier, implies that it is in itself an asymptotically efficient estimator of θ. To do this, we repeat the argument of Theorems 5.2 and 5.3 in Chapter 5 in the new context. We must restrict attention to estimates $\tilde{\theta}_n$ of θ that are regular in the full parameter $\gamma = (\theta, \eta)$. By LeCam's third lemma, this implies that $\text{Acov}\{\sqrt{n}(\theta_n - \theta), S_n\} = (I, 0)$, where I is the identity matrix of dimension d_θ and 0 is a matrix of zeros. So $\text{Acov}\{\sqrt{n}(\theta_n - \theta), S_n^*\} = \text{Acov}\{\sqrt{n}(\theta_n - \theta), S_n\} = I$ Writing

$$\sqrt{n}(\tilde{\theta}_n - \theta) = (B^*)^{-1}S_n^* + R_n.$$

as above and using the fact that $\text{var}(S_{\theta n}) = B^*$ shown above we find that

$$\text{Acov}(R_n, S_n^*) = I - (B^*)^{-1}B^* = 0,$$

so that the asymptotic variance of $\sqrt{n}(\tilde{\theta}_n - \theta)$ is minimized when $R_n = o_p(1)$ as before. □

It is often of interest to compare the best possible large-sample variance for estimating θ when η is known to that when η is unknown, that is to compare $B_{\theta\theta}^{-1}$ with $(B^*)^{-1}$. Surely, the former must be less than or equal to the latter. This is formally verified as a matrix fact in Appendix I. When $d = 2$ and $d' = 1$, so that B is a 2×2 matrix with diagonal elements a, b say and off-diagonal element c, an elementary calculation gives $B_{\theta\theta}^{-1} = 1/a \leq (a - c^2/b)^{-1}$.

In the general case, equality is seen to hold if and only if the *co-information* $B_{\theta\eta}$ is zero, that is, if the scores for θ and for η are orthogonal. In this case, the scores for η carry no information about θ. When this is so, the best estimates of θ when η is unknown have the same asymptotic behavior as the best estimates when η is known. We say that it is possible to *adapt for* η. There is no loss in asymptotic efficiency for estimation of θ if η is moved from known to unknown status. The familiar and elementary example is that a sample mean, as an estimate of the mean of a normally distributed population, has the same (exact and) large-sample behavior when the population variance σ^2 is known as when it is unknown: we can adapt for σ^2. The reverse situation is not quite so simple, but still leads to adaptation for μ; see *Exercise 9*.

Returning to the two-parameter Weibull example in Section 5.5 of Chapter 5, with α the parameter of interest, we find that we cannot adapt for β since $B_{\alpha\beta} \neq 0$. We find that

$$B^* = \frac{\beta^2}{\alpha^2}\left\{1 - \frac{\Gamma'(2)^2}{1+\Gamma''(2)}\right\} \approx 0.9020 \cdot \frac{\beta^2}{\alpha^2},$$

to be compared with $B_{\alpha\alpha} = \beta^2/\alpha^2$, the information if β were known. There is approximately a 10% loss of information about α when β is unknown, whatever the parameter values.

This concept of adaptation can be extended and recast as follows, in the spirit of its original presentation by Stein (1956). Suppose that $\hat{\theta}_n$ is asymptotically efficient for θ in a model with nuisance parameter η. Now suppose the model is enlarged by introduction of an additional nuisance parameter τ. Under what conditions will $\hat{\theta}_n$ remain asymptotically efficient? That is, can we *adapt for* τ? The necessary and sufficient condition turns out to be that $S_n^* \perp S_{n\tau}$; equivalently, $B_{\theta\tau} = B_{\theta\eta}B_{\eta\eta}^{-1}B_{\eta\tau}$ ($= 0$ if the role of η is vacuous). This can be verified by a matrix argument, verifying that the effective information for θ in the presence of both η and τ agrees with the effective information for θ when τ is known, that is, when the only nuisance parameter is η, only under the circumstances stated; see *Exercise 10*. The underlying matrix result is proved in Appendix I.

We close this section with a proposition useful in the large-sample theory of hypothesis testing (Chapter 7).

Proposition 6.1 *Suppose that the Regular Scores and Continuity of Information properties hold for the parameter* $\gamma = (\theta, \eta)$. *Suppose also that when* θ *is known,* $\tilde{\eta}_n(\theta)$ *is root-n consistent for* η *and* $\hat{\eta}_n(\theta)$ *is asymptotically efficient for* η. *Then,*

i. $S_{\theta n}\{\theta, \hat{\eta}_n(\theta)\} = S_n^*(\theta, \eta) + o_p(1)$;

ii. $S_n^*(\theta, \tilde{\eta}_n) = S_n^*(\theta, \eta) + o_p(1)$.

Proof (i) By the Regular Scores property,

$$S_{\theta n}\{\theta, \hat{\eta}_n(\theta)\} = S_{\theta n}(\theta, \eta) - B_{\theta\eta}\sqrt{n}(\hat{\eta} - \eta) + o_p(1),$$

where $\hat{\eta} = \hat{\eta}_n(\theta)$, and by the asymptotic efficiency of $\hat{\eta}$ (in η for given θ) we have

$$\sqrt{n}(\hat{\eta} - \eta) = B_{\eta\eta}^{-1} S_{\eta n}(\theta, \eta) + o_p(1),$$

so that substituting for $\sqrt{n}(\hat{\eta} - \eta)$ gives

$$S_{\theta n}(\theta, \hat{\eta}) = S_{\theta n}(\theta, \eta) - B_{\theta\eta} B_{\eta\eta}^{-1} S_{\eta n}(\theta, \eta) + o_p(1) = S_n^*(\theta, \eta) + o_p(1).$$

Here, the Continuity of Information assumption allows the argument of the information matrix $B_{\gamma\gamma}$ and its submatrices to be replaced by γ, since the regular scores assumption ensures that they are within $o_p(1)$ of γ.

(ii) By the Regular Scores property,

$$S_{\theta n}\{\theta, \tilde{\eta}_n(\theta)\} = S_{\theta n}(\theta, \eta) - B_{\theta\eta}\sqrt{n}(\tilde{\eta}_n - \eta) + o_p(1),$$

and

$$S_{\eta n}\{\theta, \tilde{\eta}_n(\theta)\} = S_{\eta n}(\theta, \eta) - B_{\eta\eta}\sqrt{n}(\tilde{\eta}_n - \eta) + o_p(1).$$

Hence

$$\begin{aligned} S_n^*(\theta, \tilde{\eta}_n) &= S_{\theta n}(\theta, \eta) - B_{\theta\eta}\sqrt{n}(\tilde{\eta}_n - \eta) - B_{\theta\eta} B_{\eta\eta}^{-1}\{S_{\theta n}(\theta, \eta) - B_{\eta\eta}\sqrt{n}(\tilde{\eta} - \eta)\} \\ &= S_n^*(\theta, \eta) \end{aligned}$$

since the terms in $\sqrt{n}(\tilde{\eta}_n - \eta)$ cancel. □

Corollary 6.1 *If $S_\theta \perp S_\eta$, so that $\hat{\theta}$ and $\hat{\eta}$ are asymptotically independent, then*

$$S_n(\theta, \tilde{\eta}_n) = S_n(\theta, \eta) + o_p(1),$$

so that we may substitute a root-n consistent estimator of the nuisance parameter into the score for the parameter of interest without affecting its asymptotic properties.

Example 6.1 *We apply the previous theory to the estimation of the effect size $\theta = \mu/\sigma$ from a sample of independent observations from a normal distribution with mean μ and standard deviation σ.*

We will assume that μ and σ are positive. We have already derived the asymptotic distribution of the obvious estimator $\hat{\theta} = \bar{X}/S_X$, the ratio of the sample mean to the sample standard deviation, in Chapter 3. Specifically, the multivariate delta method gives

$$\sqrt{n}(\hat{\theta} - \theta) \to_d N(0, 1 + \tfrac{1}{2}\theta^2).$$

But is there, perhaps, a better estimate of θ?

We rewrite the log-likelihood as a function of $\theta = \mu/\sigma$ and a nuisance parameter η. We can choose any function $\eta(\mu, \sigma)$ which makes the mapping

$(\mu, \sigma) \rightarrow_d (\theta, \eta)$ continuously differentiable and invertible, but here we will make the simple choice $\eta = \sigma$. The density for a single observation x is

$$f(x, \theta, \eta) = \frac{1}{\surd(2\pi)\eta} \exp\left\{-\frac{(x - \theta\eta)^2}{2\eta^2}\right\},$$

so that omitting the constant term $-\frac{1}{2}\log(2\pi)$, the log-density is

$$\log f(x, \theta, \eta) = -\log \eta - \frac{x^2}{2\eta^2} + \frac{\theta x}{\eta} - \frac{\theta^2}{2}.$$

The score vector per observation has components

$$s_\theta = \frac{x}{\eta} - \theta,$$

and

$$s_\eta = -\frac{1}{\eta} + \frac{x^2}{\eta^3} - \frac{\theta x}{\eta^2}. \tag{6.4}$$

Taking the negatives of the derivatives gives the elements of the observed information i:

$$i_{\theta\theta} = 1, \; i_{\theta\eta} = \frac{x}{\eta^2}, \;\; i_{\eta\eta} = -\frac{1}{\eta^2} + \frac{3x^2}{\eta^4} - \frac{2\theta x}{\eta^3}.$$

Since $E(X) = \theta\eta$ and $\text{var}(X) = \eta^2$ in this parameterization, the expected information B has elements

$$B_{\theta\theta} = 1, \;\; B_{\theta\eta} = \theta\eta, \; B_{\eta\eta} = -\frac{1}{\eta^2} + \frac{3\eta^2(1 + \theta^2)}{\eta^4} - \frac{2\theta^2}{\eta^2} = \frac{2 + \theta^2}{\eta^2}.$$

The effective information about θ is

$$B^{\theta\theta} = B_{\theta\theta} - B_{\theta\eta} B_{\eta\eta}^{-1} B_{\eta\theta} = 1 - \frac{\theta}{\eta}\left(\frac{\eta^2}{2 + \theta^2}\right)\frac{\theta}{\eta} = \frac{2}{2 + \theta^2}.$$

So the asymptotic variance of an efficient estimator of θ is $(nB^{\theta\theta})^{-1} = (1 + \frac{1}{2}\theta^2)/n$, agreeing with that of $\hat{\theta}$. This shows that there is no better estimate of θ than $\hat{\theta}$.

A general recipe for finding an asymptotically efficient estimator for θ here is to find a near-root of the score equation $S_{\eta n}(\theta, \eta) = 0$ for η as a function of θ and substitute this into the score equation for θ. Apart from a $\surd n$ multiplier, $S_{\eta n}$ has the same form as Equation (6.4) but with x, x^2 replaced by $\sum x_j$ and $\sum x_j^2$, respectively. The score equation is quadratic in η with two roots, one positive and one negative. Neither root is identical to S_X or to the maximum likelihood estimator $\hat{\sigma} = \surd\{n/(n-1)\}S_X$, but it can be shown that these are both near-roots of the score equation. Solving the score equation for θ with S_X substituted for η gives $\hat{\theta}$. An alternative approach, using the effective score S_n^* for θ, is discussed in the next section. In this example, as in many others, the theory is more useful in providing a lower bound for the asymptotic variance of an estimator than in actually computing one.

6.2 Estimating a Parametric Function

Again let γ be the d-dimensional model parameter. But now suppose that $\theta(\gamma)$ is a (smooth) d'-dimensional parametric function to be estimated, rather than just the first d' coordinates of γ. Hence, we are no longer just fitting a model, but focusing on one or more parametric functions, perhaps representing population characteristics of interest.

Let $\dot{\theta} = \dot{\theta}(\gamma)$ be the $d' \times d$ matrix of derivatives $\partial\theta/\partial\gamma$, assumed to be continuous and of full rank d'. Hence

$$\theta(\gamma_n) = \theta(\gamma) + \dot{\theta}(\gamma) \cdot (\gamma_n - \gamma) + o(\| \gamma_n - \gamma \|). \tag{6.5}$$

Writing $\gamma_n = \gamma + h/\sqrt{n}$ so that (6.5) becomes $\theta(\gamma_n) = \theta(\gamma) + \dot{\theta}(\gamma)h/\sqrt{n} + o(1/\sqrt{n})$, we have $\theta_n \equiv \theta(\gamma_n) = \theta(\gamma) + h_\theta/\sqrt{n} + o(1/\sqrt{n})$ with $h_\theta = \dot{\theta}h$.

A simple way to proceed, illustrated by the discussion of the last example in the previous section, is to fill out the transformation from γ to (θ, η), say, and apply the foregoing theory in the new parameterization. Alternatively, we can extend the theory to deal with estimation of $\theta(\gamma)$ directly. The obvious estimate of $\theta(\gamma)$ is $\hat{\theta} = \theta(\hat{\gamma})$ and the delta method shows that this estimate is asymptotically normal, with $\sqrt{n}(\hat{\theta} - \theta) \sim AN(0, \dot{\theta}B^{-1}\dot{\theta}^T)$. Here and later, we drop the argument γ from $\dot{\theta}(\gamma)$ for notational simplicity. The inverse $B^* = (\dot{\theta}B^{-1}\dot{\theta}^T)^{-1}$ of the asymptotic covariance matrix is a natural candidate for the information about $\theta(\gamma)$.

To determine the effective score for $\theta(\gamma)$, S_n^* say, extending the notation used in the previous section, we equate $\sqrt{n}(\hat{\theta}_n - \theta) = (B^*)^{-1}S_n^*$ to $\dot{\theta}\sqrt{n}(\hat{\gamma} - \gamma) = \dot{\theta}B^{-1}S_n$ obtaining $S_n^* \equiv B^*\dot{\theta}B^{-1}S_n$.

We take these formulas as definitions of B^* and S_n^* and develop the theory as before. We have $\text{var}(S_n^*) = B^*\dot{\theta}B^{-1}\text{var}(S_n)B^{-1}\dot{\theta}^T B^*$ which simplifies to $B^*\dot{\theta}B^{-1}\dot{\theta}^T B^* = B^*$ since $\text{var}(S_n) = B$.

Definition 6.2 *An estimator $\tilde{\theta}_n$ is regular for θ if it is asymptotically linear and the limiting distribution of $\tilde{\theta}_n - \theta_n$, where $\theta_n = \theta(\gamma_n)$ with $\gamma_n = \gamma + h/\sqrt{n}$, does not depend on h.*

Note If $\tilde{\gamma}_n$ is regular as an estimate of γ then $\tilde{\theta}_n = \theta(\tilde{\gamma}_n)$ is regular as an estimate of θ. However, since $d' < d$, regularity of $\tilde{\theta}_n$ for θ is a less restrictive condition than regularity of $\tilde{\gamma}_n$ for γ.

Lemma 6.1 *The asymptotically linear estimate $\tilde{\theta}_n$ is regular as an estimate of $\theta(\gamma)$ if and only if $C = \text{Acov}\{\sqrt{n}(\tilde{\theta}_n - \theta), S_n\} = \dot{\theta}(\gamma)$.*

Proof If $\tilde{\theta}_n$ is regular, $\sqrt{n}(\tilde{\theta}_n - \theta_n) \to N(0, B)$ under all local alternatives $\gamma_n = \gamma + h/\sqrt{n}$. By LeCam's third lemma, $\sqrt{n}(\tilde{\theta}_n - \theta_n) \to N(Ch, B)$ under γ, where $C = \text{Acov}\{\sqrt{n}(\tilde{\theta}_n - \theta), S_n\}$ Hence $\sqrt{n}(\tilde{\theta}_n - \theta) \to N\{(C - \dot{\theta})h, B\}$ under γ. Since the mean of this distribution must be zero for every h, we must have $C = \dot{\theta}$. To show the converse, we may reverse the steps. □

Note Since $S_n^* = B^* \dot{\theta} B^{-1} S_n$, this condition is equivalent to requiring that $\text{Acov}\{\sqrt{n}(\tilde{\theta}_n - \theta), S_n^*\} = I$. For

$$\text{Acov}\{\sqrt{n}(\tilde{\theta}_n - \theta), S_n^*\} = \text{Acov}\{\sqrt{n}(\tilde{\theta}_n - \theta), B^* \dot{\theta} B^{-1} S_n\}$$

$$= E\{\sqrt{n}(\tilde{\theta}_n - \theta) \cdot S_n B^{-1} \dot{\theta}^T B^* = \dot{\theta} B^{-1} \dot{\theta}^T B^* = (\dot{\theta} B^{-1} \dot{\theta}^T) B^* = I.$$

Theorem 6.3 *(Characterization of Asymptotically Efficient Estimates of $\theta(\gamma)$): An estimate $\hat{\theta}_n$ of $\theta(\gamma)$ is asymptotically efficient if and only if*

$$\sqrt{n}(\hat{\theta}_n - \theta) = (B^*)^{-1} S_n^* + o_p(1).$$

Proof Let $\tilde{\theta}_n$ be a regular estimate of $\theta(\gamma)$ and write

$$\sqrt{n}(\tilde{\theta}_n - \theta) = \sqrt{n}(\hat{\theta}_n - \theta) + R_n.$$

We show that $\text{Acov}\{R_n, (B^*)^{-1} S_n^*\} = 0$. Writing $R_n = \sqrt{n}(\tilde{\theta}_n - \theta) - (B^*)^{-1} S_n^*$, we see that the covariance is zero, being the difference between two terms both equalling $(B^*)^{-1}$ Hence, the asymptotic variance of $\tilde{\theta}_n$ is minimized by taking $R_n = o_p(1)$.

□

Example 6.2 *Estimation of the effect size $\theta = (\mu/\sigma)$ from a normal distribution with mean μ and standard deviation σ, continued.*

As in the previous section, we consider the estimation of $\theta = \mu/\sigma$ from a sample from $N(\mu, \sigma^2)$. The density is

$$f(x, \mu, \sigma) = \frac{1}{\sqrt{2\pi}} \exp\left\{-\frac{(x-\mu)^2}{2\sigma^2}\right\}$$

so that the score statistic per observation is

$$s = \begin{pmatrix} s_\mu \\ s_\sigma \end{pmatrix} = \begin{pmatrix} (x-\mu)/\sigma^2 \\ -1/\sigma + (x-\mu)^2/\sigma^3 \end{pmatrix}$$

with information matrix

$$B = \begin{pmatrix} 1/\sigma^2 & 0 \\ 0 & 2/\sigma^2 \end{pmatrix},$$

so that

$$B^{-1} = \begin{pmatrix} \sigma^2 & 0 \\ 0 & \frac{1}{2}\sigma^2 \end{pmatrix}.$$

We also have $\dot{\theta} = (1/\sigma, -\mu/\sigma^2)$, in this case, a row vector because $\theta(\mu, \sigma)$ is a scalar function. So

$$(B^*)^{-1} = (1/\sigma, -\mu/\sigma) \begin{pmatrix} \sigma^2 & 0 \\ 0 & \frac{1}{2}\sigma^2 \end{pmatrix} \begin{pmatrix} 1/\sigma \\ -\mu/\sigma \end{pmatrix} = 1 + \tfrac{1}{2}\mu^2/\sigma^2 = 1 + \tfrac{1}{2}\theta^2.$$

This calculation, naturally, gives the same result as in the previous discussions of this example in Chapter 3 and the previous section of this chapter.

The effective score per observation for θ is then

$$\left(\frac{1}{1+\frac{1}{2}\theta^2}\right)(1/\sigma, -\mu/\sigma^2)\begin{pmatrix}\sigma^2 & 0 \\ 0 & \sigma^2/2\end{pmatrix}\begin{pmatrix}s_\mu \\ s_\sigma\end{pmatrix} = \left(\frac{1}{1+\frac{1}{2}\theta^2}\right)(\sigma s_\mu - \tfrac{1}{2}\mu s_\sigma),$$

with the effective score S_n^* for the full sample obtained by replacing the scores per observation s_μ, s_θ by the corresponding full scores $S_{n\mu}$ and $S_{n\sigma}$, respectively. The calculation outlined in the previous section gives the same result.

Note As a final comment, if $d' = d$ so that θ is simply a transformation (reparameterization) of γ then the transformed scores are $\dot{\gamma}^T S_n$ with $\dot{\gamma} = \dot{\theta}^{-1}$, the derivative of the inverse transformation $\gamma = \gamma(\theta)$, and the transformed information is $\dot{\gamma}^T B \dot{\gamma}$ with inverse $\dot{\theta} B^{-1} \dot{\theta}^T$. See the Transformation Proposition in Chapter 4. The lower bound on the asymptotic variance of regular estimates of $\theta(\gamma)$ becomes $\dot{\theta} B^{-1} \dot{\theta}^T$.

6.3 Estimation Under Constraints on the Parameter

A problem in a sense dual to that of estimating a function $\theta(\gamma)$ is that of estimating the full parameter vector γ subject to a constraint, $\theta(\gamma) = 0$, say. Again, $\theta(\gamma)$ is a vector function of dimension $d' < d$. Our notation here follows that in the previous section, but now $\theta(\gamma)$ defines a constraint rather than a function of interest. As before $\theta(\gamma)$ is assumed to be continuously differentiable in a neighborhood of the true value of γ. We assume that there is no redundancy among the constraints and, more specifically, that the $d' \times d$ matrix $\dot{\theta} = \dot{\theta}(\gamma)$ of derivatives is of full rank d'. The definition of regularity of an estimate $\tilde{\gamma}$ of γ is the same as that in Section 6.1, except that the local alternatives $\gamma_n = \gamma + h/\sqrt{n}$ and their directions h must satisfy the constraint $\theta(\gamma_n) = 0$, so that $\dot{\theta}(\gamma)h = 0$. If $\tilde{\gamma}$ is regular in the unconstrained model, it will also be regular in the constrained model, but not conversely. The property of asymptotic efficiency will not in general carry over from the unconstrained to the constrained model, as the relaxation of the regularity condition increases the class of regular estimates under consideration.

As in the previous section, one approach, often the simplest, is to reparameterize, leading here to a new, unconstrained, parameter vector of dimension $d - d'$. Instead, we proceed directly. The standard approach to constrained maximization of a function $L(\gamma)$ is to introduce a d'-dimensional vector λ of *Lagrange multipliers* and consider the modified objective function $L(\gamma) - \theta(\gamma)^T\lambda$. Differentiation of this function in γ gives d equations. Together with the d' equations obtained by differentiating the constraint $\theta(\gamma) = 0$, these equations are solved for the $d + d'$ unknowns in (γ, λ). We write $\hat{\lambda}$ rather than

λ in the sequel to indicate the dependence of this vector of multipliers on the data.

Here, continuing to assume the usual regularity conditions, we obtain

$$S_n(\hat{\gamma}) - \dot{\theta}(\hat{\gamma})\hat{\lambda}/\sqrt{n} = 0, \quad \sqrt{n}\dot{\theta}(\hat{\gamma}) = 0,$$

the factor $\sqrt{n}$ arising because $S_n(\theta)$ is the score, not the total score. Linearizing these equations, justified as before under the Basic Assumptions, gives

$$S_n(\gamma) - B\sqrt{n}(\hat{\gamma} - \gamma) - (1/\sqrt{n})\dot{\theta}(\gamma)\hat{\lambda}/\sqrt{n} = o_p(1)$$

and

$$\dot{\theta}(\gamma)\sqrt{n}(\hat{\gamma} - \gamma) = o_p(1).$$

In matrix notation, and dropping the argument γ from $S_n(\gamma)$ and $\dot{\theta}(\gamma)$ for notational simplicity,

$$M\begin{pmatrix}\sqrt{n}(\hat{\gamma} - \gamma) \\ \hat{\lambda}/\sqrt{n}\end{pmatrix} = \begin{pmatrix}S_n \\ 0\end{pmatrix},$$

where

$$M = \begin{pmatrix}B & \dot{\theta}^T \\ \dot{\theta} & 0\end{pmatrix}.$$

Although M is not positive definite, it is invertible. In the Appendix, we verify that

$$M^{-1} = \begin{pmatrix}P & Q^T \\ Q & -B^*\end{pmatrix},$$

where the $d' \times d'$ matrix $B^* = (\dot{\theta}B^{-1}\dot{\theta}^T)^{-1}$, the $d' \times d$ matrix $Q = B^*\dot{\theta}B^{-1}$ and the $d \times d$ matrix $P = B^{-1}(I - \dot{\theta}^T Q)$. It follows that

$$\begin{pmatrix}\sqrt{n}(\hat{\gamma} - \gamma) \\ \hat{\lambda}/\sqrt{n}\end{pmatrix} = \begin{pmatrix}P & Q^T \\ Q & -B^*\end{pmatrix}\begin{pmatrix}S_n \\ 0\end{pmatrix}.$$

From the previous definitions, $Q\dot{\theta}^T$ is the identity matrix of dimension d' and $\dot{\theta}^T Q$, a $d \times d$ matrix, is idempotent, $(\dot{\theta}^T Q) \times (\dot{\theta}^T Q) = \dot{\theta}^T Q$. A little further manipulation (*Exercise 13*) shows that $PBP = P$, $QBQ^T = B^*$ and that PBQ^T and QBP are zero matrices.

It follows that

$$\begin{pmatrix}\sqrt{n}(\hat{\gamma} - \gamma) \\ \hat{\lambda}/\sqrt{n}\end{pmatrix} \sim AN(0, D),$$

where D is a block-diagonal matrix with entries P and B^*. The normalized Lagrange multiplier $\hat{\lambda}/\sqrt{n}$ is asymptotically independent of the constrained parameter estimate $\hat{\gamma}$ with asymptotic variance B^*. Interestingly, the formula for $\hat{\lambda}/\sqrt{n}$ is identical to that for the effective score S_n^* for $\theta(\gamma)$ defined in the previous section and its variance is also given by the same formula as that given earlier. The second term of P, written in full as $B^{-1}\dot{\theta}^T(\dot{\theta}B^{-1}\dot{\theta}^T)^{-1}\dot{\theta}B^{-1}$, represents the reduction in attainable asymptotic variance of $\hat{\gamma}$ due to the imposition of the constraint $\theta(\gamma) = 0$.

We now investigate the regularity and asymptotic optimality of the proposed estimate. We first give a criterion for the regularity of a constrained estimate:

Lemma 6.2 *The asymptotically linear estimate $\tilde{\gamma}_n$ is regular in the constrained model if and only if its asymptotic covariance matrix C with the full score vector S satisfies $CP = P$, where $P = B^{-1} - B^{-1}\dot{\theta}^T B^* \dot{\theta} B^{-1}$ is the symmetric matrix defined above.*

Note: In the absence of constraints, $\dot{\theta} = 0$, a zero matrix, $P = B^{-1}$ and the condition becomes $C = I$, as proved earlier. When $\dot{\theta}$ is not a zero matrix, P is not invertible, and we cannot deduce that $C = I$.

Proof We first observe that $h' = Ph$ satisfies $\dot{\theta}h' = 0$ for all h and conversely that any h' with $\dot{\theta}h' = 0$ is the image under this map of $h = Bh'$. See *Exercise 14*. Regularity under the constrained model is therefore equivalent to regularity in the full model under the restricted set of local deviations, $\gamma_n = \gamma + Ph/\sqrt{n}$. By Le Cam's third lemma, the asymptotic distribution of $\tilde{\gamma}_n - \gamma$ is $N(CPh, B)$ and so that of $\tilde{\gamma}_n - \gamma_n$ is $N(CPh - Ph, B)$. For the mean of this distribution to be zero for every h, it is necessary and sufficient that $CP = P$. □

Theorem 6.4 *The estimator $\hat{\gamma}_n = PS$ is regular in the constrained model and has smallest asymptotic variance among all regular estimators for this model.*

Proof

(i) The asymptotic covariance of $\hat{\gamma}_n$ with the full score vector S_n is $C = \text{Acov}\{\sqrt{n}(\hat{\gamma}_n - \gamma), S_n\} = \text{cov}(PS_n, S_n) = PB$, so $CP = PBP = P$, showing that $\hat{\gamma}_n$ is regular.

(ii) To show optimality of $\hat{\gamma}_n$ within the class of regular estimators we proceed as earlier by considering an alternative regular estimator $\tilde{\gamma}_n$ defined as

$$\tilde{\gamma}_n = \hat{\gamma}_n + R_n.$$

We then have

$$\text{Acov}(R_n, \gamma_n) = \text{Acov}(\tilde{\gamma}_n - PS_n, PS_n) = CP - PBP = CP - P = 0,$$

by the Lemma. It follows that

$$\sqrt{n}(\tilde{\gamma} - \gamma) = PS_n + R_n,$$

with $\text{Acov}(R_n, S_n) = 0$. The asymptotic variance of γ_n is minimized by taking $R_n = o_p(1)$ as in Theorems 6.1 and 6.2. □

Definition 6.3 *Effective Score and Information. The previous results suggest the following definitions: The effective score and effective information in the constrained model are given by*

$$S_n^{**} = BPS_n = (I - \dot{\theta}^T Q)S_n, \quad B^{**} = B - \dot{\theta}^T B^* \dot{\theta}.$$

With these definitions, it is easily seen that S_n^* and S_n^{**} are uncorrelated and hence asymptotically independent. The two score functions are not conformable, as S_n^* is indexed by the d' components of $\theta(\gamma)$ while S_n^{**} is indexed by the d components of γ. Premultiplying S_n^* by the $d \times d'$ matrix $\dot{\theta}^T$ converts it into a vector of length d compatible with S_n^{**}, allowing the additive decomposition

$$S_n = \dot{\theta}^T S_n^* + S_n^{**}$$

of the full score in γ into components orthogonal to and parallel to the constraint $\theta(\gamma) = 0$, and a corresponding decomposition of the full information as $B = \dot{\theta}^T B^* \dot{\theta} + B^{**}$.

Example 6.3 *(Continuation) We consider the dual problem of estimating the mean of a normal distribution $N(\mu, \sigma^2)$ subject to the constraint that the effect size $\theta = \mu/\sigma$, assumed positive, is known.*

Alternatively and equivalently, we may assume a known coefficient of variation $\kappa = \sigma/\mu = 1/\theta$. This version of the problem is more common in the literature, but we choose to keep our original formulation. After substituting for $\sigma = \mu/\theta$ and dropping μ-free terms from the sum, we obtain a log-likelihood function in μ from a sample of size n as

$$-n \log \mu - \frac{n\theta^2}{2\mu^2} \left\{ S_X^2 + (\bar{X}_n - \mu)^2 \right\}$$

where $\bar{X}_n$ is the sample mean and S_X^2 the sample variance, here calculated with divisor n. After setting the derivative in μ to zero and simplifying, we obtain a quadratic equation with positive root,

$$\hat{\mu} = \tfrac{1}{2}\theta^2 \left[\left\{ \bar{X}^2 + \frac{4(S_X^2 + \bar{X}^2)}{\theta^2} \right\}^{\frac{1}{2}} - \bar{X} \right]$$

a result that is computationally straightforward but not very intuitive. It can also be obtained using the Lagrange multiplier approach described above. The asymptotic variance of $\hat{\mu}$ can be calculated by the delta method or more simply from the information for μ.

For illustration, we use the general formula derived above, $P = B^{-1} - B^{-1}\dot{\theta}^T B^* \dot{\theta} B^{-1}$, where $B^* = (\dot{\theta} B^{-1} \dot{\theta}^T)^{-1}$. When, as here, the constraint is scalar, B is also scalar and may be moved outside the matrix multiplication. We also have

$$\dot{\theta}\dot{\theta}^T = \begin{pmatrix} 1/\sigma^2 & -\mu/\sigma^3 \\ -\mu/\sigma^3 & \mu^2/\sigma^4 \end{pmatrix},$$

$$B^{-1} = \begin{pmatrix} \sigma^2 & 0 \\ 0 & \frac{1}{2}\sigma^2 \end{pmatrix},$$

so that

$$P = \begin{pmatrix} \sigma^2 & 0 \\ 0 & \frac{1}{2}\sigma^2 \end{pmatrix} - \frac{\sigma^2}{1+\frac{1}{2}\theta^2}\begin{pmatrix} 1 & \frac{1}{2}\theta \\ -\frac{1}{2}\theta & \frac{1}{4}\theta^2 \end{pmatrix} = \frac{\sigma^2}{1+\frac{1}{2}\theta^2}\begin{pmatrix} \frac{1}{2}\theta^2 & -\frac{1}{2}\theta \\ -\frac{1}{2}\theta & \frac{1}{2} \end{pmatrix}.$$

The matrix is clearly of rank one, as we would expect, and immediately yields the asymptotic variances of $\hat{\mu}$ and $\hat{\sigma}$ in this constrained model. We find that

$$\text{Avar}\{\sqrt{n}(\hat{\mu}-\mu)\} = \frac{\frac{1}{2}\theta^2\sigma^2}{1+\frac{1}{2}\theta^2}. \tag{6.6}$$

An alternative approach to estimating μ, even simpler than maximum likelihood, is to combine the individual maximum likelihood estimates of μ from the sample mean $\bar{X}$ and from θ times the sample standard deviation S_X. We know that $\text{var}(\bar{X}) = \sigma^2/n$ and $\text{var}(s_X^2) = 2\sigma^4/(n-1)$ and that these estimates are independent. It follows from the delta method that $\sqrt{n}(\theta S_X - \theta\sigma)$ has asymptotic variance $\frac{1}{2}\theta^2\sigma^2$. Weighing these two independent estimates inversely according to their asymptotic variance (see Chapter 2) gives the optimal linear combination as

$$\tilde{\mu} = \frac{\bar{X} + (2/\theta)S_X}{1+2/\theta^2},$$

with asymptotic variance agreeing with the best achievable, shown in Equation (6.6). A further advantage of this approach, not pursued here, is that the estimator $\tilde{\mu}$ may be modified to be exactly unbiased for every n by multiplying S_X by an appropriate factor $a(n)$ depending on n.

6.4 Relative Efficiency

Suppose we compare a regular estimate $\tilde{\theta}_n$ with an asymptotically efficient estimate $\hat{\theta}_n$. If the former has large-sample variance $(1/n)V$ (while the latter has $(1/n)(B^*)^{-1}$), then $V \geq (B^*)^{-1}$ since $(B^*)^{-1}$ is best possible by the large-sample Cramér-Rao Inequality.

We start with the case of real θ, but allow the possibility of there being additional parameters, writing γ for the full model parameter (as in Sections 6.5 and 6.6). The ratio of asymptotic variances,

$$\frac{\text{Avar}(\hat{\theta}_n)}{\text{Avar}(\tilde{\theta}_n)} = (B^*)^{-1}V \leq 1 \tag{6.7}$$

is defined to be the *asymptotic relative efficiency.* This may vary with the model parameter γ, of course; we write $ARE(\tilde{\theta}_n;\gamma)$. It can be interpreted as the ratio of sample sizes needed, if using an efficient estimate relative to using the inefficient estimate $\tilde{\theta}_n$, to achieve the same large-sample variances (at γ). For, with sample sizes n and n' (both large) and setting

$$\text{Avar}(\hat{\theta}_n)/\text{Avar}(\tilde{\theta}_{n'}) \sim 1,$$

we find the left side of Equation (6.7) to be $(1/n)(B^*)^{-1}/\{(1/n')V\} = (n'/n)\cdot ARE$; hence, $ARE \sim (n/n')$.

Using the Orthogonal Decomposition Theorem, the large-sample covariance between $\tilde{\theta}_n$ and $\hat{\theta}_n$ is $(B^*)^{-1}$. Converting to a correlation coefficient and using 6.7, we find

Proposition 6.2 *For real θ, $ARE(\tilde{\theta}_n;\gamma) = \rho^2(\tilde{\theta}_n)$, where ρ is the large-sample correlation coefficient (assumed non-negative) between $\tilde{\theta}_n$ and any asymptotically efficient estimate $\hat{\theta}_n$.*

Not surprisingly, the asymptotic relative efficiency is identically (in γ) unity if and only if $\tilde{\theta}_n = \hat{\theta}_n + o_p(1/\sqrt{n})$.

For asymptotically linear estimates (in the random sampling case), this correlation is that between the kernel of $\tilde{\theta}_n$ and the effective score $s^*(X,\gamma)$. We discussed something similar in Chapter 3: The ratio of variances of asymptotically linear statistics is the same as their squared correlation, when one of the statistics had a kernel proportional to a score. As an example, recall that the asymptotic correlation between the sample mean and sample median when sampling from a normal distribution is $\sqrt{(2/\pi)}$ (see Chapter 3). Hence, the asymptotic relative efficiency of the sample median in this case is $2/\pi \approx 0.637 = 63.7\%$, not very high. In this example, the asymptotic relative efficiency is parameter-free, but this is not typical.

With vector parameters measures of relative efficiency will usually be directional, that is they will depend on the specific linear combination of parameters of interest. The matrix inequality $V \geq (B^*)^{-1}$ is equivalent to $a^TVa \geq a^T(B^*)^{-1}a$ for all vectors a — that is, $\text{var}(a^T\tilde{\theta}_n) \geq \text{var}(a^T\hat{\theta}_n)$ for every regular $\tilde{\theta}_n$, every asymptotically efficient $\hat{\theta}_n$ and every a. The asymptotic relative efficiency in *direction* a is

$$\text{ARE}(a^T\tilde{\theta}_n;\gamma) = \frac{a^T(B^*)^{-1}a}{a^TVa}.$$

We could get a single measure by defining the asymptotic relative efficiency to be the minimum over a of the directional asymptotic relative efficiencies; this turns out to be the smallest root λ of the determinantal equation $|(B^*)^{-1} - \lambda V| = 0$.

6.5 Brief Bibliographic Notes

Sections 6.1 and 6.2 mostly follow and elaborate on HALL & MATHIASON (1990). SILVEY (1975) and COX (2006, Section 6.8) discuss parameter estimation in the presence of constraints. Section 6.3 uses ideas expounded in MARZETTA (1993). For discussion of the use of an external source to estimate nuisance parameters, (*Exercise 12*), see JONKER & VAN DER VAART (2014).

Appendix I: Block Matrix Formulas

It is often convenient to partition an $r \times c$ matrix B into submatrices, sometimes called *blocks*, as follows:

$$B = \begin{pmatrix} B_{11} & B_{12} \\ B_{21} & B_{22} \end{pmatrix},$$

where B_{11} and B_{22} are square. It is easily verified that two square matrices A and B with the same size and partitioned into submatrices of the same dimensions may be added and multiplied using formulas analogous to the usual rules for matrix addition and multiplications, thus

$$A + B = \begin{pmatrix} A_{11} & A_{12} \\ A_{21} & A_{22} \end{pmatrix} + \begin{pmatrix} B_{11} & B_{12} \\ B_{21} & B_{22} \end{pmatrix} = \begin{pmatrix} A_{11} + B_{11} & A_{12} + B_{12} \\ A_{21} + B_{21} & A_{22} + B_{22} \end{pmatrix},$$

and

$$AB = \begin{pmatrix} A_{11}B_{11} + A_{12}B_{21} & A_{11}B_{12} + A_{12}B_{22} \\ A_{21}B_{11} + A_{22}B_{21} & A_{21}B_{12} + A_{22}B_{22} \end{pmatrix}.$$

The procedure for inverting a partitioned matrix is not as simple, but this can be accomplished using the following lemma.

Lemma A - 1 *Block Matrix Lemma: Suppose that B is a non-singular square matrix, and that B and B^{-1} are partitioned into block matrices B_{ij} and B^{ij} as follows:*

$$B = \begin{pmatrix} B_{11} & B_{12} \\ B_{21} & B_{22} \end{pmatrix} \text{ and } B^{-1} = \begin{pmatrix} B^{11} & B^{12} \\ B^{21} & B^{22} \end{pmatrix}$$

(with diagonal blocks square). If B_{11} and B_{22} are non-singular, then

$$B^{11} = \left(B_{11} - B_{12}B_{22}^{-1}B_{21}\right)^{-1}, \quad B^{22} = \left(B_{22} - B_{21}B_{11}^{-1}B_{12}\right)^{-1}$$

$$B^{12} = -B^{11}B_{12}B_{22}^{-1} \quad \text{and} \quad B^{21} = -B^{22}B_{21}B_{11}^{-1}$$

Proof We show that with the B^{ij} defined as above, $B^{-1}B = I$. The upper left block of $B^{11}B_{11}$ is

$$B^{11}B_{11} + B^{12}B_{21} = B^{11}\{B_{11} - B_{12}B_{22}^{-1}B_{21}\} = I,$$

and the upper right block is

$$B^{11}B_{12} + B^{12}B_{22} = B^{11}B_{12} - B^{11}B_{12}B_{22}^{-1}B_{22} = \mathbf{0}.$$

Here I and $\mathbf{0}$ are the identity and zero matrices of appropriate dimensions ($\mathbf{0}$ need not be square). Similar formulas apply for the lower left and right blocks, giving

$$\begin{pmatrix} B^{11} & B^{12} \\ B^{21} & B^{22} \end{pmatrix} \begin{pmatrix} B_{11} & B_{12} \\ B_{21} & B_{22} \end{pmatrix} = \begin{pmatrix} I, & \mathbf{0} \\ \mathbf{0} & I \end{pmatrix},$$

as required.

The implied non-singularities follow from identities such as

$$\begin{pmatrix} B_{11} & B_{12} \\ B_{21} & B_{22} \end{pmatrix} = \begin{pmatrix} B_{11} & 0 \\ B_{21} & B_{22} - B_{21}B_{11}^{-1}B_{12} \end{pmatrix} \begin{pmatrix} I & B_{11}^{-1}B_{12} \\ 0 & I \end{pmatrix}.$$

Hence, inversion of B only requires inversion of matrices of the orders of B_{11} and B_{22}. In particular, inversion of a 3×3 or 4×4 matrix becomes very simple. [Try it.]

Our primary interest is when B is symmetric and positive definite. Then B_{11} and B_{22} are also positive definite; therefore, the necessary inverses B_{11}^{-1} and B_{22}^{-1} exist.

Corollary A - 1 *Suppose that B is symmetric positive definite. Then $B_{11} - (B^{11})^{-1} = B_{12}(B_{22})^{-1}B_{21}$ and is positive semi-definite; hence*

$$B_{11} \geq (B^{11})^{-1} \tag{A.1}$$

with equality if and only if $B_{12} = 0$.

Corollary A - 2 *Suppose that B is symmetric positive definite. Then $B^{11} - (B_{11})^{-1} = B_{11}^{-1}B_{12}B^{22}B_{21}B_{11}^{-1}$ and is positive semi-definite.; hence*

$$B^{11} \geq B_{11}^{-1} \tag{A.2}$$

with equality if and only if $B_{12} = 0$.

Similar formulas obtain with the values 1 and 2 of the subscripts and superscripts interchanged, or with subscripts interchanged with superscripts. Recall that inequalities like (A.1) and (A.2) mean that the appropriate difference is positive semi-definite.

When B is an information matrix, the inequality in the first corollary may be interpreted as stating that information for one parameter when the other is known is at least as great as effective information when the other is unknown.

The inequality in the second corollary may be interpreted as stating that the large-sample variance when estimating a parameter when there is an unknown nuisance parameter is at least as great as the large-sample variance when the nuisance parameter is known. See Section 6.1.

Proof The formula in Equation (A.1) follows immediately from the lemma. The right side may be re-expressed as $(CB_{21})^T(CB_{21})$ with $B_{22}^{-1} = C^TC$, and hence is positive semi-definite since a symmetric matrix M is positive semi-definite if and only if $M = A^TA$ for some A of the same rank as M.

For Equation (A.2), we have $B^{11} - B_{11}^{-1} = B^{12}(B^{22})^{-1}B^{21}$ by Corollary 6.1 with an interchange of superscripts and subscripts, and this is positive semi-definite by the argument just given. Now $B^{12} = (B^{21})^T = -B_{11}^{-1}B_{12}B^{22}$ by the last formula in the Lemma, and the claimed formula follows. □

Next, we present a *Cauchy-Schwarz inequality* for random vectors.

Lemma A - 2 *(A Cauchy-Schwarz Lemma): Suppose that* $\mathbf{X}$ *and* $\mathbf{Y}$ *are random d-vectors with* $\text{var}(\mathbf{X}) = A > 0$, $\text{var}(\mathbf{Y}) = B \geq 0$, *and* $\text{cov}(\mathbf{X}, \mathbf{Y}) = C$. *Then*

$$C^TA^{-1}C \leq B \tag{A.3}$$

with equality if and only if $\mathbf{Y} = D\mathbf{X} + \mathbf{c}$ *(with probability 1) for some constant matrix D and some constant vector* $\mathbf{c}$, *in which case* $D = C^TA^{-1}$.

Recall that the meaning of (A.2) is that $B - C^TA^{-1}C$ is positive semi-definite; an equivalent statement is

$$\mathbf{u}^TC^TA^{-1}C\mathbf{u} \leq \mathbf{u}^TB\mathbf{u} \quad \text{for all} \quad d\text{-vectors } \mathbf{u}.$$

When $d = 1$, this reduces to the familiar statement that the square of a covariance is less than or equal to the product of the variances.

Proof Set $\mathbf{Z} = \mathbf{Y} - C^TA^{-1}\mathbf{X}$. Then $0 \leq \text{var}(\mathbf{Z}) = B - C^TA^{-1}C$, proving (A3). Moreover, $0 = \text{var}(\mathbf{Z})$ if and only if $\mathbf{Z}$ is constant (with probability 1); that is, $\mathbf{Y} - C^TA^{-1}\mathbf{X} = \mathbf{c}$, say. And if $\mathbf{Y} = D\mathbf{X} + \mathbf{c}$, we find $C = AD^T$ so that $D = C^TA^{-1}$, as claimed. □

Some variations of this lemma appear in *Exercises 16–17*. Next we present a lemma relevant to the discussion of constrained estimation in Section 6.8. Our treatment is adapted from that in SILVEY (1970), Appendix A.

Lemma A - 3 *(Silvey): If B is a symmetric positive definite matrix of order d and H is a* $p \times q$ *matrix of rank* d' *(so that* $d' \leq d$*) then the partitioned symmetric matrix*

$$\begin{pmatrix} B & H \\ H^T & O \end{pmatrix}$$

of order $d + d'$ *is non-singular with inverse*

$$\begin{pmatrix} P & Q \\ Q^T & R \end{pmatrix},$$

where

$$P = B^{-1} - B^{-1}H(H^TB^{-1}H)^{-1}H^TB^{-1},$$
$$Q = B^{-1}H(H^TB^{-1}H)^{-1} \quad \text{and } R = -(H^TB^{-1}H)^{-1}.$$

Proof First note that the square matrix $H^TB^{-1}H$ of order d' is invertible since H is of rank d'. The result can now be verified by checking that the product of the two matrices is the identity matrix of order $d + d'$. □

Finally, we present a matrix lemma proved by STEIN (1956), important in the theory of adaptive estimation (Section 6.1).

Lemma A - 4 *(Stein): Consider the symmetric positive definite partitioned matrix*

$$B = \begin{pmatrix} B_{11} & B_{12} & B_{13} \\ B_{21} & B_{22} & B_{23} \\ B_{31} & B_{32} & B_{33} \end{pmatrix},$$

where $B_{12} = B_{21}^T$, $B_{31} = B_{13}^T$ and $B_{23} = B_{32}^T$. Then in order for the top left submatrix B_{11}^{-1} of B^{-1} to be the same as that of the inverse of the smaller matrix

$$\begin{pmatrix} B_{11} & B_{12} \\ B_{21} & B_{22} \end{pmatrix},$$

it is necessary and sufficient that Stein's condition $B_{13} = B_{12}B_{22}^{-1}B_{23}$ hold

Note Although the result is quite general, our matrix B will be the information matrix for a regular parametric model.

Proof For $j = 1, 2, 3$ let d_j denote the dimension of the square matrix B_{jj} and let I_j denote the identity matrix of dimension d_j. From partitioned matrix formulas, the two submatrices will be identical if and only if

$$B_{11} - B_{12}B_{22}^{-1}B_{21} = B_{11} - (B_{12}, B_{13})\begin{pmatrix} B_{22} & B_{23} \\ B_{32} & B_{33} \end{pmatrix}^{-1}\begin{pmatrix} B_{21} \\ B_{31} \end{pmatrix},$$

or equivalently

$$B_{12}B_{22}^{-1}B_{21} = (B_{12}, B_{13})\begin{pmatrix} B_{22} & B_{23} \\ B_{32} & B_{33} \end{pmatrix}^{-1}\begin{pmatrix} B_{21} \\ B_{31} \end{pmatrix}.$$

Setting $H = B_{13} - B_{12}B_{22}^{-1}B_{23}$ we will show that this equation is equivalent to $H = 0$. It is equivalent to requiring that $B_{12}B_{22}^{-1}B_{21}$ equal

$$\{B_{12}B_{22}^{-1}(B_{22}, B_{23}) + (0, H)\}\begin{pmatrix} B_{22} & B_{23} \\ B_{32} & B_{33} \end{pmatrix}^{-1}\left\{\begin{pmatrix} B_{22} \\ B_{23}B_{22}^{-1}B_{21} \end{pmatrix} + \begin{pmatrix} 0 \\ H^T \end{pmatrix}\right\}, \tag{A.4}$$

since the first term in the product equals $(B_{12}, B_{12}B_{22}^{-1}B_{23} + H) = (B_{12}, B_{13})$.

By the definition of an inverse matrix,

$$\begin{pmatrix} B_{22} & B_{23} \\ B_{32} & B_{33} \end{pmatrix} \begin{pmatrix} B_{22} & B_{23} \\ B_{32} & B_{33} \end{pmatrix}^{-1} = \begin{pmatrix} I_2 & 0 \\ 0 & I_3 \end{pmatrix}.$$

from which it follows that

$$(B_{22}, B_{23}) \begin{pmatrix} B_{22} & B_{23} \\ B_{32} & B_{33} \end{pmatrix}^{-1} = (I_2, 0)$$

and

$$\begin{pmatrix} B_{22} & B_{23} \\ B_{32} & B_{33} \end{pmatrix}^{-1} \begin{pmatrix} B_{22} \\ B_{32} \end{pmatrix} = \begin{pmatrix} I_2 \\ 0 \end{pmatrix}.$$

Consider the four terms obtained by expanding out the right side of (A.4). The lead term, which does not involve H, is

$$\{B_{12}B_{22}^{-1}(B_{22}, B_{23})\} \begin{pmatrix} B_{22} & B_{23} \\ B_{32} & B_{33} \end{pmatrix}^{-1} \left\{ \begin{pmatrix} B_{22} \\ B_{23} \end{pmatrix} B_{22}^{-1} B_{21} \right\},$$

which equals

$$B_{12}B_{22}^{-1}(I_2, 0) \begin{pmatrix} B_{22} \\ B_{23} \end{pmatrix} B_{22}^{-1} B_{21} = B_{12}B_{22}^{-1}B_{21}.$$

The second and third terms are

$$B_{12}B_{22}^{-1}(B_{22}, B_{23}) \begin{pmatrix} B_{22} & B_{23} \\ B_{32} & B_{33} \end{pmatrix}^{-1} \begin{pmatrix} 0 \\ H^T \end{pmatrix} = B_{12}B_{22}^{-1}(I_2, 0) \begin{pmatrix} 0 \\ H^T \end{pmatrix} = 0,$$

and its transpose, also the zero matrix. The last term is

$$(0, H) \begin{pmatrix} B_{22} & B_{23} \\ B_{32} & B_{33} \end{pmatrix}^{-1} \begin{pmatrix} 0 \\ H^T \end{pmatrix},$$

So (A.4) is equivalent to

$$B_{12}B_{22}^{-1}B_{21} = B_{12}B_{22}^{-1}B_{21} + (0, H) \begin{pmatrix} B_{22} & B_{23} \\ B_{32} & B_{33} \end{pmatrix}^{-1} \begin{pmatrix} 0 \\ H^T \end{pmatrix},$$

Since the matrix

$$\begin{pmatrix} B_{22} & B_{23} \\ B_{32} & B_{33} \end{pmatrix}^{-1}$$

is positive definite the second term is the zero matrix if and only if $H = 0$. □

Appendix II: A Property of the Profile Log-Likelihood

In this appendix, we discuss the properties of the profile log-likelihood function, defined as

$$g(\theta) = L\{\theta, \hat{\eta}(\theta)\},$$

where

$$\hat{\eta}(\theta) = \text{argmax}\{L(\theta, \eta)\},$$

the function of θ obtained by maximizing out the nuisance parameter η. We will assume here that, for every θ, the maximizing value $\hat{\eta} = \hat{\eta}(\theta)$ is unique, and that $L(\theta, \hat{\eta})$ has continuous matrix of second derivatives of full rank at $(\theta, \hat{\eta}(\theta)$.

The key result, which motivates much of the development in the chapter, is that the second derivative

$$\frac{\partial^2 g(\theta)}{\partial \theta^2},$$

of g, regarded as a function of θ alone, is equal to

$$\frac{\partial^2 L}{\partial \theta^2} - \frac{\partial^2 L}{\partial \theta \partial \eta}\left(\frac{\partial^2 L}{\partial \eta^2}\right)^{-1}\frac{\partial^2 L}{\partial \eta \partial \theta^2}, \tag{A.5}$$

the inverse of the leading corner of the inverse of the full second derivative matrix of L. The proof is a straightforward application of the implicit function theorem and the connection between total and partial derivatives. Care is needed in keeping track of matrix transposes, recalling for example that the transpose of a product of matrices is the product of transposes in reverse order. We write

$$\frac{\partial^2 L}{\partial \theta \partial \eta} = -i_{\theta\eta} = \left(\frac{\partial^2 L}{\partial \eta \partial \theta}\right)^T = -i_{\eta\theta}^T,$$

where here i denotes the total sample information matrix.

The derivative of $g(\theta)$, which accounts for both the direct effect on $g(\theta)$ of the change in θ and indirect effect of the change in η due to the change in θ, is

$$\dot{g}(\theta) = \frac{dL}{d\theta} = \frac{\partial L}{\partial \theta} + \dot{\eta}(\theta)\frac{\partial L}{\partial \eta},$$

and the second derivative is

$$\ddot{g}(\theta) = \frac{d}{d\theta}\left\{\frac{\partial L}{\partial \theta} + \dot{\eta}(\theta)\frac{\partial L}{\partial \eta}\right\}^T = \left\{\frac{\partial}{\partial \theta} + \dot{\eta}(\theta)\frac{\partial}{\partial \eta}\right\}\left\{\left(\frac{\partial L}{\partial \theta}\right)^T + \left(\frac{\partial L}{\partial \eta}\right)^T \dot{\eta}(\theta)^T\right\}.$$

The result is a sum of five terms, namely

$$\frac{\partial^2 g(\theta)}{\partial \theta^2} + \sum_k \frac{\partial L}{\partial \eta_k}\ddot{\eta}_k(\theta) + \frac{\partial^2 L}{\partial \theta \partial \eta}\dot{\eta}(\theta)^T + \dot{\eta}(\theta)\frac{\partial^2 L}{\partial \eta \partial \theta} + \dot{\eta}(\theta)\frac{\partial^2 L}{\partial \eta^2}\dot{\eta}(\theta)^T. \tag{A.6}$$

We have written the second term as a sum involving the components η_k of η to avoid three-dimensional arrays. However, this term is zero as $\partial L/\partial \eta_k = 0$ at $\hat{\eta}$ for all k. To simplify the remaining terms we use the fact that $\eta(\theta)$ is defined implicitly by the equation

$$\frac{\partial L\{\theta, \eta(\theta)\}}{\partial \eta} = 0 \tag{A.7}$$

to give an expression for $\dot{\theta}(\eta)$.

Since Equation (A.7) holds for every θ, the total derivative of the left side in θ must be zero. Hence

$$\left\{\frac{\partial}{\partial \theta} + \dot{\eta}(\theta)\frac{\partial}{\partial \eta}\right\}\frac{\partial L}{\partial \eta} = 0,$$

yielding

$$\frac{\partial^2 L}{\partial \theta \partial \eta} + \dot{\eta}(\theta)\frac{\partial^2 L}{\partial \eta^2} = 0,$$

so that

$$\dot{\eta}(\theta) = -\frac{\partial^2 L}{\partial \theta \partial \eta}\left(\frac{\partial^2 L}{\partial \eta^2}\right)^{-1}.$$

We find on substitution that the third and fourth terms and the negative of the last term in Equation (A.6) all equal the second term in Equation (A.5), proving the result.

Example 6.4 *We return to the Weibull distribution of the previous chapter, with density*

$$f(x, \theta, \eta) = \theta \eta^\theta x^{\theta - 1} \exp\{-(\eta x)^\theta\},$$

where the index parameter θ is of interest and the scale parameter is a nuisance parameter.

The log-likelihood from a sample of size n is

$$L = n \log \theta + n\theta \log \eta + (\theta - 1)\sum \log x_i - \sum (\eta x_i)^\theta.$$

So

$$\frac{\partial L}{\partial \eta} = \frac{n\theta}{\eta} - \frac{\theta}{\eta}\sum (\eta x_i)^\theta, \tag{A.8}$$

and

$$\frac{\partial L}{\partial \theta} = \frac{n}{\eta} + n \log \eta - \sum \log x_i - \sum (\eta x_i)^\theta \log(\eta x_i).$$

The second derivatives are

$$\frac{\partial^2 L}{\partial \eta^2} = -n\theta\eta^2 - \theta(\theta - 1)\eta^2 \sum (\eta x_i)^\theta,$$

$$\frac{\partial^2 L}{\partial\eta\partial\theta} = \frac{n}{\eta} - \frac{1}{\eta}\sum(\eta x_i)^\theta - \frac{\theta}{\eta}\sum(\eta x_i)^\theta \log(\eta x_i).$$

$$\frac{\partial^2 L}{\partial\theta^2} = -\frac{n}{\theta^2} - \sum(\eta x_i)^\theta\{\log(\eta x_i)\}^2.$$

As was shown in Chapter 5, in this model the maximizing value of $\hat\eta$ can be calculated explicitly as a function of θ by solving (A.8). We find that $\sum(\hat\eta x_i)^\theta = n$, so that $\hat\eta = (n/\sum x_i^\theta)^{1/\theta}$. When this relation holds the first two formulas simplify, giving

$$\frac{\partial^2 L}{\partial\eta\partial\theta} = -\frac{\theta}{\hat\eta}\sum(\hat\eta x_i)^\theta \log(\hat\eta x_i),$$

and

$$\frac{\partial^2 L}{\partial\eta^2} = -\frac{n\theta^2}{\hat\eta^2}.$$

With $A_j = \sum(\hat\eta x_i)^\theta\{\log(\hat\eta x_i)\}^j$ for $j = 1, 2$ the matrix of negative second derivatives is

$$\begin{pmatrix} n/\theta^2 + A_2 & \theta A_1/\hat\eta \\ . & n\theta^2/\hat\eta^2 \end{pmatrix}$$

so that the block matrix formula, which reduces to the ordinary inversion formula for the leading element of this 2×2 matrix, gives

$$(i^{\theta\theta})^{-1} = n/\theta^2 + A_2 - A_1^2/n. \tag{A.9}$$

The profile log-likelihood function $g(\theta)$ is obtained by substitution for $\hat\eta(\theta)$ in $L(\theta)$, giving

$$g(\theta) = n\log\theta - n\log(\sum x_i^\theta) + n\log n + (\theta - 1)\sum \log x_i - n.$$

The derivative in θ is

$$\dot g(\theta) = \frac{n}{\theta} - n\frac{\sum x_i^\theta \log x_i}{\sum x_i^\theta} + \sum \log x_i.$$

and the second derivative is

$$\ddot g(\theta) = -\frac{n}{\theta^2} - n\frac{\sum x_i^\theta (\log x_i)^2}{\sum x_i^\theta} + n\left(\frac{\sum x_i^\theta \log x_i}{\sum x_i^\theta}\right)^2.$$

To compare this expression with (A.9) above, note that $\sum x^\theta = n/\hat\eta^\theta$ so that

$$\ddot g(\theta) = -\frac{n}{\theta^2} - n\sum(\hat\eta x_i)^\theta(\log x_i)^2 + n\left\{\sum(\hat\eta x_i)^\theta \log x_i\right\}^2.$$

The second and third terms combined are the negative of the variance of the $\log x_i$ when individual i is selected from the finite population $i = 1, \ldots, n$

with probability $n^{-1}(\hat{\eta}x_i)^\theta$. This variance is unchanged when the $\log x_i$ are replaced by $\log(\hat{\eta}x_i)$, since this transformation just adds $\log\hat{\eta}$ to each value. So $\ddot{g}(\theta) = -(i^{\theta\theta})^{-1}$ as asserted.

We conclude with a few general comments about the use of profile likelihood:

1. The example is special in that an explicit formula is available for $\hat{\eta}$. More commonly $g(\theta)$ has to be computed numerically, requiring a second level of iteration.
2. The use of the profile log likelihood avoids the need to invert the full matrix $i_{\gamma\gamma}$, a useful feature when d'' is large. Contrariwise, computation can sometimes be eased by the introduction of artificial nuisance parameters if this leads to a simpler form for the log likelihood.
3. The argument holds for every pair $\{\theta, \hat{\eta}(\theta)\}$, not just at the overall maximum likelihood estimate $\{\hat{\theta}, \hat{\eta}(\hat{\theta})\}$. However, it does not hold away from the curve $\{\theta, \hat{\eta}(\theta)\}$. In fact the log-likelihood function for the Weibull example just discussed may not be concave away from this curve. Avoidance of possible regions of noncavity of the full log-likelihood function is another advantage of the profile log-likelihood approach.
4. This Appendix does not use any distributional assumptions or probabilistic reasoning. However, it provides intuitive support for the approach discussed in Section 6.1 of replacing the nuisance parameter η in the effective score function S_n^* for θ by an asymptotically efficient estimate.

Exercises

1. Consider sampling from a *trinomial distribution* with parameter $\gamma = (p, q)$; that is, each observation $X = (U, V)$ has density

$$f_{p,q}(u, v) = p^u q^v (1 - p - q)^{1-u-v}$$

where $(u, v) = (0, 0)$, $(0, 1)$, or $(1, 0)$ and $p > 0$, $q > 0$ and $p + q < 1$. Find the *effective score* for p and the *effective information* for p (i.e., $\theta = p$ and $\eta = q$). Note that they are identical to the score and information in the Bernoulli (binomial) model in which only the u's are observed, i.e., the model is

$$f_p(u) = p^u (1 - p)^{1-u}, u = 0 \text{ or } 1.$$

Hence inference about the probability p of the first category, can be based just on the counts in the first category and those in the other two categories combined. This seems right (does it not?). [*Note:* The problem becomes much simpler if we reparameterize by setting $h_1 = p$ and $h_2 = q/(1-p)$, for then the likelihood factorizes into functions of h_1 and h_2 The idea extends to multiple categories in an obvious way].

2. Show that the effective score is the residual from large-sample-regression of $S_{\theta n}$ on $S_{\eta n}$, that is $S_n^* = S_{\theta n} - E(S_{\theta n}|S_{\eta n})$ using the large-sample distribution to compute the conditional expectation.

3. Extend the profile likelihood argument in Appendix II to the censored Weibull example shown in *Exercise 23* of Chapter 5. Using the data from that example derive the profile likelihood function as a function β and plot it.

4. An alternative definition of the effective score per observation in a random sampling model with parameter $\gamma = (\theta, \eta)$ is

$$s_n^o(x, \theta) = s_\theta\{x, \theta, \hat{\eta}_n(\theta)\}$$

where $s_\theta(x, \theta, \eta)$ is the score for θ (per observation) and $\hat{\eta}_n(\theta)$ is an asymptotically efficient estimate of η in the random sampling model with θ known. The *average effective score* is then the sample average of the $s_n^o(x_j, \theta)$'s. Show that this average effective score is asymptotically normaL, $AN\{0, (1/n)B^*\}$. You may assume that both θ and η are real.

5. Consider a random sampling model with parameter $\gamma = (\theta, \eta)$. The *effective log likelihood*, for a single observation x, written $l^*(x, \theta)$ is defined as the likelihood for a single x with η replaced by an asymptotically efficient estimate $\hat{\eta}_n(\theta)$ under the assumption that θ is known. Dependence of l^* on n and on $\mathbf{x}_n$ is not made explicit in the notation. Show that the effective score for θ, as defined in the previous problem, is (to a first order of approximation) the derivative of the effective log likelihood. Note the connection with the profile log-likelihood function discussed in Appendix II.

6. Suppose that we have data in the form of a random sample of observations from a density $f(x, \theta, \eta)$, where θ is of interest and η is a nuisance parameter. Let $(\hat{\theta}, \hat{\eta})$ be an efficient estimator of (θ, η). Now suppose that we are told that $\eta = 0$. Show, by direct calculation using partitioned matrices, that, when $\eta = 0$,

$$\hat{\theta}' = \hat{\theta} - B_{\theta\theta}^{-1} B_{\theta\eta} \hat{\eta},$$

has

$$\sqrt{n}(\hat{\theta}' - \theta) \sim AN(0, B_{\theta\theta}^{-1}).$$

and comment on this result (see Cox (2006), Section 6.8 for related discussion).

7. Suppose that $\gamma = (\theta, \eta)$ and $S_\theta \perp S_\eta$. Let $\tilde{\eta}_n$ be a root-n consistent estimate of η. Show, under assumptions, that a near root of $S_{\theta n}(\theta, \tilde{\eta}_n)$ is asymptotically efficient for θ. Hence, it is sufficient to iterate only the equations for θ rather than those for both θ and η, having substituted $\tilde{\eta}_n$ for η. (Note also that $\tilde{\eta}_n$ may be allowed to depend on θ.)
8. A partial converse to Proposition 6.4. Suppose that $B_{\theta\eta} \neq 0$ and that $\tilde{\eta}$ is a regular but inefficient estimator of η when θ is known. Show that substituting $\tilde{\eta}$ for η into the score equation $S_{n\theta} = 0$ for θ yields a regular but inefficient estimator for θ.
9. Consider random sampling from a $N(\mu, \sigma^2)$ distribution. Show that it is possible, when estimating σ^2, to adapt for μ, that is, the large-sample distribution of an asymptotically efficient estimate of σ^2 is the same whether or not μ is known. (Note, however, that this is only a large-sample result: the exact distribution of the uniform minimum variance unbiased estimator estimate of σ^2 does depend on whether or not μ is known, in contrast to the case of estimating μ.)
10. Consider a model with parameter (θ, η, τ) and a reduced model with τ known to be τ_o.

    (a) Show that the effective information for θ is the same in each model if and only if $B_{\theta\tau} = B_{\theta\eta} B_{\eta\eta}^{-1} B_{\eta\tau}$. Use Stein's theorem described in Appendix 1.

    (b) Suppose that data are available from two independent samples of sizes n and np respectively from densities $f_0(x, \theta, \eta_1)$ and $f_0(x, \theta, \eta_2)$ respectively where f_0 is a known density function, θ is the parameter of interest and η is a nuisance parameter. So the parameter of interest θ has the same value for the two samples but the values of the nuisance parameter may differ. Reparameterizing with $\eta_1 = \eta$ and $\eta_2 = \eta + \tau$, show that when $\eta_1 = \eta_2$ the information matrix in (θ, η, τ) from the combined sample is

    $$\begin{pmatrix} (1+p)B_{0\theta\theta} & (1+p)B_{0\theta\eta} & pB_{0\theta\eta} \\ \cdots & (1+p)B_{0\eta\eta} & pB_{0\theta\eta} \\ \cdots & \cdots & pB_{0\eta\eta} \end{pmatrix}$$

    where

    $$\begin{pmatrix} B_{0\theta\theta} & B_{0\theta\eta} \\ \cdots & B_{0\eta\eta} \end{pmatrix}$$

    is the information matrix for f_0.

    Show that the condition in (a) applies here, so that we can adapt for τ. When the values of the nuisance parameter in the two samples are the same, there is no penalty (asymptotically!) for a cautious estimation strategy which allows them to differ. The situation is analogous to that of the usual two-sample t-test, where the two population variances may be assumed to be equal (or to have known ratio) or may be allowed to differ. However, for the t-test, the mean and variance parameters are orthogonal, which need not be the case here.

11. (*See the discussion of location-scale families in Chapter 4*). Consider the location-scale model with $f(x;\theta,\eta) = f_0\{(x-\theta)/\eta\}$, where $f_0(\cdot)$ is a known function, not necessarily symmetric. Consider the modified location parameter $\alpha_\lambda = \alpha + \lambda\eta$, where λ is to be specified. Calculate the asymptotic variance of $\hat{\alpha}_\lambda$ as a function of λ and show that it always has a unique minimum in λ and that, for the minimizing value of λ, α_λ and η are asymptotically independent. Show that if $f_0(\cdot)$ is symmetric about zero, $f_0(x) = f_0(-x)$ for all x, then $\hat{\alpha}$ and $\hat{\eta}$ are asymptotically independent and the minimizing value of λ is zero.

12. *Estimation of the nuisance parameter from an external source.* Let θ be the parameter of interest and η a nuisance parameter. Suppose that observations $X_1, \ldots, X_n$ are sampled from a density $f_X(x,\theta,\eta)$ depending on the full parameter vector and and a further n independent observations $Y_1, \ldots, Y_n$ from the density $f_Y(y,\eta)$ depending only on the nuisance parameter. Let $B_X(\theta,\eta)$ and $B_Y(\eta)$ denote the information matrices from the two densities, S_X and S_Y the two score vectors. Let (θ_0,η_0) denote the true value of (θ,η).

    (a) Supposing that both parameters are scalar, consider the estimation of (θ,η) (i) from the data on X alone and (ii) from the combined data on (X,Y). Calculate the asymptotic relative efficiency of strategy (i) compared with strategy (ii), and show that it is always less than unity except when $\theta \perp \eta$, but it always exceeds the corresponding asymptotic efficiency for the estimates of η.

    (b) A third strategy, which may be useful when η is high-dimensional and $f_Y(y,\eta)$ has a simple form, is to estimate η from the data on Y alone and substitute this estimate $\tilde{\eta}$ say, for the true value η_0 in the score equation for θ from the data on X. Show that the resulting estimate $\tilde{\theta}$ has $\sqrt{n}(\tilde{\theta}-\theta_0) \sim AN(0,V)$, where

    $$V = B_{X\theta\theta}^{-1} + B_{X\theta\theta}^{-1} B_{X\theta\eta} B_{Y\eta\eta}^{-1} B_{X\eta\theta} B_{X\theta\theta}^{-1}.$$

    Note: (i) The result extends easily to samples of different sizes $X_1, \ldots, X_m$ and $Y_1, \ldots Y_n$ say, as long as m and n increase at the same rate, $m/n = O(1)$.

13. Verify the following assertions in Section 6.3, using the notation defined there:

    (a) $Q\dot{\theta}^T$ is the identity matrix of dimension d';

(b) $\dot\theta^T Q$, is idempotent, $(\dot\theta^T Q) \times (\dot\theta^T Q) = \dot\theta^T Q$;

(c) $PBP = P$, $QBQ^T = B^*$, PBQ^T and QBP are zero matrices.

14. With the notation of Section 6.3, verify that for any h, $h' = Ph$ satisfies $\dot\theta h' = 0$ and that any h' with $\dot\theta h' = 0$ is the image under this map of $h = Bh'$.

15. Verify that the formulas for effective scores and effective information in Sections 6.8 and 6.9 reduce correctly when $\theta(\gamma) = (I, 0)\gamma$, where I is the identity matrix of size d' and 0 a $d' \times d - d'$ zero matrix, so that $\theta(\gamma)$ just picks out the first d' components of γ.

16. Consider the model described in *Exercise 27* of Chapter 2, where, for $i = 1, \ldots, n$, the $\{X_i\}$, $\{Y_i\}$ and $\{Z_i\}$ have independent Poisson distributions with means $\mu_X = \mu \exp(-\beta)$, $\mu_Y = \mu$ and $\mu_Z = \mu \exp(\beta)$ respectively.

    (a) Derive the joint log likelihood in (β, μ) and find the asymptotic information about β (i) when μ is known and (ii) when it is unknown.

    (b) Hence show that the weighted estimator derived in that Exercise is asymptotically efficient when μ is unknown.

    (c) Formulate this model as a three-parameter model subject to the constraint $\mu_X - 2\mu_Y + \mu_Z = 0$ on the three Poisson means and derive the corresponding constrained likelihood estimator and its (singular) covariance matrix using the formulas in Section 6.3.

17. Prove an alternative Cauchy-Schwarz Lemma, as in Appendix I but with $A = E(\mathbf{X}\mathbf{X}^T)$, $B = E(\mathbf{Y}\mathbf{Y}^T)$, $C = E(\mathbf{X}\mathbf{Y}^T)$ and $\mathbf{c} = \mathbf{0}$.

18. Extend the Cauchy-Schwarz Lemma of Appendix I to allow $A \geq 0$, with A^{-1} replaced by any *generalized inverse* A^- for which $A^- A A^- = A^-$, for example the *Moore generalized inverse.* (For information about such g-inverses, see Chapter 9 or RAO (1973).)

19. (a) Show that if the $m \times m$ matrix A is invertible and b, c are column vectors of length m then $(A + bc^T)$ is invertible with

$$(A + bc^T)^{-1} = A^{-1} - A^{-1}\frac{(A^{-1}b)(c^T A^{-1})}{1 + c^T A^{-1} b}.$$

    Note that $c^T A^{-1} b$ is a scalar quantity, and that division is interpreted elementwise.

    (b) Suppose that $(V_1, \ldots, V_m)$ follow a multinomial distribution with index n and probabilities $p_1, \ldots, p_m$ with $\sum p_j = 1$ and the first $m - 1$ values of the p_j are taken as parameters with $p_m = 1 - (p_1 + \cdots + p_{m-1})$. Derive the log-likelihood function in $p_1, \ldots, p_{m-1}$ and the corresponding score functions in these p_j. Show that the maximum likelihood estimates of the p_j are

$$\hat p_j = V_j / n$$

and that the covariance matrix of these estimates is $n^{-1}\Sigma$ where Σ has diagonal elements $\Sigma_{jj} = p_j(1-p_j)$ and off-diagonal elements $\Sigma_{jk} = -p_jp_k$. Show also that the information matrix is nB where B has diagonal elements $B_{jj} = 1/p_j + 1/p_m$ and off-diagonal elements $B_{jk} = 1/p_m$

(c) Use part (a) to verify that $\Sigma^{-1} = B$. Take A to be the diagonal matrix with entries $(p_1, \ldots, p_{m-1})$ and $b = -c = (p_1, \ldots, p_{m-1})^T$.

[*Note:* See, for example, RAO (1973) Exercise 2.8. for part (a). We return to the multinomial distribution in Chapter 9.]

20. Investigate the form of the profile log-likelihood function $g(\theta)$ for the effect size parameter $\theta = \mu/\sigma$ from a normally distributed random sample. Verify that $g''(\theta)$ is given by equation (A.5).

21. A simplified form of a model of BOX & COX (1968). Suppose that, for some unknown non-negative value of λ, the $X_1, X_2, \ldots, X_n$ are such that

$$Y_i^\lambda = \begin{cases} (X_i^\lambda - 1)/\lambda, & \text{if } \lambda > 0; \\ \log X_i, & \text{if } \lambda = 0. \end{cases}$$

are independent and identically distributed with normal distribution with mean $\mu > 0$ and variance $\sigma^2 > 0$. It may be assumed that μ is sufficiently large compared to σ for the possibility of negative values may be ignored.

(a) Find an expression for the profile log-likelihood function in λ.

(b) Suppose that λ is restricted to a finite set of possible values, for example $\{0, \frac{1}{2}, 1\}$. Show that the true value of λ can be identified with probability approaching unity as $n \to \infty$ and that the asymptotic distribution of $(\hat{\mu}, \hat{\lambda})$ is the same as it would have been if λ were known.

(c) Using the approach outlined in the Appendix to Chapter 5, show how (b) may be extended to the case where λ is known only to be rational.

7

Large Sample Hypothesis Testing and Confidence Sets

We present an approach to define and derive asymptotically efficient tests. It requires assumptions similar to those made in large-sample estimation, in particular, a requirement that tests be constructed from asymptotically normal test statistics. But instead of a lower bound on the large-sample variance of estimates, with conditions for equality characterizing efficient estimates, we develop an upper bound on the large-sample power of tests having a specified large-sample significance level, with conditions for equality characterizing asymptotically efficient tests.

Section 7.1 reviews the basic concepts of hypothesis testing in finite samples, including the Neyman-Pearson lemma, existence of uniformly most powerful tests and the notion of an unbiased test. Emphasis is placed on tests constructed from normally distributed variables.

In Section 7.2, we present an asymptotic theory for one- and two-sided testing problems in a model with a single (real) parameter. In Section 7.3, we extend it to models with nuisance parameters, but still with hypotheses about the real parameter of interest. In the first case, we find that any asymptotically efficient test statistic must be asymptotically equivalent to the *standardized score*; this includes both the *Rao (score) test statistic* and the *Wald test statistic*. With nuisance parameters, any asymptotically efficient test statistic is found to be asymptotically equivalent to the *standardized effective score*; such tests require estimation of the nuisance parameter. They include *effective score tests*, as well as *score tests* and *Wald tests*. Indeed, all of these tests are shown to be equivalent, to the first order of approximation used here.

In Section 7.4, we consider tests about a multidimensional parameter of interest, allowing for additional nuisance parameters, and characterize the asymptotically efficient tests among those based on quadratic forms in asymptotically normal statistics. Section 7.5 considers *likelihood-ratio tests*, showing them to be asymptotically equivalent to the other asymptotically efficient tests of Sections 7.2–7.4.

Along the way, in Section 7.4, brief attention is given to relative efficiency of alternative tests and to the cost of not knowing nuisance parameters. These results are very similar to those of Chapter 6. Power approximations and large-sample p-values are also considered.

 DOI: 10.1201/9780429160080-7

Having developed efficiency theory and methods for tests, we show in Section 7.6 how to invert a family of tests to obtain confidence intervals and sets, and briefly discuss the corresponding efficiency properties. Finally, Section 7.7 extends the theory and methods of Sections 7.2–7.6 to tests and confidence sets about a parametric function $\theta = \theta(\gamma)$ say, of the full parameter vector γ.

Appendix I contains some material on a normal-theory hypothesis-testing problem, without asymptotics, that the asymptotic theory parallels. In Appendix II, some of the assumptions imposed in the development of the asymptotic theory are removed; we show that the asymptotically efficient tests for the one-sided testing problems in Sections 7.1 and 7.2 are *asymptotically uniformly most powerful*, quite generally—without restriction to tests constructed from asymptotically normal test statistics; similarly, the two-sided tests are *asymptotically uniformly most powerful unbiased* and the d-dimensional tests are *asymptotically uniformly most powerful invariant*, quite generally. (For concepts of *unbiased tests* see Section 1.2. Invariant tests are discussed more fully in LEHMANN & ROMANO (2005)).

Throughout, we tacitly assume *Local Asymptotic Normality*, and sometimes additionally *Uniform Local Asymptotic Normality*, and/or *Continuity of Information*, see Chapter 4). And we use concepts of *effective scores* and *effective information*, introduced in Section 6.5 of Chapter 6, in order to allow for nuisance parameters. As always, we emphasize the random sampling case, but carry along greater generality so as to cover models such as two-sample, k-sample, regression, and censoring models; examples of some of these are included.

7.1 Hypothesis Testing in Finite Samples

The classical formulation of a hypothesis testing problem postulates a decision problem, whether, on the basis of available data, to accept or reject a hypothesis concerning that data. For example, we may want to use data from a study of a new drug to reduce blood pressure to decide whether or not to recommend the drug for general use.

In the simplest form of the problem, testing a point null hypothesis $\theta = \theta_0$ versus a point alternative $\theta = \theta_1$, we are asked to decide between two possible values of the unknown parameter θ. We are not allowed to defer a decision, or to decide that neither value is consistent with the data, although either of these might sometimes be reasonable courses of action. We are forced to make a choice and would like this choice to be correct. We are allowed to observe data $\mathbf{X}$, from a density $f(\mathbf{x}, \theta)$ depending on θ. Boldface notation is used here to indicate that data are usually available from a sample $\mathbf{X} = (X_1, \cdots, X_n)$ rather than just a single value of X. Then $f(\mathbf{x})$ denotes the joint density of

$(X_1, \ldots, X_n)$. We can specify the test as a decision rule $\psi(\mathbf{x})$, a function of the data $\mathbf{x}$, so that we choose $\theta = \theta_0$ when $\psi(\mathbf{x}) = 0$ and $\theta = \theta_1$ when $\psi(\mathbf{x}) = 1$. The set $\{\mathbf{x} : \psi(\mathbf{x}) = 1\}$ is called the rejection region of the test. With data generated from a discrete distribution, we sometimes allow $0 < \psi(\mathbf{x}) < 1$, for certain values of $\mathbf{x}$ on the boundary of the rejection region, corresponding to choosing $\theta = \theta_0$ with probability $\psi(\mathbf{x})$ and $\theta = \theta_1$ with probability $1 - \psi(\mathbf{x})$, both positive. In any case $0 \leq \psi(\mathbf{x}) \leq 1$ for all $\mathbf{x}$.

We say that $\theta = \theta_0$ is the *null hypothesis*. The probability $\alpha = \mathrm{pr}\{\psi(\mathbf{X}) = 1; \theta_0\}$ of rejecting the null hypothesis when it is actually true is call the *alpha-level* or just *level*, *Type I error* or *size* of the test. The probability $\beta = \mathrm{pr}\{\psi(\mathbf{X}) = 0; \theta_1\}$ of accepting the null hypothesis when it is actually false is called the *beta error* or *Type II error* of the test. The probability $\phi = 1 - \beta$ of rejecting the null hypothesis, that is of accepting the alternative hypothesis, when $\theta = \theta_1$ is called the *power* of the test. The terminology suggests that the two possible values of θ are not on an equal footing, usually the null value θ_0 corresponds to lack of effect, for example that a proposed new treatment for a disease is equivalent to placebo treatment, whereas the value θ_1 under the alternative hypothesis corresponds to a difference in effectiveness of the two treatments.

There is a trade-off between the two objectives, each desirable, of minimizing the Type I and Type II errors of a test, as can be seen by the extreme cases $\psi(\mathbf{x}) \equiv 1$ (always reject the null hypothesis) and $\psi(\mathbf{x}) \equiv 0$ (never reject the null hypothesis). A sensible strategy is to seek the test which maximizes the power ϕ for a specified level α. The Neyman-Pearson lemma proved below gives a simple and appealing solution to this problem. It states that among all tests with same level α the test with maximum power is based on the likelihood-ratio statistic $L(\mathbf{x}) = f_1(\mathbf{x})/f_0(\mathbf{x})$, where, for $j = 0, 1$ we write $f_j(\mathbf{x}) = f(\mathbf{x}, \theta_j)$. Specifically, we write $\psi(\mathbf{x}) = \mathbf{1}[L(\mathbf{x}) > c]$, where c is chosen so that $\int_{\mathbf{x}: L(\mathbf{x}) \geq c} f_0(\mathbf{x}) d\mathbf{x} = \alpha$. When $L(\mathbf{x})$ takes discrete values, achieving an exact size α may require use of a randomized decision rule. Here is the result stated formally.

Theorem 7.1 *(Neyman-Pearson lemma): Suppose that* $\mathbf{X} \sim f(\mathbf{x}, \theta)$. *Let* $\psi(\mathbf{x})$ *denote a likelihood-ratio test* $\psi(\mathbf{x}) = \mathbf{1}[f_1(\mathbf{x})/f_0(\mathbf{x}) \geq c]$ *with size* α. *Let* $\tilde{\psi}(\mathbf{x})$ *denote any other test of size* α. *Let* $\phi = 1 - \beta$ *and* $\tilde{\phi} = 1 - \tilde{\beta}$ *denote the powers of the two tests. Then* $\phi \geq \tilde{\phi}$.

Proof This is very straightforward. Consider the expression

$$\int \{\psi(\mathbf{x}) - \tilde{\psi}(\mathbf{x})\}\{f_1(\mathbf{x}) - c f_0(\mathbf{x})\} d\mathbf{x}$$

where the integral is over all values of $\mathbf{x}$ with either density positive. Since $0 \leq \tilde{\psi}(\mathbf{x}) \leq 1$ for all $\mathbf{x}$ and $\psi(\mathbf{x}) = \mathbf{1}[f_1(\mathbf{x}) > c f_0(x)]$ both terms are non-negative if $f_1(\mathbf{x}) > c f_0(\mathbf{x})$ and both non-positive if $f_1(\mathbf{x}) \leq c f_0(\mathbf{x})$. Hence the integrand is non-negative everywhere. Since

$$\int \psi(\mathbf{x}) f_0(\mathbf{x}) d\mathbf{x} = \int \tilde{\psi}(\mathbf{x}) f_0(\mathbf{x}) d\mathbf{x} = \alpha$$

it follows that

$$\int \psi(\mathbf{x}) f_1(\mathbf{x}) d\mathbf{x} \geq \int \tilde{\psi}(\mathbf{x}) f_1(\mathbf{x}) d\mathbf{x},$$

which completes the proof. □

The proof also applies to discrete distributions after substituting sums for integrals with the proviso that achieving an exact size α may require that a value of $\mathbf{x}$ on the boundary of the rejection region result in a randomized decision with $0 < \psi(\mathbf{x}) < 1$.

It is instructive to reformulate the Neyman-Pearson lemma as the solution to a constrained optimization problem, to find the function $\psi(\mathbf{x})$ satisfying $0 \leq \psi(\mathbf{x}) \leq 1$ that maximizes the power of the test

$$\phi = \int \psi(\mathbf{x}) f_1(\mathbf{x}) d\mathbf{x} \tag{7.1}$$

subject to the constraint

$$\int \psi(\mathbf{x}) f_0(\mathbf{x}) d\mathbf{x} = \alpha \tag{7.2}$$

on its level. The standard approach to such problems is to introduce a "Lagrange multiplier" λ and examine the function

$$\phi - \lambda \int \psi(\mathbf{x}) f_0(\mathbf{x}) d\mathbf{x} = \int \psi(\mathbf{x}) \{f_1(\mathbf{x}) - \lambda f_0(\mathbf{x})\} d\mathbf{x}. \tag{7.3}$$

Since the second term equals α when the constraint is satisfied, maximization of Equation (7.1) subject to Equation (7.2) is equivalent to maximization of Equation (7.3) subject to the same constraint. If it should happen that the unconstrained maximization of Equation (7.3) is such that the constraint is satisfied, then it will also be the maximum of the same function subject to the constraint and therefore the solution to the original constrained maximization problem. We can ensure this happy circumstance by choosing λ to make it so.

Clearly, once λ is specified, the unconstrained maximum of Equation (7.3) occurs when $\psi(x) = 1$ if $f_1(\mathbf{x})/f_0(\mathbf{x}) > \lambda$ and $\psi(\mathbf{x}) = 0$ if $f_1(\mathbf{x})/f_0(\mathbf{x}) < \lambda$; its value when $f_1(\mathbf{x})/f_0(\mathbf{x}) = \lambda$ can be arbitrary. The relevant value of λ is just that which ensures that the test has level α.

7.1.1 Uniformly Most Powerful Tests

Write $\psi_c(\mathbf{x})$ for the test which rejects H_0 when $f(\mathbf{x}, \theta_1) > cf(\mathbf{x}, \theta_0)$. In general, the form of this likelihood-ratio test and the corresponding critical region, the set of values of $\mathbf{x}$ for which $\psi(\mathbf{x}) = 1$, leading to rejection of the null hypothesis, will depend on both θ_0 and θ_1. In an important special case, however we obtain the same test statistic for any $\theta_1 > \theta_0$.

Definition 7.1 *The class of densities $f(\mathbf{x}, \theta)$ has monotone likelihood ratio, if, for any θ_0, θ_1 with $\theta_1 > \theta_0$, $f(\mathbf{x}, \theta_1)/f(\mathbf{x}, \theta_0)$ is an increasing function of some scalar function $T(\mathbf{x})$. Typically, $T(\mathbf{x})$ will be a sufficient statistic for θ in this model.*

We could of course replace "increasing" by "decreasing" in this definition by replacing $T(\mathbf{x})$ by $-T(\mathbf{x})$. The notion of monotone likelihood ratio was introduced in Chapter 1. *Exercise 1.33* proved that the monotone likelihood property of the density implies stochastic ordering of the distributions, so that, for any c, $\text{pr}\{T(\mathbf{x}) > c\}$ is an increasing function of θ.

Example 7.1 *With a single observation $X \sim N(\theta, 1)$, the condition is clearly satisfied with $T(x) = x$, since $f_1(x)/f_0(x) = \exp[(\theta_1 - \theta_0)x - \frac{1}{2}(\theta_1^2 - \theta_0^2)]$ which is clearly increasing in x if $\theta_1 > \theta_0$. A similar form obtains for the ratios of the two densities of the mean $\bar{x}$ of a random sample from $N(\theta, 1)$.*

Example 7.2 *Suppose now that $X \sim N(0, \sigma^2)$. Then the condition is no longer satisfied with $T(x) = x$ but it is satisfied with $T(x) = x^2$. For a random sample, take $T(x) = \sum x^2$.*

For distributions with monotone likelihood ratio, the rejection region for the likelihood-ratio test of θ_0 versus any $\theta_1 > \theta_0$ will be of the form $\{\mathbf{x} : T(\mathbf{x}) > c\}$. The value of c is determined by the size of the test, which is calculated using only the distribution of $\mathbf{x}$ under the null hypothesis $\theta = \theta_0$. The resulting test $\psi(\mathbf{x})$ will be the same for all $\theta_1 > \theta_0$. It is said to be *uniformly most powerful* for testing $\theta = \theta_0$ against $\theta = \theta_1 > \theta_0$. The actual power of the test depends on θ_1: for θ_1 close to θ_0 it will be close to α, for $\theta_1 \gg \theta_0$ it will be close to unity. The power can be calculated for $\theta_1 < \theta_0$. It will be less than α in this region, usually vanishingly small.

7.1.2 Composite Null Hypotheses

Suppose now that we have a random sample $X_1, \ldots, X_n$ with $X_i \sim N(\theta, 1)$ as before, but now interest centers on whether $\theta \leq \theta_0$ or $\theta > \theta_0$. This is a subtle change from the previous formulation where the point null hypothesis $\theta = \theta_0$ was of interest, now we are concerned with the composite null hypothesis $\theta \leq \theta_0$. This formulation is relevant when the point null hypothesis is unrealistic, and we would like to assess the evidence that $\theta - \theta_0$ is positive rather than negative. We would still like to control the size of the test and use the most powerful test with that size. Specifically, we require the Type I error to be less than or equal to α for every $\theta \in H_0$ so that

$$\sup_{\theta \leq \theta_0} E_\theta \psi(\mathbf{x}) \leq \alpha, \tag{7.4}$$

the *level* of the test $\psi(\mathbf{x})$.

We wish to choose $\psi(\mathbf{x})$ to maximize the power $E_{\theta_1}\{\psi(\mathbf{X})\}$ at each $\theta_1 > \theta_0$ subject to the constraint (7.4). This leads to the same test statistic as before. To see this, note that, for any $\theta_1 > \theta_0$, the likelihood-ratio test $\psi_0(\mathbf{x}) = \mathbf{1}[T(\mathbf{x}) > c]$ maximizes the power $E_{\theta_1}\psi(\mathbf{X})$ subject to the constraint

$$E_{\theta_0}\{\psi(\mathbf{X})\} \leq \alpha$$

This constraint is weaker than the constraint (7.4) which requires the inequality to hold for all $\theta \leq \theta_0$. Inclusion of the additional constraints on $\psi(\mathbf{x})$ cannot increase the achievable power. However, since the monotone likelihood-ratio property implies stochastic ordering, the test $\psi_0(\mathbf{x})$ satisfies the larger class of constraints implied by Equation (7.4) and so solves the broader optimization problem. In summary, we have

Theorem 7.2 *Suppose that the joint density of the data $f(\mathbf{x}, \theta)$ has the monotone likelihood-ratio property with summarizing statistic $T(\mathbf{x})$. Then, for any $\theta_0' < \theta_0 < \theta_1$, the test $\psi(\mathbf{x}) = \mathbf{1}[T(\mathbf{x}) > c]$ is uniformly most powerful of its size for testing $\theta = \theta_0'$ against $\theta = \theta_1$.*

7.1.3 Two-Sided Tests, Unbiased Tests

We can reverse the inequalities in the previous discussion and consider tests of the null hypothesis $\theta = \theta_0$ versus alternatives of the form $\theta < \theta_0$. Such tests, like that above, of a point or interval null hypothesis against all values of the parameter on one side of the null value(s) are called one-sided tests. Sometimes however we may wish to test whether $\theta = \theta_0$ against $\theta \neq \theta_0$, a two-sided alternative hypothesis. We want to detect differences in either direction between the effects of two treatments rather than just whether the second is better than the first.

One approach to specify a two-sided test is to combine two one-sided tests, rejecting the null hypothesis $\theta = \theta_0$ if either $T(\mathbf{x}) - \theta_0 > c_1$ or $T(\mathbf{x}) - \theta_0 < -c_2$, where usually c_1 and c_2 are both positive. For simplicity, we consider the case that $X \sim N(\theta, 1)$, a distribution that is symmetric about θ, and $T(\mathbf{x}) = x$. We first prove a simple

Lemma 7.1 *For $c > 0$ the total area $A = \Phi(x + c) - \Phi(x - c)$ under the standard normal curve within the interval $(x - c, x + c)$ the standard normal is maximized as a function of x when $x = 0$.*

Proof The derivative of A is

$$\varphi(x + c) - \varphi(x - c) = \varphi(c)\exp(-\tfrac{1}{2}x^2)\{\exp(-cx) - \exp(cx)\}$$

which is positive for $x < 0$, negative for $x > 0$ and zero only when $x = 0$. □

Corollary For $c_1, c_2 > 0$ the expression $\Phi(x + c_2) + 1 - \Phi(x - c_1)$ corresponding to the area under the interval $(x - c_1, x + c_2)$ is maximized when $x = \frac{1}{2}(c_1 - c_2)$

Proof Set $c = \frac{1}{2}(c_2 - c_1)$ and $x' = x - c$ in Lemma 7.1. □

The power of the proposed combined test against the hypothesis $\theta = \theta_0 + \delta$ is $\Phi(x + c_2) - \Phi(x - c_1)$. As a consequence of the corollary, unless $c_1 - c_2 = 0$, we have the strange result that the power of the test for a range of values within the alternative hypothesis $\theta_1 \neq \theta_0$ is less than its power when $\theta_1 = \theta_0$, i.e. is less than its level. On occasion, when deviations from the null hypothesis in one direction are viewed as more important than deviations in the other direction, this may not matter, but if deviations in both directions are viewed as equally important then such behavior of the test is undesirable. We can rule it out by restricting attention to tests that are *unbiased.*

Definition 7.2 *A two-sided test $\psi(\mathbf{x})$ is unbiased for the hypothesis H_0 if $\sup_{\theta \in H_0} E_\theta \psi(\mathbf{x}) \leq \inf_{\theta \notin H_0} E_\theta \psi(\mathbf{x})$ so that the power of the test when null hypothesis is false always exceeds its level.*

Sometimes we can find an optimal, that is most powerful, test among the smaller class of unbiased tests with size α when there is no optimal test among the larger class of all tests with size α.

Definition 7.3 *A test is uniformly most powerful unbiased if it is unbiased and is the most powerful test of any $\theta \in H_0$ among all unbiased tests of the same level.*

We illustrate with the previous example, a single observation $X \sim N(\theta, \sigma^2)$, with σ^2 known. For testing the mean of a random sample of n observations replace σ^2 by σ^2/n. We seek a most powerful test unbiased test of level α of the null hypothesis $\theta = \theta_0$ against the alternative hypothesis $\theta = \theta_1$, where θ_1 is any specified value of θ greater or less than θ_0. Unbiasedness requires that the power $\phi(\theta)$, viewed as function of θ, is never less than $\phi(\theta_0)$, so that $\phi(\theta)$ achieves its minimum value at $\theta = \theta_0$. Now the power at θ is

$$\phi(\theta) = \int \psi(x) f(x, \theta) dx = \frac{1}{\sqrt{(2\pi)}\sigma} \int \psi(x) \exp\Big\{-\frac{(x-\theta)^2}{2\sigma^2}\Big\} dx$$

Differentiation under the integral sign gives

$$\phi'(\theta) = -\int \psi(x) \Big\{\frac{x-\theta}{\sigma^2}\Big\} f(x, \theta) dx.$$

A necessary condition for $\psi(x)$ to be unbiased is that $\phi'(\theta_0) = 0$, giving

$$\int (x - \theta_0) \psi(x) f(x, \theta_0) dx = 0, \tag{7.5}$$

and we must also have

$$\int \psi(x) f(x, \theta_0) dx = \alpha. \tag{7.6}$$

We now seek an optimal test within the class of tests satisfying these two constraints, a potentially larger class than the class of all unbiased tests of size α. We will verify that this test *is* unbiased, and therefore the most powerful test of θ against the specified θ_1 among the smaller class of unbiased tests. Finally, we will note that the resulting test $\psi_0(x)$ does not depend on θ_1, even on whether $\theta_1 < \theta_0$ or $\theta_1 > \theta_0$, showing that $\psi_0(x)$ is uniformly the most powerful among all unbiased tests for testing $\theta = \theta_0$ against $\theta \neq \theta_0$.

We use the Lagrange multiplier approach described earlier but now with two constraints (7.6) and (7.5) and corresponding multipliers λ_2 and λ_1. With $f_1(x) = f(x, \theta_1)$ this requires us to maximize

$$\begin{aligned} \phi(\theta_1) \quad &- \quad \lambda_1 \left\{ \int \psi(x) f_0(x) dx - \alpha \right\} - \lambda_2 \int (x - \theta_0) \psi(x) f_0(x) dx \\ &= \quad \int \psi(x) [f_1(x) - \{\lambda_1 + \lambda_2 (x - \theta_0)\} f_0(x)] dx + \lambda_1 \alpha \end{aligned}$$

among all functions $\psi(x)$ satisfying $0 \leq \psi(x) \leq 1$. Once λ_1 and λ_2 are specified, the solution is clear. We should take

$$\psi(x) = \mathbf{1}[f_1(x)/f_0(x) \geq \lambda_1 + \lambda_2 (x - \theta_0)].$$

Now

$$\frac{f_1(x)}{f_0(x)} = \frac{f(x, \theta_1)}{f(x, \theta_0)} = \exp\left\{ \frac{(\theta_1 - \theta_0)x}{\sigma^2} - \frac{\theta_1^2 - \theta_0^2}{2\sigma^2} \right\}. \tag{7.7}$$

As functions of x, the left and right sides of Equation (7.7) are exponential and linear, respectively. So they can be equal for at most two values of x. Clearly $\psi(-\infty) = \psi(\infty) = 1$. Therefore, the solution must be of the form $\psi(x) = \mathbf{1}[\{x \leq a\} \cup \{x \geq b\}$ where a and b are functions of θ_1, λ_1 and λ_2 and satisfy $-\infty \leq a \leq b \leq \infty$. The possibilities that a or b may be infinite or that $a = b$ are not (yet) excluded.

We must now identify the impact of the constraints (7.6) and (7.5) on the choice of λ_1 and λ_2. We have

$$\int_a^b f(x, \theta_0) dx = 1 - \alpha \quad \text{and} \quad \int_a^b (x - \theta_0) f(x, \theta_0) dx = 0.$$

Since the normal density $f_0(x) = f(x - \theta_0)$ is symmetric about θ_0 the second constraint implies that $b - \theta_0 = \theta_0 - a$, so that the rejection region is symmetric around θ_0. The first condition determines the common value as σ times the $1 - \frac{1}{2}\alpha$'th quantile of the standard normal distribution. The resulting test is unbiased of size α and most powerful among all such tests of $\theta = \theta_0$ against $\theta = \theta_1$. Since a and b do not depend on θ_1 the test is uniformly most powerful among all unbiased tests of $\theta = \theta_0$ versus $\theta \neq \theta_0$ that are of size α.

To complete the proof, it is necessary to show that, for any θ_1, there exist values of λ_1 and λ_2 for which the interval defined by $f_1(x)/f_0(x) \leq \lambda_1 + \lambda_2(x - \theta_0)$ coincides with $(\theta_0 - a, \theta_0 + a)$. This is straightforward, see

Exercise 2. These values do depend on θ_1, but do not affect the choice of a, as indicated above.

The approach and much of the detailed calculation extends to tests based on a random sample $\mathbf{X} = (X_1, \ldots, X_n)$ when the X_i follow a one parameter exponential family with natural parameter θ and sufficient statistic $T(\mathbf{x})$ that has a continuous distribution. We again find that a test which rejects $\theta = \theta_0$ when $T(\mathbf{x}) \notin (a, b)$ is uniformly most powerful of size α against the two-sided alternative $\theta \neq \theta_0$ for suitable choice of a and b. With $\psi(X) = \mathbf{1}[T(\mathbf{x}) \notin (a, b)]$, the conditions on a and b are that $E\{\psi(X)\} = \alpha$ and $E\{T(\mathbf{X})\psi(\mathbf{X})\} = \alpha E\{T(\mathbf{X})\}$. For details see, for example, LEHMANN & ROMANO (2005). When the distribution of $T(\mathbf{X})$ is not symmetric, we may find that under the null hypothesis $\text{pr}\{T(\mathbf{X}) \leq a\} \neq \text{pr}\{T(\mathbf{X}) \geq b\}$ so that the two parts of the rejection region do not have equal probability under the null hypothesis. See *Exercise 1* for a rather extreme example.

7.2 Asymptotic Theory for Models with a Single Real Parameter

We now develop the asymptotic theory of testing a fixed value θ_0 of a parameter θ against local alternatives of the form $\theta = \theta_0 + h/\sqrt{n}$, based on data $\mathbf{X}_n = (X_1, \ldots, X_n)$ from a sample of size n. The motivation for considering alternative hypotheses of this form is the same as that for estimating a real parameter. With a fixed alternative hypothesis $\theta = \theta_1$, say, it becomes possible as $n \to \infty$ to discriminate between the two possible values with increasing certainty as $n \to \infty$, resulting in an uninformative theory.

7.2.1 One-Sided Tests

We begin with the simple case of a model for the potential data $\mathbf{X}_n$ depending only on a single (real) parameter θ. We first consider the *one-sided testing problem*, with hypotheses

$$H_0 : \theta \leq \theta_0 \quad \text{versus} \quad H_A : \theta > \theta_0 \tag{7.8}$$

for a specified θ_0 interior to the parameter space. Since with large samples, we can distinguish with near certainty between values that are far apart—as noted in Chapter 4—we confine attention to the performance of a statistical test locally to θ_0. With $\theta_n = \theta_0 + h/\sqrt{n}$, we focus on the corresponding *local hypotheses*, with $\theta = \theta_0$ and

$$H_0' : h \leq 0 \quad \text{versus} \quad H_A' : h > 0. \tag{7.9}$$

For motivation, first consider testing $\theta = \theta_0$ versus a point local alternative $\theta_n = \theta_0 + h/\sqrt{n}$ ($h > 0$). The Neyman-Pearson lemma characterizes the most powerful test of specified significance level as one that rejects θ_0 for suitably large values of the likelihood ratio. Assuming local asymptotic normality, as we do throughout, and ignoring terms that are $o_p(1)$, we see that the log likelihood-ratio is increasing in the score $S_n(\mathbf{x}_n, \theta_0)$. But $S_n \sim AN(0, \sigma^2)$, where $\sigma^2 = B(\theta_0)$, under θ_0, so a test that rejects the null hypothesis whenever the *standardized score* S_n/σ exceeds z_α (with $\bar{\Phi}(z_\alpha) = \alpha$) is most powerful at θ_n among tests of significance level α—based on this large-sample approximation. Since $S_n/\sigma \sim AN(\sigma h, 1)$ under θ_n, the claim remains true when allowing for $h < 0$ in H_0'; indeed, such a test is uniformly most powerful against $h > 0$. It seems likely that these results should have large-sample legitimacy—that is, that the ignoring of $o_p(1)$ terms and the use of large-sample approximations can be justified. That this is so is shown formally in Appendix II. Here, instead, we take a simpler approach and use this simply to motivate our confining attention to so-called 'regular' tests. Throughout, $\alpha \in (0, 1)$.

We now adapt these concepts to the large-sample setting.

Definition 7.4 *(Asymptotic Level): A test $\psi_n = \psi(\mathbf{x}_n)$ has <u>asymptotic level</u> α if*

$$\lim_{n\to\infty} E_{\theta_n}\{\psi_n(\mathbf{X}_n)\} \leq \alpha \quad \text{for all} \quad h \in H_0' \quad \text{and} \quad \theta_n = \theta_0 + h/\sqrt{n}$$

To allow for the possibility that the limit does not exist, we could replace 'lim' by 'lim sup' in this definition.

Definition 7.5 *Regular Test. A test statistic T_n and the associated test $\psi_n(\mathbf{X}_n) = \mathbf{1}[T_n > z]$ are <u>regular</u> if, under θ_0, T_n is asymptotically normal jointly with the score $S_n(\mathbf{X}_n, \theta_0)$, and if the test $\psi_n(\mathbf{X}_n)$ has an asymptotic level (strictly between zero and one) for each z.*

Without loss of generality, we can and do assume that $T_n \sim AN(0, 1)$ under θ_0. And, as we noted, local asymptotic normality is assumed throughout, thereby permitting consideration of scores, information, and associated facts. We shall find that regularity plays a role similar to its role in estimation theory, in particular, assuring *local unbiasedness* of tests. As seen in later sections, this enables the construction of optimal tests about a parameter when there are nuisance parameters, just as a requirement that a test be unbiased sometimes enables the construction of uniformly most powerful tests, see LEHMANN & ROMANO (2005).

The joint asymptotic normality with the scores permits use of LeCam's third lemma. Specifically, as shown in Chapter 4, $T_n \to_{\theta_n} N(\omega h, 1)$ where $\omega = \text{Acov}(T_n, S_n)$ the large-sample covariance between T_n and the score S_n, all evaluated at θ_0. It follows that, for each h,

$$\lim_{n\to\infty} P_{\theta_n}(T_n \geq z) = \bar{\Phi}(z - \omega h) = \Phi(\omega h - z). \tag{7.10}$$

Now confine attention to tests of *asymptotic level* α, for a specified $\alpha \in (0, 1)$—and so strictly less than unity—that is, we require

$$\lim_{n\to\infty} P_{\theta_n}(T_n \geq z) \leq \alpha \quad \text{for} \ \ h \in H_0', \tag{7.11}$$

(i.e., $h \leq 0$). It follows from Equations (7.10) and (7.11) that a regular test must have $\omega \geq 0$, for otherwise multiplying by a negative h of large magnitude would yield a number close to unity on the right side in Equation (7.10), contradicting (7.11). It also follows, by considering $h = 0$, that $z \geq z_\alpha$, with z_α defined as before by $\bar{\Phi}(z_\alpha) = \alpha$.

We define *local power* as

$$\text{pow}(h) \equiv \lim_{n\to\infty} P_{\theta_n}(T_n \geq z) \quad \text{for} \ \ h \in H_A' \tag{7.12}$$

(i.e., $h > 0$). We would like to choose a regular statistic T_n for which this power is as large as possible, for each $h > 0$. Replacing the covariance ω in Equation (7.12) by $\rho\sigma$, with $\rho = \text{Acorr}(T_n, S_n)$ and $\sigma^2 = B(\theta_0)$, the information, and recalling that the large-sample variance of T_n is unity, we see that the local power in 7.12 is $\Phi(\rho\sigma h - z)$. Whatever the value of ρ and h, this local power is maximized by choosing $z = z_\alpha$. Since $h > 0$, further maximization occurs with $\rho = 1$.

Now denote $T_n^0 = S_n/\sigma$, evaluated at θ_0, here termed the *standardized score*. For T_n^0, $\rho = 1$; in fact, $\rho = 1$ only if $T_n = T_n^0 + o_p(1)$, since (T_n, T_n^0) converging to bivariate normal with means zero and unit variances and covariance implies that $T_n - T_n^0 = o_p(1)$.

We therefore have

Proposition 7.1 *(Power Bound): For any regular test of the local hypotheses (7.9) of asymptotic level α, the local power satisfies*

$$\text{pow}(h) = \Phi(\rho\sigma h - z_\alpha) \leq \Phi(\sigma h - z_\alpha) \quad \text{for} \quad h > 0$$

with equality holding for every such h if and only if the regular test statistic T_n differs from the standardized score T_n^0 by $o_p(1)$.

Here, $\sigma = B(\theta_0)^{1/2}$, the asymptotic standard deviation of S_n, and $\rho = \text{Acorr}(T_n, S_n)$. We refer to any regular test achieving the bound, or to the corresponding test statistic, simply as *asymptotically efficient.* In alternative terminology, an asymptotically efficient test is *asymptotically uniformly most powerful* with respect to $h \in H_A'$ among regular tests of *asymptotic level* α (for $h \in H_0'$).

A test based on $T_R \equiv T_n^0$, rejecting H_0 in (1) whenever $T_R \geq z_\alpha$, is called a *Rao* or *score test.* But it is not the only asymptotically efficient test. For example we may construct a test from an asymptotically efficient estimate $\hat{\theta}_n$ of θ, rejecting H_0 whenever it is 'significantly larger' than θ_0. An asymptotically efficient estimate has large-sample variance $B^{-1}/n = 1/(n\sigma^2)$, leading to the test statistic

$$T_W \equiv \sqrt{n}\sigma \cdot (\hat{\theta}_n - \theta_0).$$

Recall, from the *Characterization Theorem for Asymptotically Efficient Estimates* in Chapter 5, that $\sqrt{n}(\hat{\theta}_n - \theta_0) = B(\theta_0)^{-1}S_n(\theta_0) + o_p(1)$. It follows that $T_W = T_n^0 + o_p(1)$, and is therefore asymptotically efficient. A test based on T_W is called a *Wald test*, or sometimes a maximum likelihood test when, as often happens the efficient estimate θ_n is actually the maximum likelihood estimate. If σ is replaced by a consistent estimate $\tilde{\sigma}_n$, rather than the null value, the resulting test is asymptotically equivalent and may still be called a Wald test. A test constructed from an inefficient estimate of θ will not be asymptotically efficient so is usually called a *Wald-type test*. See *Exercises 20 and 21.*

We take up *likelihood-ratio tests* in Section 7.6, finding them also to be asymptotically efficient.

7.2.2 Two-Sided Tests

We now turn to the *two-sided testing problem*, with hypotheses

$$H_0 : \theta = \theta_0 \quad \text{versus} \quad H_A : \theta \neq \theta_0,$$

having local form

$$H_0'' : h = 0 \quad \text{versus} \quad H_A'' : h \neq 0. \tag{7.13}$$

We modify the regularity definition for two-sided tests by considering only test statistics $|T_n|$ and tests of the form $\mathbf{1}[|T_n| \geq z]$ for positive z, again assuming T_n to be asymptotically standard normal under θ_0. In this way we force our tests to be locally most powerful unbiased. See Section 2.3 and Appendix II. Equivalently, we could use T_n^2 as a test statistic, required to be asymptotically distributed as chi-squared with 1 degree of freedom.

Corresponding to Equation (7.10) we have

$$\lim_{n\to\infty} P_{\theta_n}(T_n| \geq z) = \Phi(\omega h - z) + \Phi(-\omega h - z). \tag{7.14}$$

For asymptotic level α, we only need to consider (7.14) with $h = 0$, leading to $z \geq z_{\alpha/2}$. The local power $\text{pow}(h)$ is given by Equation (7.14) for $h \neq 0$. Whatever the value of ω and h, this is bounded above by replacing z by $z_{\alpha/2}$, and depends only on $|\omega h|$. Now consider $g(x) \equiv \Phi(x - z) + \Phi(-x - z)$, defined for $x \geq 0$ and $z > 0$. For $x > 0$ this is an increasing function of x by Lemma 7.1.

We therefore conclude that (7.14) is increasing in $|\omega h|$, and hence, with $\omega = \rho\sigma$ as before, reaches its maximum when $|\rho| = 1$. Summarizing, we have

Proposition 7.2 *Two-Sided Power Bound: For any regular test of local hypotheses (7.13) of asymptotic level* α*, the local power satisfies*

$$\text{pow}(h) = \Phi(\rho\sigma h - z_{\alpha/2}) + \Phi(-\rho\sigma h - z_{\alpha/2}) \leq \Phi(\sigma h - z_{\alpha/2}) + \Phi(-\sigma h - z_{\alpha/2}), h \neq 0,$$

with equality holding for every such h *if and only if the regular test statistic* $|T_n|$ *differs from* $|T_n^0|$ *by* $o_p(1)$.

We call tests achieving this bound *asymptotically efficient two-sided tests.* The test statistic $|T_R| = |T_n^0|$ is the *two-sided score (Rao) test statistic*, and $|T_W|$ is the *two-sided Wald test statistic* (verified as in the one-sided case).

As a simple example, consider sampling from the distribution $N(\theta, \tau^2)$ with τ known, testing $\theta \le \theta_0$ versus $\theta > \theta_0$. The score is found to be $\sqrt{n}(\bar{X}_n - \theta_0)/\tau^2$ with information $B = 1/\tau^2$. Here T_W and T_R are identical and equal to the *uniformly most powerful* test statistic $\sqrt{n}(\bar{X}_n - \theta_0)/\tau$ derived in the previous section.

For approximations to the power, relative efficiency considerations, and large-sample p-values, see Section 7.3. We close with another example:

Example 7.3 *Testing the exponential rate parameter.*

Consider testing hypotheses about the scale parameter θ when sampling from an exponential distribution with survival function $\exp(-\theta x)$ on $\mathcal{R}^+$. Suppose that for specified θ_0 $H_0 = \{\theta \ge \theta_0\}$ and $H_A = \{\theta < \theta_0\}$. Since the inequalities are in the opposite direction to those in Equation (7.8), we need to change the sign of θ, to reparameterize by its reciprocal, or to reject when $T_n \le -z_\alpha$; we choose the last of these.

The standardized score is found to be $T_n^0 = \sqrt{n}(1 - \theta_0 \bar{X}_n)$, and so the score test rejects H_0 whenever this is at most $-z_\alpha$. The Wald test statistic is found to be $\sqrt{n}(1 - \theta_0 \bar{X}_n)/(\theta_0 \bar{X}_n)$, which differs from T_n^0 by $o_p(1)$ since the denominator converges to unity in probability. See *Exercise 21.*

This is a one-parameter exponential family testing problem, and hence the Neyman-Pearson lemma provides a uniformly most powerful test $\psi(\mathbf{x}) = \mathbf{1}[\bar{x} \ge c_\alpha]$ of size α where c_α is determined from gamma-distribution considerations so that the test has size α. By central limit theorem $\bar{X}_n \sim AN\{1/\theta_0, 1/(n\theta_0^2)\}$ when $\theta = \theta_0$, and so $c_\alpha \approx (1 + z_\alpha/\sqrt{n})/\theta_0$. Therefore, ψ agrees with T_n^0 asymptotically.

Example 7.4 *(Testing exponentiality within a Weibull model).*

Consider a random sample from a Weibull distribution with shape parameter θ and scale parameter η, the density of a single observation being

$$f_{\theta,\eta}(x) = \theta \eta^\theta x^{\theta-1} \exp\{-(\eta x)^\theta\} \mathbf{1}[x > 0].$$

We consider testing the hypothesis that $\theta \le 1$ versus the alternative that $\theta > 1$, assuming for now that the scale parameter η is known, say equal to unity. This is a one-parameter testing problem involving a distribution not of exponential family form.

If $\theta = 1$, the distribution is exponential; if $\theta < 1$, the density is unbounded at the origin, whereas if $\theta > 1$, the density vanishes at the origin. The *hazard rate* (the density divided by the survival function), is decreasing in x when $\theta < 1$, constant when $\theta = 1$ and increasing when $\theta > 1$. Hence, the hypotheses are parametric forms of non-increasing or increasing hazard functions. A two-sided version, testing $\theta = 1$ versus $\theta \neq 1$, is more precisely a test of exponentiality against a Weibull alternative.

Referring back to Section 4.2 of Chapter 4, we find that $s(x, \theta) = (1/\theta) + (1 - x^\theta)\log x$ and $B \approx 1.8237/\theta^2 = (1.350/\theta)^2$. Hence, an asymptotically efficient test statistic is

$$T_n^0 = T_R = \frac{1}{1.350\sqrt{n}} \sum_{i=1}^{n} \{1 + (1 - X_i)\log X_i\}.$$

A Wald test statistic is $T_W = 1.350\sqrt{n}(\hat{\theta}_n - 1)$, with $\hat{\theta}$ an asymptotically efficient estimate; see Chapter 5.

7.3 Testing a Real Parameter with Nuisance Parameters

We extend the developments in Section 7.3 to a model with parameter $\gamma = (\theta, \eta)$ with θ the (real) parameter of interest and η a nuisance parameter, of arbitrary dimension, say d. The hypotheses are as in Equation (7.8), with $\eta \in \mathcal{R}^d$ added to both H_0 and H_A.

We focus on a neighborhood of $\gamma = \gamma_0 \equiv (\theta_0, \eta)$ with η unknown, writing $\gamma_n = \gamma_0 + h/\sqrt{n}$ with $h = (h_\theta, h_\eta)$. Corresponding local hypotheses are now

$$H_0' : h_\theta \leq 0,\ h_\eta \in \mathcal{R}^d \quad \text{versus} \quad H_A' : h_\theta > 0\ h_\eta \in \mathcal{R}^d, \tag{7.15}$$

with η unknown.

The definitions of asymptotic level (7.11) and local power (7.12) are as before but with the subscript θ_n replaced by γ_n. The definition of regularity is unchanged, but now recognizes that the joint asymptotic normality is with both components $S_{\theta n}$ and $S_{\eta n}$ of the full score vector $S_n = (S_{\theta n}, S_{\eta n})$.

Again we assume $T_n \sim AN(0, 1)$ under $\gamma_0 = (\theta_0, \eta)$, for every η. Now $T_n \to_{\gamma_n} N(\omega h, 1)$ with $\omega h = \omega_\theta h_\theta + \omega_\eta h_\eta$, $\omega_\theta = \text{Acov}(T_n, S_{\theta n})$ and $\omega_\eta = \text{Acov}(T_n, S_{\eta n})$, a row vector of length d_η.

We again use formulas (7.10–7.12), now re-labelled $(10' - 12')$ upon replacing θ_n by γ_n. The right side of $(10')$ becomes $\Phi(\omega_\theta h_\theta + \omega_\eta h_\eta - z)$. Requirement $(11')$ now implies $\omega_\eta = 0$, for otherwise there would exist h_η to make $\omega_\eta h_\eta$ so large that the right side of $(10')$ exceeds α. Hence, $T_n \perp S_{\eta n}$—*a regular test statistic T_n must be asymptotically uncorrelated with the score for η*. This orthogonality parallels that found in Chapter 6 for regular estimates.

With $\omega_\eta = 0$, the same argument as in Section 7.1 implies that $\omega_\theta \geq 0$ and $z \geq z_\alpha$. Now the local power (12') is maximal, as before, if $z = z_\alpha$—whatever the regular test statistic T_n—yielding $\text{pow}(h) = \Phi(\omega_\theta h_\theta - z_\alpha)$. Here, $\omega_\theta = \text{Acov}(T_n, S_{\theta n}) = \text{Acov}(T_n, S_n^*)$ with S_n^* the *effective score for* θ (see Section 5.5 of Chapter 5); this is so since $T_n \perp S_{\eta n}$. Hence, $\omega_\theta = \rho^* \sigma^*$ with

$\rho^* = \text{Acorr}(T_n, S_n^*)$ and ${\sigma^*}^2 = B^*$, the *effective information* (all evaluated at γ_0). Proceeding as in Section 7.1, and defining

$$T_n^0(\eta) \equiv S_n^*/\sigma^*, \quad \text{evaluated at } \gamma_0$$

to be the *standardized effective score*, we have

Proposition 7.3 *(Power Bound with Nuisance Parameters): For any regular test of local hypotheses (7.15) of asymptotic level α, the local power satisfies*

$$\text{pow}(h) = \Phi(\rho^*\sigma^* h_\theta - z_\alpha) \le \Phi(\sigma^* h_\theta - z_\alpha) \quad \text{for} \quad \text{h} \in H_A' \quad \text{in} \quad (7.15)$$

with equality holding for every such h if and only if the regular test statistic T_n differs from the standardized effective score $T_n^0(\eta)$ by $o_p(1)$.

Here, $\rho^* = \text{Acorr}(T_n, S_n^*)$. Any test achieving the power bound is said to be asymptotically efficient.

Note We could have derived $\Phi(\sigma h_\theta - z_\alpha)$ as a power bound, by writing $\omega_\theta = \rho\sigma$, using characteristics of the asymptotic distribution of $(T_n, S_{\theta n})$. But this bound typically exceeds that given in Proposition 7.3 since $\sigma \ge \sigma^*$, and only the smaller bound is worth pursuing.

Typically, the standardized effective score depends on the unknown parameter η. Accordingly, we replace η by a root-n consistent estimator η_n and invoke Proposition 5.3 in Chapter 5. This states that, provided the Regular Scores and Continuity of Information properties hold, $S_n^*(\theta_0, \tilde{\eta}_n) = S_n^*(\theta_0, \eta) + o_p(1)$ when $\tilde{\eta}_n$ is root-n consistent under γ_0.

Letting $\tilde{\sigma}_n^*$ be any consistent estimate of σ^*, (consistency under γ_0 is sufficient) we therefore find that

$$T_{NR} \equiv S_n^*(\theta_0, \tilde{\eta}_n)/\tilde{\sigma}_n^*,$$

the *effective score* (or *Neyman-Rao) test statistic*, is asymptotically efficient (assuming Regular Scores and Continuity of Information). The null hypothesis is to be rejected whenever $T_{NR} \ge z_\alpha$.

But why use an inefficient estimate of η? If an asymptotically efficient estimate $\hat{\eta}(\theta_0)$ is used—asymptotically efficient under γ_0—it is a near-root of the score for η (at θ_0) and hence $S_n^*\{\theta_0, \hat{\eta}_n(\theta_0)\} = S_{\theta n}\{\theta_0, \hat{\eta}_n(\theta_0)\} + o_p(1) = S_n^*(\theta_0, \eta) + o_p(1)$, the latter equality another consequence of Proposition 5.3 of Chapter 5 (assuming Regular Scores). We therefore find that

$$T_R \equiv S_{\theta n}\{\theta_0, \hat{\eta}_n(\theta_0)\}/\tilde{\sigma}_n^*$$

differs from $T_n^0(\eta)$ by $o_p(1)$. This statistic is called the *score* or *Rao test statistic*, and is also asymptotically efficient (under the Regular Scores assumption).

Note The denominator in T_R is a consistent estimate of σ^*, not of σ. Thus, it is best to think of the numerator in the Rao statistic as an estimate of the effective score for θ, which just happens to simplify an estimate of the score for θ. Therefore standardization is by an estimate of the standard deviation of the effective score. If the scores are orthogonal, $S_n^* = S_{\theta n}$ and T_{NR} is like T_R but with a possibly inefficient estimate $\tilde{\eta}_n$.

How about a Wald test statistic? Since $\sqrt{n}(\hat{\theta}_n - \theta_0) = S_n^*/(\sigma^*)^2 + o_p(1)$ for an asymptotically efficient estimate $\hat{\theta}_n$, we find again that the *Wald test statistic*, now defined as

$$T_W \equiv \sqrt{n}\tilde{\sigma}_n^* \cdot (\hat{\theta}_n - \theta_0)$$

is asymptotically efficient.

Since all asymptotically efficient test statistics differ by at most $o_p(1)$ from the standardized effective score, they are all approximately equal to each other:

$$T_{NR} \approx T_R \approx T_W.$$

Also, to the first order of approximation considered here, they each have the same large-sample ('local') power; second-order approximations (not developed here), or case-specific numerical investigations, are needed to distinguish among them.

To test $\theta = \theta_0$ against a two-sided alternative when there is a nuisance parameter η, the developments in Section 7.1 can be extended in a straightforward manner. Regularity is modified as before. We arrive at the power bound

$$\text{pow}(h) \leq \Phi(\sigma^* h_\theta - z_{\alpha/2}) + \Phi(-\sigma^* h_\theta - z_{\alpha/2}), \quad h_\theta \neq 0, \, h_\eta \in \mathcal{R}^d \quad (7.16)$$

with equality holding for every such h if and only if the regular test statistic $|T_n|$ differs from $|T_n^0(\eta)|$ by $o_p(1)$. The *two-sided effective score, score* and *Wald tests*—based on $|T_{NR}|$, $T_R|$ and $|T_W|$, respectively—are each asymptotically efficient. For an inefficient test, local power is given by Equation (7.16) with ρ^* inserted in front of σ^*.

Example 7.5 *Example 7.4 (continued) Testing Exponentiality.*

Consider a random sample from a Weibull distribution with shape parameter θ but now with the scale parameter η being an unknown nuisance parameter. We consider testing the hypothesis $\theta \leq 1$ versus $\theta > 1$, as before.

Scores and information were derived in Section 6.1 of Chapter 6. The effective score for θ, per observation, is found to be

$$s^*(x, \theta, \eta) = s_\theta(x) - (B_{\theta\eta}/B_{\eta\eta})s_\eta(x) = \frac{1}{\theta}[1 + \{1 - (\eta x)^\theta\}\{\theta \log(\eta x) - \Gamma'(2)\}]$$

and the effective information is

$$\sigma^*(\theta, \eta)^2 = \text{var}_{\theta,\eta} s^*(X, \theta, \eta) = B_{\theta\theta} - B_{\theta\eta}^2/B_{\eta\eta} = \frac{1}{\theta^2}\{1 + \Gamma''(2) - \Gamma'(2)^2\}$$

Evaluating these at $\theta = 1$ yields

$$s^*(x) \equiv s^*(x, 1, \eta) = 1 + \{1 - \eta x\}\{\log(\eta x) - \Gamma'(2)\},$$

$$(\sigma^*)^2 = 1 + \Gamma''(2) - \Gamma'(2)^2 \approx 1 + 0.8237 - (0.4228)^2 \approx 1.6449 \approx (1.283)^2. \tag{7.17}$$

The effective information, 1.6449, is of course less than the information when η is known, namely 1.8237.

The standardized effective score is $T_n^0(\eta) = S_n^*/\sigma^* = \sum_{j=1}^n s^*(X_j) /(\sqrt{n}\sigma^*)$. The Neyman-Rao test statistic is obtained by substituting any root-n consistent estimate for the unknown η in s^*; it is sufficient to assume exponentiality. A quantile estimate is $\tilde{\eta}_n = \log 2/\text{med}$ where med is the sample median. In this example, there is no need to estimate σ^* since it was found to be parameter-free (7.17).

If we use an asymptotically efficient estimate of η (asymptotically efficient when $\theta = 1$), namely $\hat{\eta}_n = 1/\bar{X}_n$, we obtain the Rao (score) test statistic

$$T_R = \frac{1}{1.283\sqrt{n}} \sum_{j=1}^n \Big[1 + \big(1 - \frac{X_j}{\bar{X}_n}\big)\Big\{\log\big(\frac{X_j}{\bar{X}_n}\big) - \Gamma'(2)\Big\}\Big]$$
$$= \frac{1}{1.283\sqrt{n}} \Big[n - \frac{1}{\bar{X}_n} \sum_{j=1}^n \{(X_j - \bar{X}_n)\log(X_j/\bar{X}_n)\}\Big]$$

(the $\Gamma'(2)$ term dropping out). The two-sided score test statistic, for testing exponentiality against any Weibull alternative, is $|T_R|$.

The Wald test statistic is $T_W = 1.283\sqrt{n}(\hat{\theta}_n - 1)$ requiring asymptotically efficient estimation of (θ, η) under the Weibull model, as described in Chapter 5.

We now modify the example by introducing right-censoring. We are given a sequence of positive numbers $c_1, c_2, \ldots, c_n, \ldots$ and the distribution of the jth observation X_j is *right-censored at* c_j. That is, the survival function of X_j is assumed to be $\exp\{-(\eta x)^\theta\}1[x \le c_j]$. This distribution is of mixed-type, being absolutely continuous on the interval $(0, c_j)$ (with density as in the uncensored case) and with a discrete mass point at c_j with probability mass equal to $p_j = \exp\{-(\eta c_j)^\theta\}$.

Such models are common in studies in clinical medicine. Suppose heart-disease patients are recruited into a study, given a treatment, and then observed until a particular date to see whether and when a heart attack occurs. The assumed model is that the elapsed time from treatment until a (first) heart attack occurs follows a Weibull distribution. Suppose the recruitment date for patient j is r_j and the termination date of the study is t days beyond the initial date of the study, and let $c_j = t - r_j$, the observation time for patient j. If patient j has a heart attack within this observation time, it will be observed during the study; if not, it will only be observed that no attack had yet occurred at the end of observation time c_j, which occurs with probability p_j.

Under minor conditions on the c_j's, the local asymptotic normality condition continues to hold. Each component of the score vector is again proportional to the sum of scores per observation, but the scores are no longer identical or identically distributed, the jth one depending on c_j as well as x_j. And the elements of the information matrix are limits (assumed to exist) of averages of the expected information per observation. Details are left to the reader. However, even if the c_j are all known, which is often not the case, the expected information per observation cannot be computed except by numerical integration. Hence, it is more convenient to estimate B using the sample information matrix, which is easily calculated as a function of (θ, η) followed by substitution of $\theta_0 = 1$ and the corresponding restricted maximum likelihood estimate for η. See Chapter 6 for more details of the calculations. The Neyman-Rao and Rao test statistics may then be readily obtained.

Note An alternative approach is to force the model into a random sampling form, by treating the potential observation times, the c_j, as a random sample, independent of the true failure times, those times at which a heart attack occurs. Then the data consist of a random sample of the following data from each subject: (i) the smaller of the failure time and the potential observation (censoring) time, and (ii) an indicator of whether (i) was a failure time. If we act as if the distribution $G(c) = \text{pr}(C \leq c)$ say, of the c_j's is known we find that it does not enter the score and observed information for the Weibull parameters (η, θ) of the distribution of the T_i, so that we may proceed using these quantities as before. Full justification of this approach requires semiparametric theory—outside the scope of this book.

Example 7.6 *(Pairs of independent Cauchy variables):.*

Consider a random sample of pairs (X, Y), where X and Y are independent Cauchy random variables with distributions centered at μ and ν, respectively, and with common scale parameter τ. The joint density is therefore

$$f_\gamma(x, y) = \frac{\tau^2}{\pi^2} \cdot \frac{1}{\tau^2 + (x - \mu)^2} \cdot \frac{1}{\tau^2 + (y - \nu)^2} \quad \text{with } \gamma = (\tau, \mu, \nu).$$

We consider testing hypotheses about τ:

$$H_0 : \tau \leq \tau_0 \qquad \text{versus} \qquad H_A : \tau > \tau_0,$$

with μ and ν unknown nuisance parameters.

Such a model could arise as follows: A light source emits photons in random directions. An infinite screen, a distance τ away, captures the photons which hit it, n in number. On the screen there are coordinate axes, and (x_j, y_j) $(j = 1, \ldots, n)$ are the positions where the photons landed. The point (μ, ν) is the closest point to the light source. A little geometric reasoning leads to the conclusion that the (X_j, Y_j) have the distribution stated above: The tangent of the angle from the light source, between the perpendicular from (μ, ν) along the x-direction to (x, ν), is $(x - \mu)/\tau$; this angle is uniformly distributed on $(-\frac{\pi}{2}, \frac{\pi}{2})$, implying that $(X - \mu)/\tau$ has a standard Cauchy distribution; and similarly for $(Y - \nu)/\tau$, independently. Here, we are interested in the distance of the light source from the screen. Later, we will consider this example again, with interest in the location of (μ, ν).

The following additional notation (with dependence on parameters suppressed) is useful:

$$u = \frac{x-\mu}{\tau}, \quad w = \frac{1}{1+u^2}, \quad v = \frac{y-\nu}{\tau}, \quad z = \frac{1}{1+v^2}.$$

Using this notation, and the facts that $u^2 w = 1 - w$ and $v^2 z = 1 - z$, the scores for τ, μ and ν are found to be

$$s_\tau = \frac{2}{\tau}(1 - w - z), \quad s_\mu = \frac{2}{\tau}uw, \quad s_\nu = \frac{2}{\tau}vz.$$

To compute information, we need some integrals:

$$\frac{1}{\pi}\int w(u)^r du = 1, \tfrac{1}{2}, \tfrac{3}{8} \quad \text{for} \quad r = 1, 2, 3 \quad \text{and} \quad \int uw(u)^r du = 0 \quad \text{for} \quad r > \tfrac{3}{2}$$

(derivations omitted). Then the information matrix B is found to be diagonal with diagonal elements $B_{\tau\tau} = 1/\tau^2$ and $B_{\mu\mu} = B_{\nu\nu} = 1/(2\tau^2)$.

Since the information matrix is diagonal, the effective score and effective information for τ are just the ordinary score and information for τ. Hence, the standardized effective score is

$$T_n^0(\mu, \nu) = \frac{2}{\sqrt{n}}\sum_{j=1}^{n}(1 - w_j - z_j)$$

where $w_j = 1/(1+u_j^2)$, $u_j = (x_j - \mu)/\tau_0$ and similarly for z_j in terms of v_j, y_j and ν. The effective score test statistic T_{NR} is obtained by replacing μ and ν by root-n consistent estimates, for example the sample medians of the x's and y's, respectively (since $S_n^* = S_{\theta n}$ by orthogonality of the scores). Or, asymptotically efficient estimates may be used—roots of $\sum u_j w_j = 0$ and of $\sum v_j z_j = 0$ (with $\tau = \tau_0$) leading to T_R. A Wald test requires an asymptotically efficient estimate of τ in the 3-parameter model.

7.4 Power Approximation, Relative Efficiency and *p*-values

We conclude our discussion of testing of a single real parameter by describing applications of the asymptotic theory to approximating the power of a finite sample test, efficiency properties of tests and the use of *p*-values

7.4.1 Power Approximation

Since the limiting value of the power at a fixed value of $\theta \in H_A$ as the sample size $n \to \infty$ is typically unity we base our approximations on local power. To approximate the power of an asymptotically efficient test at a particular value θ_A, set $h_\theta = \sqrt{n}(\theta_A - \theta_0)$, and substitute into the appropriate power bound. Thus, for a one-sided asymptotically efficient test with nuisance parameter,

$$\text{power of } T_n \text{ at } (\theta_A, \eta) \;\approx\; \Phi\{\sigma^* \sqrt{n}(\theta_A - \theta_0) - z_\alpha\}. \tag{7.18}$$

This increases smoothly from α at $\theta_A = \theta_0$ towards 1 as θ_A increases. For an inefficient regular test, insert ρ^* in front of σ^* in Equation (7.18). Dependence on η is through σ^* (and ρ^* if present).

In Example 7.3 above, the power of an asymptotically efficient test of asymptotic level α of $\tau \leq \tau_0$, at a value of τ close to τ_0, is approximated by $\Phi(\sqrt{n}\frac{\tau-\tau_0}{\tau_0} - z_\alpha)$; this is free of dependence on the nuisance parameters (μ, ν) since $\sigma^* = 1/\tau_0$—a fact specific to this example. The same holds true in the two-parameter version of Example 7.2. Often, power approximation also depends on η, thereby requiring a preliminary guess at η to be fully useful.

7.4.2 Relative Efficiency

The asymptotic correlation coefficient ρ^*—or ρ if there is no nuisance parameter—in the local power formula provides a basis for quantifying asymptotic efficiency. The square of ρ^* is called the *asymptotic relative efficiency (ARE)*, and has the following interpretation (so long as ρ^* is positive!):

When comparing a regular test statistic $T'_{n'}$, with n' the index of the amount of data (usually sample size), with an asymptotically efficient test statistic T_n, the corresponding powers will be equal if $\rho^* \sqrt{n'} \approx \sqrt{n}$, or

$$ARE \equiv (\rho^*)^2 \sim n/n'.$$

Thus, if $(\rho^*)^2 = 0.5$, an asymptotically efficient test (based on T_n) with half the sample size ($n = \frac{1}{2}n'$) would have the same approximate power at θ_A as the inefficient test (based on $T'_{n'}$). Note that, conveniently, this quantity does not depend on α nor on θ_A; it does of course depend on θ_0, and typically also on η through ρ^*.

The effect of not knowing η on the power of asymptotically efficient tests may be seen in Equation (7.18), since if η were known σ^* would be replaced by σ, the asymptotic standard deviation of $S_{n\theta}$ instead of S_n^*. As in estimation (Chapter 6), the ratio of these variances enables a measure of the cost of not knowing η:

$$(\sigma^*)^2/\sigma^2 \leq 1. \tag{7.19}$$

This may also be given a sample-size interpretation, setting $\sigma^* \sqrt{n^*} \sim \sigma\sqrt{n}$. The ratio in Equation (7.19) is then n/n^*. For example, if this ratio is 0.9, so that $n = 0.9n^*$, a sample size 90% of that to be used when η is unknown would be sufficient if η were known. Put another way, the sample size would need to be increased by $100\times(\frac{1}{.9}-1)\%$ ($\approx 11\%$) from an η-known value n to an η-unknown value n^* in order to achieve the same power.

7.4.3 *p*-values

Use of a particular significance level α requires a critical value z_α (or $z_{\alpha/2}$) in all of these power bounds and formulas, and puts the test in accept-reject form. In practice, *observed significance levels*, or *p-values*, are more useful, still, these quantities, being data-dependent, cannot replace the role of α in power evaluations.

The *large-sample p-value* associated with a one-sided regular test statistic T_n is defined to be

$$p_n = \lim_{n\to\infty} P_{\theta_0,\eta}(T_n \geq t)\big|_{t=t_n^*} = \bar{\Phi}(t_n^*),$$

with t_n^* taking the observed value of T_n. No maximization over a local neighborhood within H_0 is needed, since the maximum occurs at $h_\theta = 0$ and the probability is free of η. In words, assuming that the null hypothesis to be true (and maximizing locally over it), it is the probability—or rather a large-sample approximation thereto—of obtaining a test statistic T_n as large or larger than the one actually observed. It provides a quantitative measure of 'statistical significance', with small values—say less than 5% or 1% or even 0.1%—indicating evidence, or strong evidence, against the null hypothesis, while larger values— say those greater than 20%—show little or no evidence against the null hypothesis. Note that the statements $p_n \leq \alpha$ and $T_n \geq z_\alpha$ are equivalent.

For two-sided tests, replace T_n and t_n^* by their absolute values, yielding $p_n = 2\bar{\Phi}(|t_n^*|)$—representing the large-sample probability, under the local null hypothesis, of a value of $|T_n|$ as large or larger than the value actually observed.

7.5 Multidimensional Hypotheses

We now turn to tests about a d_θ-dimensional θ, with a d_η-dimensional nuisance parameter η. The hypotheses are

$$H_0 : \theta = \theta_0,\ \eta \in \mathcal{R}^{d_\eta} \quad \text{versus} \quad H_A : \theta \neq \theta_0,\ \eta \in \mathcal{R}^{d_\eta}, \tag{7.20}$$

with local form

$$H_0'' : h_\theta = 0,\ h_\eta \in \mathcal{R}^{d_\eta} \quad \text{versus} \quad H_A'' : h_\theta \neq 0,\ h_\eta \in \mathcal{R}^{d_\eta}, \tag{7.21}$$

with θ_0 specified and η unknown—just as in the two-sided case with a nuisance parameter but with θ now d_θ-dimensional. Now $\gamma_n = (\theta_n, \eta_n) = (\theta_0, \eta) + h/\sqrt{n}$ and $h = (h_\theta, h_\eta)$.

We extend regularity as follows:

Definition 7.6 *Regular Test (Multidimensional Case): A test statistic Q_n is regular if $Q_n = T_n^T \Sigma^{-1} T_n + o_p(1)$ for some d_θ-dimensional T_n which, under γ_0, is asymptotically normal, with mean-vector 0 and variance matrix $\Sigma(\gamma_0)$ for T_n, jointly with the score S_n, and the test $\psi_n(\mathbf{x}) = \mathbf{1}[Q_n \geq w]$ has an asymptotic level for each positive w. Any such test is also said to be regular.*

Thus, Q_n is analogous to $T_n^2/\text{Avar}(T_n)$ in the one-dimensional case.

Notes

1. We could require that Σ be the identity matrix I—in closer analogy with the two-sided one-dimensional case—by making a suitable linear transformation of T_n.
2. The restriction to quadratic forms can be justified by an appeal to invariance theory. Appendix II gives some discussion.

The *continuous mapping theorem* implies that the quadratic form $Q_n = T_n^T \Sigma^{-1} T_n$ converges in distribution (under γ_0) to chi-squared with d_θ degrees of freedom (see Chapter 2). Arguing as in Section 7.1, LeCam's third lemma implies that $T_n \sim AN(\Omega h, \Sigma)$ under the local alternative γ_n with Ω ($d_\theta \times d$) the asymptotic covariance matrix of T_n and S_n. Hence Q_n converges in distribution under γ_n to noncentral chi-squared with d_θ degrees of freedom and noncentrality parameter $\delta^2 = h^T(\Omega^T \Sigma^{-1} \Omega)h$, again by the continuous mapping theorem. Write $\Omega = (\Omega_\theta, \Omega_\eta)$ with $\Omega_\theta = \text{Acov}(T_n, S_{\theta n})$ ($d_\theta \times d_\theta$) and $\Omega_\eta = \text{Acov}(T_n, S_{\eta n})$ ($d_\theta \times d_\eta$). The role of Equation (7.3) is now replaced by

$$\lim_{n\to\infty} P_{\gamma_n}(Q_n \geq w) = P\{\chi^2_{d_\theta}(\delta^2) \geq w\}. \tag{7.22}$$

We now show that regularity requires $\Omega_\eta = 0$; that is, $T_n \perp S_{\eta n}$, as in Section 7.2. Recall that noncentral chi-squared is *stochastically ordered* in its noncentrality δ^2 (see Section 1.4 of Chapter 1); that is, the right side in Equation (7.22) is increasing in δ^2. In the local null hypothesis, $h_\theta = 0$ so that δ^2 reduces to $h_\eta^T \Omega_\eta^T \Sigma^{-1} \Omega_\eta h_\eta$. In order that an asymptotic level be maintained for all h_η, this δ^2 must vanish, and hence Ω_η must be 0. Then $w \geq w_\alpha$, defined by $P\{\chi^2_{d_\theta}(0) \geq w_\alpha\} = \alpha$, will assure asymptotic level α. (The dependence of w_α on the degrees of freedom, d_θ, is suppressed.)

The local power of such a test is then

$$\text{pow}(h) \equiv \lim_{n\to\infty} P_{\gamma_n}(Q_n \geq w) = P\{\chi^2_{d_\theta}(\delta^2) \geq w) \tag{7.23}$$

with

$$\delta^2 = h_\theta^T \Omega_\theta^T \Sigma^{-1} \Omega_\theta h_\theta$$

(to be compared with (7.22). For Equation (7.23) to be maximal, $w = w_\alpha$. As in Equation (7.22), the resulting local power is increasing in δ^2. Hence, to bound the local power, among all regular tests of asymptotic level α, we maximize δ^2 by choice of the d_θ-dimensional statistic T_n. Lemma 7.1 below—an immediate corollary to the *Cauchy-Schwarz Lemma* of Appendix II of Chapter 5—asserts that this maximization, uniformly in h_θ (and h_η), is achieved if and only if $T_n = DS_n^* + o_p(1)$ for some (constant) non-singular matrix D, in which case the resulting Q_n is of the form $Q_n^0(\eta) + o_p(1)$, where $Q_n^0(\eta)$ is the *effective score quadratic form*:

$$Q_n^0(\eta) \equiv S_n^{*T}(B^*)^{-1}S_n^*, \text{ evaluated at } \gamma_0 = (\theta_0, \eta); \tag{7.24}$$

the resulting δ^2 is $\delta_0^2 \equiv h_\theta^T B^* h_\theta$. (It would be sufficient to take $D = (B^*)^{-1/2}$ so that, up to $o_p(1)$, $T_n = (B^*)^{-1/2}S_n^*$, a 'standardized effective score'. But the resulting $Q_n^0(\eta)$ is unchanged.)

Summarizing, we have

Proposition 7.4 *(Power Bound for the Multidimensional Case): For any regular test of local hypotheses (21) of asymptotic level α, the local power satisfies*

$$\text{pow}(h) = P\{\chi^2_{d_\theta}(\delta^2) \geq w_\alpha\} \leq P\{\chi^2_{d_\theta}(\delta_0^2) \geq w_\alpha\}$$

with δ^2 in (23) and $\delta_0^2 = h_\theta^T B^*(\gamma_0)h_\theta$. Equality holds for every $h \in H_A''$ in (21) if and only if the regular test statistic Q_n differs from the effective score quadratic form $Q_n^0(\eta)$ by $o_p(1)$ — equivalently, if $T_n = DS_n^* + o_p(1)$ for some non-singular D.

Any such test achieving the power bound is said to be asymptotically efficient. It is noteworthy that just as when $d_\theta = 1$, the dimension of the nuisance parameter η plays no role. The dependence on η is only through the non-centrality parameter.

Here is the needed Cauchy-Schwarz result:

Corollary 7.1 *(Cauchy-Schwarz Extension): Suppose that X and Y are d-vectors with means zero, variance matrices* $\text{var}(X) = A > 0$, $\text{var}(Y) = B > 0$, *and covariance matrix* $\text{cov}(X,Y) = C$, *and let $\delta^2(u) = u^T C^T A^{-1} C u$, defined for all $u \in \mathcal{R}^d$. Then*

$$\delta^2(u) \leq u^T B u \quad \text{for every } d \text{ vector } u, \tag{7.25}$$

with equality if and only if $X = KY$ for some constant non-singular matrix K, in which case $K = A(C^T)^{-1}$ and $X^T A^{-1} X = Y^T B^{-1} Y$.

Expressed in matrix form, (7.25) is: $C^T A^{-1} C \leq B$, which reduces when $d = 1$ to the familiar $\{\text{cov}(X,Y)\}^2 \leq \text{var}(X) \cdot \text{var}(Y)$.

Proof The inequality is precisely the *Cauchy-Schwarz Inequality* in Appendix I of Chapter 6. There, the conditions for equality were that $Y = C^T A^{-1} X$, and since we have assumed $B > 0$, we can equivalently write $X = A(C^T)^{-1}Y$ (equality implies that C must be non-singular); and $B = C^T A^{-1} C$ implies

$B^{-1} = C^{-1}A(C^T)^{-1}$. It follows that $Y^T B^{-1} Y$ reduces to $X^T A^{-1} X$, as claimed. □

We apply this to $T_n = X$, $S_n^* = Y$ and $B^* = B$. (It can be shown but is not proved here that the condition $X = KY$ is necessary as well as sufficient for equality of the quadratic forms, so that if the quadratic form test statistic differs from Q_n^0 by only $o_p(1)$, then the corresponding statistic T_n is essentially a linear transformation of the effective scores.)

The *effective score (Neyman-Rao) test* has the quadratic form test statistic

$$Q_{NR} \equiv \tilde{S}_n^{*T}(\tilde{B}_n^*)^{-1}\tilde{S}_n^*,$$

rejecting H_0 in Equation (7.20) whenever $Q_{NR} \geq w_\alpha$, where $\tilde{S}_n^* = S_n^*(\theta_0, \tilde{\eta}_n)$, $\tilde{\eta}_n$ is a root-n consistent estimate of η, and $\tilde{B}_n^*$ is consistent for $B^*(\gamma_0)$, either under γ_0, or generally. Therefore, under the assumptions of Regular Scores and Continuity of Information, it is an asymptotically efficient test by Proposition 6.1 in Chapter 6. If an estimate $\hat{\eta}_n(\theta_0)$ that is efficient under H_0 is used in S_n^*, then $\tilde{S}_n^*$ reduces to $S_{\theta n}\{\theta_0, \hat{\eta}_n(\theta_0)\} = \hat{S}_{\theta n}$, say (again assuming Regular Scores), and the resulting test is a *score (Rao) test* with test statistic

$$Q_R \equiv \hat{S}_{\theta n}^T(\tilde{B}_n^*)^{-1}\hat{S}_{\theta n}.$$

Note that it uses $\tilde{B}_n^*$ and not $\tilde{B}_{\theta n}$, as $\hat{S}_{\theta n}$ estimates S_n^*.

A *Wald test* uses the quadratic form

$$Q_W \equiv n(\hat{\theta}_n - \theta_0)^T \tilde{B}_n^*(\hat{\theta}_n - \theta_0)$$

where $\hat{\theta}_n$ is asymptotically efficient and, as before, $\tilde{B}_n^*$ is consistent for $B^*(\gamma_0)$. It too is equivalent to $Q_n^0(\eta)$ [why?], and hence also asymptotically efficient. In fact, all three test statistics differ from the effective score quadratic form by $o_p(1)$ (*Exercise 13*), and therefore

$$Q_{NR} \approx Q_R \approx Q_W.$$

The power at θ_A for any asymptotically efficient test is approximated by the bound in Proposition 7.4, with $\delta_0^2 = n(\theta_A - \theta_0)^T B^*(\theta_A - \theta_0)$. Calculations require quantiles of the noncentral chi-squared distribution. These are available in R and other software, see Chapter 1. Note that the degrees of freedom for these test statistics is determined solely by the dimension d_θ of θ, whatever the dimension (if any) of η; the non-centrality parameter $\delta_0^2 = h_\theta^T B^* h_\theta = \delta_0^2(\eta)$ reflects dependence on the nuisance parameter η. If there is no nuisance parameter, S_n^* is just S_n and B^* just B.

There is no simple measure of relative efficiency in this multidimensional case. However, for each linear combination of the coordinates of h_θ, the ratio of non-centralities δ_0^2/δ^2 can be calculated giving directional measure, very much as for multidimensional estimation in Section 5.8 of Chapter 5.

For any asymptotically efficient test, a corresponding large-sample p-value is $\bar{G}_{d_\theta}(Q_n^*)$ where $\bar{G}_d$ is the survivor function of central chi-squared with d

degrees of freedom and Q_n^* is the observed value of the asymptotically efficient test statistic. It represents the large-sample probability, under the local null hypothesis, of obtaining a value of Q_n^* as large or larger than that observed. Again, the dimension of any nuisance parameter is irrelevant.

Note that when θ is scalar everything in this section reduces to results for the two-sided case in the previous sections, but expressed here in terms of a chi-squared distribution with one degree of freedom rather than the two 'tails' of a normal distribution.

Example 7.7 *Example 7.6 (continued) Pairs of independent Cauchy variables*

We now consider testing hypotheses about the location (μ, ν) of the random sample of pairs of Cauchy variables, with the common scale parameter τ now being a nuisance parameter. The model was described in Example 7.3 of Section 7.2. Briefly stated, the hypotheses considered now are: $H_0: \ \mu = \nu = 0$ versus $H_A: \ \text{not} \ \ H_0$.

The effective score quadratic form is found to be

$$Q_n^0(\tau) = 2\tau^2(S_{\mu n}^2 + S_{\nu n}^2) = \frac{8}{n}\Big\{\Big(\sum_{i=1}^n u_i w_i\Big)^2 + \Big(\sum_{i=1}^n v_i z_i\Big)^2\Big\}$$

with the u_i's, w_i's, v_i's and z_i's evaluated at $\mu = \nu = 0$. There remains dependence on the nuisance parameter τ, through the u_j's and v_j's.

The effective score test statistic Q_{NR} is obtained by replacing τ by a root-n consistent estimate. A simple example is one-fourth of the sum of the two sample-specific *iqr*'s, since the iqr of a Cauchy distribution with scale paramater τ is 2τ). Alternatively, an asymptotically efficient estimate $\hat{\tau}_n^0$ can be obtained by finding a near root of the score equation for τ when $\mu = \nu = 0$—located near the simple estimate just defined—leading to Q_R. The resulting test statistic, whether Q_{NR} or Q_R, may be expressed as

$$\frac{8}{n}\left(\left[\{\sum_{i=1}^n \{\tilde{x}_i/(1+\tilde{x}_i^2)\}\right]^2 + \left[(\sum_{i=1}^n \{\tilde{y}_i/(1+\tilde{y}_i^2)\}\right]^2\right)$$

where $\tilde{x}_i = x_i/\tau$ and $\tilde{y}_i = y_i/\tau$ with τ appropriately estimated; it is asymptotically distributed as chi-squared with two degrees of freedom. The noncentrality parameter is found to be $(h_\mu^2 + h_\nu^2)/(2\tau^2) \approx n(\mu_A^2 + \nu_A^2)/(2\tau^2)$.

A Wald test requires simultaneous efficient estimation of (τ, μ, ν). Then $Q_W = n(\hat{\mu}_n^2 + \hat{\nu}_n^2)/(2\hat{\tau}_n^2)$. It is also treated as chi-squared with two degrees of freedom.

More examples will be given later.

7.6 Likelihood-Ratio Tests

We continue with consideration of hypotheses of the form Equations (7.20) and (7.15) for the one-sided case when θ is real. We allow nuisance parameters η, the special case without η being implicitly included.

Define $\hat{L}_n$ as the log likelihood-ratio with γ in the numerator replaced by a generally asymptotically efficient estimate $\hat{\gamma}_n$ and γ in the denominator replaced by $(\theta_0, \hat{\eta}_n(\theta_0))$ where $\hat{\eta}_n(\theta_0)$ is an asymptotically efficient estimate of η when θ is known to be its hypothesized value θ_0:

$$\hat{L}_n \equiv \log\left[p_{n,\hat{\gamma}_n}(\mathbf{x}_n)/p_{n,\hat{\gamma}_n(\theta_0)}(\mathbf{x}_n)\right]. \qquad (7.26)$$

Note that $\hat{L}_n$ is typically positive, since $\hat{\gamma}_n$ yields a (local) maximum of the likelihood. (To avoid difficulties, redefine $\hat{L}_n$ as 0 when negative.)

The statistic $\hat{L}_n$ has a long history as a potential test statistic, having been introduced by NEYMAN & PEARSON (1928). However, traditionally, the prescription is *(i)* the likelihood in the numerator is taken to be the likelihood maximized over the whole parameter space, and *(ii)* the likelihood in the denominator is that maximized over the nuisance parameters (if any) under the assumption of the null hypothesis. However, since unrestricted maximization of likelihoods can occasionally lead to difficulties (Chapter 5), we restrict attention to local maxima, yielding the statistic (7.26).

The *likelihood-ratio test statistic* for hypotheses (7.20) or (7.21) is now defined as

$$Q_{LR} \equiv 2\hat{L}_n.$$

When θ is real, the *one-sided likelihood-ratio test statistic* for hypotheses (1) (with $\eta \in \mathcal{R}^{d_\eta}$ inserted) or (8) is

$$T_{LR} \equiv \pm\sqrt{(2\hat{L}_n)}$$

with the sign chosen to match that of $\hat{\theta}_n - \theta_0$; the two-sided version is $|T_{LR}| = \sqrt{Q_{LR}} = \sqrt{2\hat{L}_n}$.

We show below that each of these versions differs from the corresponding effective score quadratic form or the standardized effective score by $o_p(1)$, and hence that each provides an asymptotically efficient test. The resulting likelihood-ratio test is thus asymptotically equivalent to the effective score (Neyman-Rao), score (Rao) and Wald tests. This equivalency, in contrast to all that has gone before, requires the uniform version of Local Asymptotic Normality. For the purposes of determining critical values or p-values, Q_{LR} is treated as (central) chi-squared with d_θ degrees of freedom while T_{LR} is treated as standard normal. Power is approximated as for the other asymptotically efficient tests.

Recall from Uniform Local Asymptotic Normality (Equation (4.9) in Chapter 4) that the log likelihood-ratio, γ_n to γ_0, is $L_n(\gamma_0, h) = S_n(\gamma_0)^T h - \frac{1}{2}h^T Bh$ (ignoring $o_p(1)$ (ignoring $o_p(1)$ terms, here and in what follows), and the uniformity of the error term allows replacing h by $\sqrt{n}(\hat{\gamma}_n - \gamma_0)$, which $= B^{-1}S_n(\gamma_0)$ by (8) of Chapter 5. Hence, substituting $B^{-1}S_n$ for h yields

Lemma 7.2 *Assuming Uniform Local Asymptotic Normality, and with $\hat{\gamma}_n$ an asymptotically efficient estimate of γ, we have*

$$2L_n(\gamma_0, \hat{h}_n)\Big|_{\hat{h}_n=\sqrt{n}(\hat{\gamma}_n-\gamma_0)} = S_n^T B^{-1} S_n + o_p(1).$$

This lemma enables proof that $Q_{LR} \approx Q_n^0$, the effective score quadratic form in Equation (7.24):

Proposition 7.5 *Assume Uniform Local Asymptotic Normality. Consider Q_{LR} with $\hat{\gamma}_n$ asymptotically efficient for $\gamma = (\theta, \eta)$ and $\hat{\eta}_n(\theta_0)$ asymptotically efficient for η when θ_0 is specified. Then*

$$Q_{LR} = Q_n^0(\eta) + o_p(1) \qquad \text{for each } \eta.$$

When θ is real, $T_{LR} = T_n^0(\eta) + o_p(1)$ for each η.

Proof Applying Lemma 7.2 twice, first as stated and second when θ is known to be θ_0 and $\hat{h}_n = \hat{h}_n(\theta_0) \equiv \{0, \sqrt{n}(\hat{\eta}_n(\theta_0) - \eta)\}$, we have (ignoring $o_p(1)$ terms)

$$\begin{aligned} Q_{LR} = 2\hat{L}_n &= 2L_n(\gamma_0, \hat{h}_n) - 2L_n\{\gamma_0, \hat{h}_n(\theta_0)\} \\ &= S_n^T B^{-1} S_n - S_{\eta n}^T B_{\eta\eta}^{-1} S_{\eta n} \\ &= S_{\theta n}^T B^{\theta\theta} S_{\theta n} + 2S_{\theta n}^T B^{\theta\eta} S_{\eta n} + S_{\eta n}^T (B^{\eta\eta} - B_{\eta\eta}^{-1}) S_{\eta n}. \end{aligned}$$

But

$$B^{\eta\eta} - B_{\eta\eta}^{-1} = B_{\eta\eta}^{-1} B_{\eta\theta} B^{\theta\theta} B_{\theta\eta} B_{\eta\eta}^{-1} \qquad \text{and} \qquad B^{\theta\eta} = -B^{\theta\theta} B_{\theta\eta} B_{\eta\eta}^{-1}$$

(see Appendix II of Chapter 5). The resulting expression for Q_{LR} reduces to $S_n^{*T} {B^*}^{-1} S_n^* = Q_n^0(\eta)$.

When θ is real, we therefore have $\sqrt{(2\hat{L}_n)} = |S_n^*|/\sigma^* = |T_n^0(\eta)|$. The sign to be applied in T_{LR} is that of $(\hat{\theta}_n - \theta_0)$; but $S_n^* \propto \sqrt{n}(\hat{\theta}_n - \theta_0)$ (still ignoring $o_p(1)$ terms), and so the sign is that of S_n^*. Therefore, $T_{LR} = T_n^0(\eta)$. □

In contrast to other asymptotically efficient tests, no estimate of information is required to use the likelihood-ratio test. However, this advantage is typically illusory in that efficient estimation of unknown parameters is required for the likelihood-ratio test and this generally requires estimation of information for iterative solution of the score equations. With nuisance parameters, the estimations must be performed under both null and alternative hypotheses; consequently, the likelihood-ratio test may require greater computation than its competitors.

We now re-visit the three examples of Sections 7.2 and 7.4.

Example 7.8 *Example 7.3 (continued) Testing the exponential rate parameter.*

To test $\theta = \theta_0$ versus $\theta \neq \theta_0$, we find that $|T_{LR}| = \sqrt{(2\hat{L}_n)} = \sqrt{[2n\{-\log(\theta_0\bar{X}_n) - 1 + \theta_0\bar{X}_n\}]}$. To compare this with the standardized score $|T_R|$, write $\epsilon = 1 - \theta_0\bar{X}_n = o_p(1)$ and $-\log(1-\epsilon) \approx \epsilon + \frac{1}{2}\epsilon^2$, yielding $|T_{LR}| = \sqrt{n}|1 - \theta_0\bar{X}_n| + o_p(1)$, as expected.

Example 7.9 *Example 7.5 (continued) Testing exponentiality.*

The likelihood-ratio test of exponentiality within a Weibull model (that the shape parameter $\theta = 1$) is obtained by efficiently estimating (θ, η) jointly, and then efficiently estimating η when $\theta = 1$. The latter yields $\hat{\eta}_n(1) = 1/\bar{X}_n$, but the former requires iterative solution of the two score equations, starting from root-n consistent estimates or solution of the profile likelihood equation (Chapter 6) We then find that

$$\hat{L}_n/n = 1 + \log\bar{X}_n + \log\hat{\theta} + \hat{\theta}\log\hat{\eta} + (\hat{\theta}-1)\frac{1}{n}\sum\log(X_j) - \hat{\eta}^{\hat{\theta}}\cdot\frac{1}{n}\sum X_j^{\hat{\theta}}.$$

The two-sided test statistic is $|T_{LR}| = \sqrt{(2\hat{L}_n)}$. The one-sided test statistic is positive or negative according to whether $\hat{\theta}$ is larger or smaller than unity.

Example 7.10 *Example 7.7 (continued) Pairs of independent Cauchy variables.*

To test the two-dimensional hypothesis $(\mu, \nu) = (0, 0)$, we first solve two efficient estimation problems, finding an asymptotically efficient $(\hat{\mu}, \hat{\nu}, \hat{\tau})$ and also $\hat{\tau}(0,0) = \hat{\tau}^0$, say, the latter being an asymptotically efficient estimate of τ when assuming $(\mu, \nu) = (0, 0)$. Then

$$Q_{LR} = -2\sum_{j=1}^{n}\Big[\log\Big\{1 + \Big(\frac{x_j - \hat{\mu}}{\hat{\tau}}\Big)^2\Big\} + \log\Big\{1 + \Big(\frac{y_j - \hat{\nu}}{\hat{\tau}}\Big)^2\Big\}$$
$$- \log\Big\{1 + \Big(\frac{x_j}{\hat{\tau}^0}\Big)^2\Big\} - \log\Big\{1 + \Big(\frac{y_j}{\hat{\tau}^0}\Big)^2\Big\}\Big] - 2n\log(\hat{\tau}/\hat{\tau}^0),$$

to be treated as having a chi-squared distribution with 2 degrees of freedom. For the non-centrality parameter δ_0^2, see the end of Section 7.4. This looks different from Q_R given earlier, but they differ by $o_p(1)$.

7.7 Large-Sample Confidence Sets

Recall that a confidence interval for a real parameter θ, based on potential data $\mathbf{X}$ from a model with parameter θ, is defined by a data-based interval $I(\mathbf{X})$ for which

$$P_\theta\{\theta \in I(\mathbf{X})\} \geq 1 - \alpha \quad \text{for each} \quad \theta$$

for some specified *confidence coefficient* $1-\alpha$, such as 0.95 or 0.90. The endpoints of the interval I may be interpreted as 'lower' and 'upper' estimates of θ, say $I(\mathbf{X}) = \big(\tilde{\theta}_L(\mathbf{X}), \tilde{\theta}_U(\mathbf{X})\big)$. The statement should be interpreted as: "the lower estimate is smaller than θ and the upper estimate is larger than θ"; it is *not* a probability statement about θ but a statement about the random interval I, or equivalently about its random endpoints. The probability statement is correct whatever the true but unknown value of the parameter; that is crucial to the concept: We have a data-based interval that may or may not have 'captured' the parameter, but its chances of having done so are known (or bounded below).

Sometimes only an upper, or only a lower, estimate is desired, leading to a *confidence bound*, a semi-infinite interval.

More generally, and without regard to the dimension of θ, nor whether the random set I is an interval, nor whether there is a nuisance parameter η, a data-based set $C(\mathbf{X})$ in the range of θ for which

$$P_\gamma\{\theta \in C(\mathbf{X})\} \geq 1-\alpha \quad \text{for each} \quad \gamma = (\theta, \eta)$$

is a *confidence set for* θ, with *confidence coefficient* $1-\alpha$. Again, the statement is that the random set $C(\mathbf{X})$ contains the true parameter value θ within it; it is the set that is random, not the parameter.

There is a 1-1 correspondence between a confidence set and a family of tests. Consider, for each θ_0 in the range of θ, a (non-randomized) test $\psi(\mathbf{X}, \theta_0)$ of significance level α for the hypothesis $\theta = \theta_0$, η unspecified, against some alternative. The significance level requirement (since ψ takes on only the values 0 and 1, being non-randomized) is equivalent to

$$P_{\theta_0,\eta}\{\psi(\mathbf{X}, \theta_0) = 0\} = 1 - E_{\theta_0,\eta}\{\psi(\mathbf{X}, \theta_0)\} \geq 1-\alpha \quad \text{for every} \quad \eta \qquad (7.27)$$

(since $E\psi = 1 \cdot P(\psi = 1) + 0 \cdot P(\psi = 0)$ and $P(\psi = 1) = 1 - P(\psi = 0)$). Now consider the set $C = C(\mathbf{x})$ in the range of θ defined by $C(\mathbf{x}) = \{\theta_0 | \psi(\mathbf{x}, \theta_0) = 0\}$—the set of 'acceptable' values of θ_0 upon observing $\mathbf{x}$. Then we have the logical equivalence:

$$\theta_0 \in C(\mathbf{X}) \quad \text{if and only if} \quad \psi(\mathbf{X}, \theta_0) = 0.$$

The two sides in this display have the same probability, and since (7.27) holds for each θ_0, we conclude—upon dropping the subscript on θ—that $C(\mathbf{X})$ is a $(1-\alpha)$-confidence set for θ. We say that C is determined by *inverting the family of tests.* The converse also readily follows: that is, a confidence set for θ defines a family of tests about θ—one test for each value of $\theta = \theta_0$.

These concepts and this equivalence extend directly to the large-sample setting. We consider a model for potential data $\mathbf{X}_n$ with parameter $\gamma = (\theta, \eta)$:

Definition 7.7 *A confidence set for θ, with asymptotic confidence coefficient $1-\alpha$, is a set $C_n(\mathbf{X}_n)$ in the range of θ for which*

$$\lim_{n\to\infty} P_{\gamma_n}\{\theta_n \in C_n(\mathbf{X}_n)\} \geq 1-\alpha \qquad (7.28)$$

for each $\gamma_n = (\theta_n, \eta_n) = (\theta, \eta) + (h_\theta, h_\eta)/\sqrt{n}$ in the parameter space.

Note Since the requirement holds for *every* γ_n, it also holds for (θ, η_n)—that is, the asymptotic probability of θ being in the confidence set when θ is the true parameter value is at least $1 - \alpha$. We could insert a 'lim inf' in Equation (7.28), but retain the simpler form.

The equivalence between confidence sets and tests yields:

Proposition 7.6 *Suppose that, for each θ_0 in the range of θ, $\psi_n(\mathbf{X}_n, \theta_0)$ is a regular test of asymptotic level α of the local hypothesis $h_\theta = 0$ and that h_η arbitrary, with $\theta = \theta_0$ and η unspecified, and let $C_n(\mathbf{x}_n) = \{\theta_0 | \psi_n(\mathbf{x}_n, \theta_0) = 0\}$. Then $C_n(\mathbf{X}_n)$ is a confidence set for θ with asymptotic confidence coefficient $1 - \alpha$.*

A confidence set is the set of (asymptotically) acceptable values of θ; moreover, as in the non-asymptotic case, the nature of the alternative hypotheses plays no role. (The converse of the proposition is also true.)

It should be noted that the quality of the large-sample approximation to be used in (21) is allowed to vary with (θ, η). An alternative definition would require the limit in (21) to be uniform in (θ, η). This could be accommodated in the theory presented here if Local Asymptotic Normality was assumed or established with this kind of uniformity, rather than just for the true (but possibly unknown) value of (θ, η). We proceed with the simpler definition, however.

If the component tests are asymptotically efficient when testing against $\theta \neq \theta_0$ for each θ_0, we say the corresponding confidence sets are *asymptotically efficient.* If θ is real and the alternatives are one-sided, and if the resulting confidence sets are semi-infinite intervals, the defining endpoints of the intervals may be termed *asymptotically efficient confidence bounds.*

An interpretation of asymptotically efficient in the context of confidence sets is briefly as follows—again a carry-over of corresponding non-asymptotic concepts as developed in LEHMANN & ROMANO (2005). We do so in the context of inverting tests of $\theta = \theta_0$, for real θ_0, versus a (local) one-sided alternative, allowing a nuisance parameter η. If the tests are regular, we say the confidence set is *regular.* When comparing two regular confidence sets, say C_n and C_n', we say C_n is *asymptotically more accurate at* $\gamma_n = (\theta_n, \eta_n)$ than C_n' if the asymptotic probability, when the true parameter is (θ, η_n), of the confidence set including θ_n with $h_\theta > 0$ is smaller using C_n than when using C_n'. Thus, the probability of including an incorrect (local) value is smaller. The set C_n is *asymptotically uniformly most accurate* if asymptotically more accurate than any competing regular confidence set, i.e. with the same asymptotic confidence coefficient whatever the parameter values considered. Since greater accuracy of a confidence set corresponds precisely to greater power of a test, it may be concluded that inversion of asymptotically efficient tests leads to asymptotically uniformly most accurate confidence sets—or simply *asymptotically efficient confidence sets.* We omit the details and the extensions to other cases.

Now let us consider the form of asymptotically efficient confidence sets corresponding to various asymptotically efficient tests. We start with the simplest case, the inversion of Wald tests. In the θ-real case, a two-sided Wald test accepts a null hypothesis $\theta = \theta_0$ whenever $\sqrt{n}\tilde{\sigma}_n^*|\hat{\theta}_n - \theta_0| \leq z_{\alpha/2}$ where $\hat{\theta}_n$ is asymptotically efficient and $\tilde{\sigma}_n^*$ is consistent for σ^* (at least under (θ_0, η)). If we choose a $\tilde{\sigma}_n^*$ that is θ_0-free, the corresponding confidence set is the interval with endpoints

$$\hat{\theta}_n \mp z_{\alpha/2}/(\sqrt{n}\tilde{\sigma}_n^*). \tag{7.29}$$

In higher dimensions, the Wald test is based on a quadratic form in the error-of-estimation; inverting it (when B^* is estimated in a θ_0-free way) yields an ellipse, or ellipsoid, as the Wald confidence set:

$$\left\{\theta|(\theta - \hat{\theta}_n)^T \tilde{B}_n^*(\theta - \hat{\theta}_n) \leq w_\alpha/n\right\}$$

with w_α determined from a central chi-squared distribution with d_θ degrees of freedom (whether or not there is a nuisance parameter). However, especially when there is no nuisance parameter, it would seem more natural to use an estimate of effective information that varies with θ_0. The resulting confidence set is not so readily made explicit; for example when in θ is real, the set is $\{\theta_0 | |\theta_0 - \hat{\theta}_n| \leq z_{\alpha/2}/[\sqrt{n}\tilde{\sigma}_n^*(\theta_0)]\}$.

A Bernoulli-sampling example may clarify this. If the Bernoulli parameter is p, the information is $\sigma(p)^2 = 1/(pq)$ with $q = 1 - p$. A Wald test of $p = p_0$ versus a two-sided alternative rejects when the sample proportion $\hat{p}$ differs from p_0 by more than $z_{\alpha/2}/(\sqrt{n}\sigma) = z_{\alpha/2}\sqrt{(p_0 q_0/n)}$. When inverting to obtain a confidence interval for p, common practice is to replace p_0 and q_0 in σ by sample proportions $\hat{p}_n$ and $\hat{q}_n$. Inverting then yields the standard Wald confidence interval $\hat{p}_n \mp z_{\alpha/2}\sqrt{(\hat{p}_n\hat{q}_n/n)}$. Alternatively, retaining p_0 in the information and inverting requires solving a quadratic equation in p_0, upon squaring both sides of $|p_0 - \hat{p}_n| \leq z_{\alpha/2}\sqrt{(p_0 q_0/n)}$, yielding the *Wilson confidence interval* (WILSON, 1927)

$$\frac{\hat{p}_n + \frac{1}{2}\epsilon_n^2 \pm \epsilon_n(\hat{p}_n\hat{q}_n + \frac{1}{4}\epsilon_n^2)^{1/2}}{1 + \epsilon_n^2}$$

with $\epsilon_n = z_{\alpha/2}/\sqrt{n}$. This turns out to be more reliable (see BROWN, ET AL. (2002)), and only slightly more complicated — but yet uncommon in texts and in applied work. Also, it more often has interval endpoints falling inside the restricted range $(0, 1)$ of the parameter p. See *Exercise 23* for the slightly simpler derivation of the Wilson interval for a Poisson mean

An alternative way of assuring that a Wald confidence interval for real θ falls inside the parameter space, when the range of θ is not the whole real line, is to transform θ monotonically so that its range *is* the full line, obtain a confidence interval for the transformed parameter, and back-transform the endpoints of the interval to obtain a confidence interval in the original parametrization. For a variance parameter, we might use a log transform. In the Bernoulli case, we might transform from p to $\text{logit}(p) \equiv \log\{p/(1-p)\}$, or possibly use the variance-stabilizing arc sine root transformation $\gamma \equiv 2\sin^{-1}(\sqrt{p})$, with inverse $p = \sin^2(\frac{1}{2}\gamma)$ discussed in Chapter 2. The large-sample variance of $\hat{\gamma}_n = \gamma(\hat{p}_n)$ is found to be, by the δ-method, $1/n$, parameter-free and thereby 'stable', and the resulting information for γ is unity. Hence, a Wald confidence interval for γ is $\hat{\gamma}_n \mp z_{\alpha/2}/\sqrt{n}$. Back transforming yields, as a confidence interval for p, $\sin^2[\sin^{-1}\{\sqrt{(\hat{p}_n)}\} \mp z_{\alpha/2}/(2\sqrt{n})]$.

In continuous-data problems, such transformations often improve the accuracy of large-sample approximations, but discreteness, as in the Bernoulli example, may degrade the quality of approximations in unforeseen ways (see BROWN, ET AL. (2002)).

We now turn to the inversion of *effective score tests*, *score tests* and *likelihood-ratio tests*. For the first of these, and when θ is real, the asymptotically efficient confidence set is the set of θ-values for which

$$\left|S_n^*\right| \leq z_{\alpha/2}\tilde{\sigma}_n^*, \tag{7.30}$$

where the effective information B_n^* is evaluated at $(\theta, \tilde{\eta}_n)$ for some root-n consistent $\tilde{\eta}_n$ and some consistent $\tilde{\sigma}_n^*$; dependence of $\tilde{\eta}_n$ or of $\tilde{\sigma}_n^*$ on θ is allowed. The resulting set may or may not be an interval. For a higher-dimensional θ, the confidence set is defined by

$$S_n^{*T}\tilde{B}_n^{*-1}S_n^* \leq w_\alpha$$

with the dependence on θ as before. Score-based confidence sets are similar, with S_n^* replaced by $S_{\theta n}$ and using an estimate $\hat{\eta}_n(\theta)$ of η which is asymptotically efficient when θ is specified.

A likelihood-ratio test accepts θ_0 whenever $2\hat{L}_n \leq w_\alpha$, with $\hat{L}_n$ the log likelihood-ratio with $\gamma = \hat{\gamma}_n$ in the numerator and $\hat{\gamma}_n(\theta_0)$ in the denominator—the former being asymptotically efficient for $\gamma = (\theta, \eta)$ and the latter being $(\theta_0, \hat{\eta}_n(\theta_0))$ with $\hat{\eta}_n(\theta_0)$ asymptotically efficient for η when θ_0 is specified. So the likelihood-ratio confidence set consists of those θ-values for which

$$\log p_{n,\gamma}(\mathbf{x}_n)\big|_{\gamma=\hat{\gamma}_n(\theta)} \geq \log p_{n,\gamma}(\mathbf{x}_n)\big|_{\gamma=\hat{\gamma}_n} - \tfrac{1}{2}w_\alpha, \tag{7.31}$$

which reduces, when there is no η, to

$$\left\{\theta \middle| \log p_{n,\theta}(\mathbf{x}_n) \geq \log p_{n,\hat{\theta}_n}(\mathbf{x}_n) - \tfrac{1}{2}w_\alpha\right\}. \tag{7.32}$$

The various types of asymptotically efficient confidence sets can be readily visualized in the case where θ is real and there is no nuisance parameter. Consider a plot of the log likelihood as a function of θ, evaluated at the observed data $\mathbf{x}_n^*$, and assume that this has a unique maximum, identifying $\hat{\theta}$. The Wald interval (7.29) extends an equal distance on either side of $\hat{\theta}$. The Rao (score) interval (7.30) (but without the asterisks, and when $\tilde{\sigma}_n$ is θ-free) is determined by the set of values of θ, including $\hat{\theta}$, where the slope

$(\partial/\partial\theta)\log p_{n,\theta}(\mathbf{x}_n^*) = \surd n S_n(\theta)$ is small in magnitude. The likelihood-ratio interval (7.32) is defined by the interval of values of θ for which $\log p_{n,\theta}(\mathbf{x}_n^*)$ is within $\frac{1}{2}w_\alpha$ of its maximum value. It is easily shown that, if the log likelihood is exactly parabolic near $\hat{\theta}$, as when sampling from $N(\theta,\eta^2)$, these three confidence intervals coincide.

To extend to allow a nuisance parameter η, the *profile log likelihood (effective log likelihood)* can be plotted instead, plotting $\log p_{n,\gamma}(\mathbf{x}_n^*)$ at $\gamma = (\theta, \hat{\eta}_n(\theta))$ against θ. See Appendix III to Chapter 6 for discussion. To extend to a two-dimensional parameter θ of interest, a contour plot can be used. A Wald confidence set (using a θ-free estimate of B^*) is an ellipse centered at $\hat{\theta}_n$. A *likelihood-ratio confidence set* is bounded by the contour which is a distance $\frac{1}{2}w_\alpha$ below the peak. A score-based confidence set is harder to visualize, but is determined by the orientations of the tangents to the surface near the peak. A disadvantage of Wald intervals compared with the others is that they are quite dependent — at least for moderate n — on the parameterization chosen.

7.8 Tests and Confidence Sets for a Parametric Function

Now suppose that hypotheses are stated in terms of some parametric function $\theta(\gamma)$, rather than through a coordinate of γ. For a one-sided test, θ must be one-dimensional; otherwise, θ is of dimension $d_\theta \geq 1$. One-sided hypotheses are $\theta(\gamma) \leq \theta_0$ versus $\theta(\gamma) > \theta_0$ and two-sided and multidimensional hypotheses are $\theta(\gamma) = \theta_0$ versus $\theta(\gamma) \neq \theta_0$, for specified θ_0.

There are two ways to develop asymptotically efficient tests. One approach is to transform from γ to (θ,η) by introduction of a suitable function $\eta = \eta(\gamma)$. Then the problem fits into one of the forms considered in Sections 7.1–7.5. It turns out that the choice of 1-1 parameterization of γ into (θ,η) does not matter.

An alternative direct approach is to extend the development in all of the preceding sections to allow for this level of generality. We sketch this briefly here.

Refer to Section 6.2 in Chapter 6. There we introduced the effective score and effective information for $\theta(\gamma)$, namely:

$$S_n^* = B^*\dot{\theta}B^{-1}S_n \quad \text{and} \quad B^* = \left(\dot{\theta}B^{-1}\dot{\theta}^T\right)^{-1}. \tag{7.33}$$

Here, $\dot{\theta}$ is the $d_\theta \times d$ matrix of partial derivatives of θ with respect to the d coordinates of γ.

The standardized effective score (when θ is real) is therefore

$$T_n^0(\gamma) = \frac{\dot{\theta}B^{-1}S_n}{\surd(\dot{\theta}B^{-1}\dot{\theta}^T)}. \tag{7.34}$$

In the multidimensional case, the effective score quadratic form is

$$Q_n^0(\gamma) = S_n^T B^{-1}\dot{\theta}^T\big(\dot{\theta}B^{-1}\dot{\theta}^T\big)^{-1}\dot{\theta}B^{-1}S_n. \tag{7.35}$$

In each of these, γ is evaluated at any value for which $\theta(\gamma) = \theta_0$. Then effective score and score tests are constructed by substituting inefficient or efficient estimates of γ under the constraint that $\theta = \theta_0$, and consistent estimates of the necessary information terms. A Wald test is constructed by estimating γ efficiently, without the null hypothesis constraint, and consistently estimating the effective information B^*. In the one-dimensional case, this leads to

$$T_W = \sqrt{n}\cdot\tilde{\sigma}_n^*\{\theta(\hat{\gamma}_n) - \theta_0\}$$

or its absolute value, and, in the multidimensional case, to

$$Q_W = n\{\theta(\hat{\gamma}_n) - \theta_0\}^T\tilde{B}_n^*\{\theta(\hat{\gamma}_n) - \theta_0\}.$$

A likelihood-ratio test requires both constrained and unconstrained efficient estimation of γ. Thus, the Neyman-Rao test can use either constrained (efficient under H_0) or unconstrained (inefficient) estimates of γ, the Wald test requires unconstrained efficient estimates, and the likelihood-ratio test requires both efficient estimates under both the constrained and unconstrained model.

Confidence sets for $\theta(\gamma)$ are obtained by inverting families of tests. In particular, a Wald confidence set for θ when considering two-sided alternatives for a real θ, and estimating $\tilde{\sigma}_n^*$ consistently in a θ-free manner, is

$$\theta(\hat{\gamma}_n) \mp z_{\alpha/2}/(\sqrt{n}\tilde{\sigma}_n^*).$$

A likelihood-ratio confidence set for a θ of arbitrary dimension is the set of values in the range of θ for which (7.31) holds, with $\hat{\gamma}_n(\theta)$ a constrained asymptotically efficient estimate.

Important special cases are two-sample and multi-sample settings, when a hypothesis of interest is whether or not the sampled populations agree in specified characteristics. Specifically, when sampling from two normal populations, we may wish to test whether the variances are equal, or, when assuming the variances are equal, whether the means are equal. A natural reformulation in the former is to test whether the ratio of variances is unity, and in the latter to test whether the difference between means is zero. These may be done by using a natural parametrization and employing the formulas above, or by re-parameterizing—e.g., $\theta = \sigma_2^2/\sigma_1^2$ (or $\theta = \log(\sigma_1^2/\sigma_2^2)$) and $\eta = (\sigma_1^2, \mu_1, \mu_2)$ in the former problem and $\theta = \mu_2 - \mu_1$ and $\eta = (\mu_1, \sigma_1^2, \sigma_2^2)$ in the latter (in obvious notation). Fortunately, the choice of how to re-parameterize is inconsequential, except for possible mathematical convenience; in particular, achieving some orthogonality may be worth the effort.

A third related example would be testing whether the two normal populations were identical, a two-dimensional null hypothesis. There are related multi-sample examples of popular interest—e.g., a large-sample treatment of *analysis of variance* problems.

Example 7.11 *(Two Weibull samples):*

Suppose that we have two independent random samples, of sizes n_1 and n_2, from Weibull distributions with scale parameters α_1 and α_2 and shape parameters β_1 and β_2, respectively—each parameterized as in Section 4.2 of Chapter 4.

We first consider the hypothesis of equal shapes: $\beta_2 = \beta_1$. To re-parameterize, it may be suitable to choose θ as the ratio of shapes, or the logarithm thereof, with three nuisance parameters $(\beta_1, \alpha_1, \alpha_2)$. We will instead employ the natural parametrization and use the formulas above for testing the hypothesis $\theta(\alpha_1, \beta_1, \alpha_2, \beta_2) \equiv \log(\beta_2/\beta_1) = 0$.

Let $n = n_1 + n_2$ and assume $n_j/n \sim p_j$ $(p_1 + p_2 = 1)$. (We use n_j/n and its limit p_j interchangeably in what follows.) In a two-sample problem, the score for a parameter appearing only in sample j is $\sqrt{np_j}\bar{s}_j \approx \sqrt{n_j} \cdot \sqrt{p_j}\bar{s}_j$ with $\bar{s}_j$ the average score for the parameter, averaged over the observations in sample j. (See *Exercise 15* in Chapter 4.) Information for such a parameter—the asymptotic variance of the score—is $np_j^2 B_j/n_j = p_j B_j$ with B_j the information (per observation in sample j) for this parameter. Co-information for a parameter that appears only in one sample with one that appears only in the other sample is zero.

The scores and information in a single Weibull sample appeared in Section 4.2 of Chapter 4. The scores and information for this 2-sample problem are therefore $S_n = (S_{\alpha_1 n}, S_{\beta_1 n}, S_{\alpha_2 n}, S_{\beta_2 n})^T$ with

$$S_{\alpha_j n} = \sqrt{n}\frac{p_j \beta_j}{\alpha_j} \cdot \frac{1}{n_j}\sum_{i=1}^{n_j}\{1 - (\alpha_j X_{ij})^{\beta_j}\},$$

$$S_{\beta_j n} = \sqrt{n}\frac{p_j}{\beta_j} \cdot \frac{1}{n_j}\sum_{i=1}^{n_j}\left[1 + \{1 - (\alpha_j X_{ij})^{\beta_j}\}\beta_j \log(\alpha_j X_{ij})\right]$$

and a 4×4 information matrix with non-zero entries

$$B_{\alpha_j \alpha_j} = p_j \frac{\beta_j^2}{\alpha_j^2}, B_{\alpha_j \beta_j} = p_j \frac{\Gamma'(2)}{\alpha_j}, B_{\beta_j \beta_j} = p_j \frac{1 + \Gamma''(2)}{\beta_j^2} (j = 1, 2).$$

The non-zero entries in B^{-1} are found to be

$$B^{\alpha_j \alpha_j} = B_{\beta_j \beta_j}/\Delta_j, \quad B^{\alpha_j \beta_j} = -B_{\alpha_j \beta_j}/\Delta_j, \quad B^{\beta_j \beta_j} = B_{\alpha_j \alpha_j}/\Delta_j \quad (j = 1, 2)$$

with the determinant $\Delta_j = B_{\alpha_j \alpha_j} B_{\beta_j \beta_j} - B^2_{\alpha_j \beta_j}$, which simplifies to $\Delta_j = p_j^2 \zeta/\alpha_j^2$ where $\zeta \equiv 1 + \Gamma''(2) - \Gamma'(2)^2 = \frac{1}{6}\pi^2$. (Refer back to Section 4.2 of Chapter 4.) Also, $\dot{\theta} = (0, -1/\beta_1, 0, 1/\beta_2)$. Hence,

$$\dot{\theta} B^{-1} = \frac{1}{\zeta} \cdot \left(\frac{\Gamma'(2)\alpha_1}{p_1 \beta_1}, -\frac{\beta_1}{p_1}, -\frac{\Gamma'(2)\alpha_2}{p_2 \beta_2}, \frac{\beta_2}{p_2} \right). \tag{7.36}$$

Then, from Equations (7.33) and (7.36), $B^* = (\sigma^*)^2$ is found to be $\zeta[(1/p_1)+(1/p_2)]^{-1} = \zeta p_1 p_2$. The standardized effective score (7.34) can now be calculated; we find (details omitted) that

$$T_n^0 = \sigma^* \dot{\theta} B^{-1} S_n \tag{7.37}$$

$$= \sqrt{(np_1p_2/\zeta)} \sum_{j=1}^{2} (-1)^{j-1} \frac{1}{n_j} \sum_{i=1}^{n_j} \{\Gamma'(2) - \beta_j \log(\alpha_j X_{ij})\}\{1 - (\alpha_j X_{ij})^{\beta_j}\}. \tag{7.38}$$

Substitution of root-n consistent estimates for the four parameters yields the effective score test statistic for the null hypothesis $\theta = 0$. To derive asymptotically efficient estimates under the null hypothesis would require solving the three score equations in the three-parameter problem with two α's and a common β, or equivalently solving a constrained maximization problem with a Lagrange multiplier for the constraint $\log(\beta_2/\beta_1) = 0$. However, asymptotically efficient estimates without the null hypothesis constraint are easier to derive, and they remain root-n consistent under the null hypothesis.

The Wald test statistic requires unconstrained asymptotically efficient estimates for the four parameters, as derived in Chapter 5, and derivation of the same $(\sigma^*)^2 = \zeta p_1 p_2$. A Wald confidence interval for θ is quite simple; see Equation (7.29), with $\hat{\theta}_n = \log(\hat{\beta}_2/\hat{\beta}_1)$ with the $\hat{\beta}_i$'s being unconstrained asymptotically efficient estimates. A Neyman Rao (effective score) confidence interval requires extensive computation, repeating computation of Equation (7.30) for a sequence of null values of θ (instead of just for $\theta = 0$), to find the range of values for which the statistic in Equation (7.30) is at most $z_{\alpha/2}$, as in Equation (7.30). The likelihood-ratio confidence interval is defined in Equation (7.31).

Finally, consider the null hypothesis that the two Weibull distributions are identical: $\alpha_2 = \alpha_1$ and $\beta_2 = \beta_1$. Let $\theta_1 = \log(\beta_2/\beta_1)$ (formerly just θ) and $\theta_2 = \log(\alpha_2/\alpha_1)$. We apply the methodology to test $(\theta_1, \theta_2) = (0, 0)$. Then $\dot{\theta}_2 = (-1/\alpha_1, 0, 1/\alpha_2, 0)$ and

$$\dot{\theta}_2 B^{-1} = \frac{1}{\zeta} \cdot \left[-\frac{\{1+\Gamma''(2)\}\alpha_1}{p_1\beta_1^2}, \frac{\Gamma'(2)}{p_1}, \frac{\{1+\Gamma''(2)\}\alpha_2}{p_2\beta_2^2}, -\frac{\Gamma'(2)}{p_2} \right].$$

After dropping subscripts on α and β, this leads to

$$B^* = p_1 p_2 \begin{pmatrix} 1+\Gamma''(2) & -\Gamma'(2)\beta \\ -\Gamma'(2)\beta & \beta^2 \end{pmatrix}$$

(details omitted). The test statistic, to be treated as chi-squared with two degrees of freedom, is then given by Equation (7.35) with common α's and common β's, estimated efficiently from the pooled sample — see the end of Section 5.4 in Chapter 5.

7.9 Bibliographic Notes

The first rigorous development of the large-sample efficiency of tests was by Abraham Wald in 1943; he proved that a test based on Q_W is *asymptotically most stringent*—that is, the difference between its asymptotic power and the asymptotic power of a Neyman-Pearson (most powerful) test against a simple alternative $h_A = (h_{\theta A}, h_{\eta A})$ is everywhere a minimum. He also proved that it has greatest power among tests with large-sample power that is constant on certain surfaces in the parameter space, and that the likelihood-ratio test is approximately equal to it. ROUSSAS (1972) developed large-sample properties of tests without nuisance parameters, using the methods of LeCam and Hájek. That the one-sided Wald test with nuisance parameters (and all the equivalent tests) is asymptotically uniformly most powerful was proved by CHOI (1989); see also WEFELMEYER (1987) and CHOI, HALL & SCHICK (1996). The asymptotically uniformly most powerful *unbiased* property of two-sided tests is described in STRASSER (1985); see also CHOI, HALL & SCHICK (1996).

Likelihood-ratio tests were introduced by NEYMAN & PEARSON (1928); their large-sample power was developed by WILKS (1938). Rao tests were introduced by RAO (1948), and Neyman-Rao tests by MORAN (1970), and more extensively by MATHIASON (1982) and HALL & MATHIASON (1990). CHOI, HALL & SCHICK (1996) present an invariance argument to support confining attention to regular (quadratic form) tests in the multidimensional case. This is somewhat analogous to the 'constant on certain surfaces' argument of WALD (1943).

SILVEY (1959) introduced a "Lagrange multiplier" test, widely cited in the economics literature, that turns out to be identical to the Neyman-Rao test. See our discussion in Section 6.3 of Chapter 6. The papers by BERA & BILIAS (2001) and RAO (2001) in a special issue of the Journal of Statistical Planning and Inference provide additional background. See also Boos (1992).

Appendix I: Some Normal-Theory Test Problems

In this appendix, we consider some normal-theory test problems with nuisance parameters and derive *uniformly most powerful tests*. These problems are analogous to the large-sample test problems considered in this chapter, but here there are no asymptotics. In a second appendix, we consider the large-sample analogs.

A one-sided test problem, with nuisance parameters, is considered in Section A.I.1, followed by two-sided and multidimensional test problems in Section A.I.2. In Section A.I.2, we need to restrict attention to *unbiased tests* or to *invariant tests*. Throughout, we confine attention to tests with a fixed *significance level* α.

A.I.1 A One-Sided Normal-Theory Test Problem

We start with a bivariate normal testing problem. The potential data are (X, Y) with a bivariate normal distribution with known variances σ^2 and τ^2, respectively, and known correlation coefficient ρ. Let B be the corresponding 2×2 variance matrix. The means of X and Y are each special linear combinations of two unknown parameters θ and η; specifically, $E(X,Y)^T = B \cdot (\theta, \eta)^T$. Upon observing (X, Y), we want to test the hypotheses

$$H_0: \quad \theta \leq 0,\ \eta \in \mathcal{R} \qquad \text{versus} \quad H_A: \quad \theta > 0,\ \eta \in \mathcal{R} \tag{A.1}$$

with B known.

This model may appear peculiar, but it parallels the large-sample fact that scores for θ and η are asymptotically normal with variance B and mean vector $B \cdot (h_\theta, h_\eta)^T$. A more natural form of the model treated here can be obtained by letting $\Sigma = B^{-1}$ and considering potential data $(U, V)^T = \Sigma(X, Y)^T$, which is bivariate normal with mean vector $(\theta, \eta)^T$ and known variance $\Sigma = B^{-1}$ with B defined above. The results below apply to this model as well.

(Also, we could introduce a random sample of (X, Y) values as our potential data, rather than a single pair, but then a sufficient statistic would be the sums of the X's and Y's, and these sums would have the same model as above except for a factor n on B. Hence, we may as well consider just a single (X, Y) pair.)

What kind of optimal test might we seek? A one-sided testing problem about a single parameter might well have a uniformly most powerful (α) test available (*uniformly most powerful of significance level* α), but the hypotheses (A.1) have a nuisance parameter η. There are very few examples in the literature of uniformly most powerful tests when there is a nuisance parameter. To deal with nuisance parameters, even in normal-theory models, it is usually necessary to confine attention to *unbiased tests*, or to *invariant tests*, and to seek uniform maximization of power within these restricted families. For example, the t-test of whether the mean of a normal distribution is at most zero or is positive, with the standard deviation σ as a nuisance parameter, is uniformly most powerful only within the class of unbiased tests (Section 7.1) or within the class of tests which are *scale-invariant*. See LEHMANN & ROMANO (2005). But we shall show that there does exist a uniformly most powerful test of size α for the hypotheses in (A.1); specifically:

Proposition A - 3 *Let* $X^* \equiv X - \frac{\rho\sigma}{\tau} Y$ *and* $\sigma^* \equiv \sigma \cdot \sqrt{(1-\rho^2)}$. *Then* $X^* \perp Y$ *and the test* $\psi(X, Y) \equiv 1[X^* \geq \sigma^* z_\alpha]$ *is uniformly most powerful of size* α *for the hypotheses in Equation (A.1).*

Proof We first re-formulate the testing problem in terms of the transformed data (X^*, Y). We find that X^* is $N(\theta', {\sigma^*}^2)$ with $\theta' = (\sigma^*)^2\theta$ and Y is $N(\eta', \tau^2)$ (as before) with $\eta' = \rho\sigma\tau\theta + \tau^2\eta$; moreover, X^* and Y are independent. The hypotheses (A.1) can be equivalently written as

$$H_0': \theta' \leq 0,\ \eta' \in \mathcal{R} \quad \text{versus} \quad H_A': \theta' > 0,\ \eta' \in \mathcal{R}. \tag{A.2}$$

The test ψ in the proposition is known to be uniformly most powerful (α) for testing $\theta' \leq 0$ versus $\theta' > 0$ among tests of size α based solely on X^*. Application of Lemma A.1 below completes the proof. □

There was no need to restrict Y and η to be real-valued; only X and θ need be real: Suppose that (X, Y) is multinormal with mean vector $B(\gamma)$, where $\gamma^T = (\theta^T, \eta^T)$ and variance matrix B. The proposition holds with $X^* = X - B_{\theta\eta}B_{\eta\eta}^{-1}Y$ and $(\sigma^*)^2 = B_{\theta\theta} - B_{\theta\eta}B_{\eta\eta}^{-1}B_{\eta\theta}$ with subscripts on B denoting blocks therein.

Now for the lemma, which asserts the intuitive fact that extra independent data, with parameter unrelated to that of the primary data, can safely be ignored when constructing optimal tests. We present it for the case of real η, but the dimension plays no role.

Lemma A - 5 *Suppose that when observing* X *with density* $f_\theta(x)$, $\theta \in \mathcal{R}$, *that* $\psi^\circ(X)$ *is uniformly most powerful of size* α *for testing* $H_0'' : \theta \leq 0$ *versus* $H_A'' : \theta > 0$. *Suppose that* Y *has density* $g_\eta(y)$, $\eta \in \mathcal{R}$, *and that* X *and* Y *are independent. Then, when observing* (X, Y), $\psi^\circ(X)$ *is uniformly most powerful of size* α *for testing the hypotheses in Equation (A.1).*

Proof Let $\mathcal{T}$ represent the set of all tests based on (X, Y) which have significance level α for H_0 in Equation (A.1). We will show: (i) that $\psi^\circ(X) \in \mathcal{T}$, and (ii) that whatever the choice of $\gamma_A = (\theta_A, \eta_A)$ in H_A in Equation (A.1), $\psi^\circ(X)$ is most powerful at γ_A among all tests in $\mathcal{T}$. Then, since ψ° does not depend on the choice of γ_A, it is uniformly most powerful among tests in $\mathcal{T}$.

That (i) holds is immediate since ψ° does not depend on Y and has size α for H_0''. Now consider the class $\mathcal{T}(\eta_A)$ of all tests of significance level α for the simple null hypothesis $(0, \eta_A)$, and consider the member of $\mathcal{T}(\eta_A)$, say $\psi_{\gamma_A}(X, Y)$, which is most powerful against γ_A among all tests in $\mathcal{T}(\eta_A)$. According to the *Neyman-Pearson Lemma*, this test is determined by the likelihood ratio of these two simple hypotheses, which is identical to the likelihood ratio for $\theta = \theta_A$ to $\theta = 0$ when observing only X; hence $\psi_{\gamma_A}(X, Y) = \psi^\circ(X)$. So $\psi^\circ(X)$ is most powerful at γ_A among tests in $\mathcal{T}(\eta_A)$. But $\mathcal{T}$ is contained in $\mathcal{T}(\eta_A)$ and $\psi^\circ \in \mathcal{T}$; so ψ° is most powerful at γ_A among tests in $\mathcal{T}$. □

The reader may skip to Appendix II for the analogous large-sample problem.

A.I.2 A Two-Sided and a Multidimensional Normal-Theory Test Problem

We start with a two-sided version of the testing problem in the previous section, considering hypotheses

$$H_{02}: \quad \theta = 0, \ \eta \in \mathcal{R} \quad \text{versus} \quad H_{A2}: \quad \theta \neq 0, \ \eta \in \mathcal{R} \tag{A.3}$$

about the bivariate normal model for (X, Y) with known variance matrix B and with $E(X, Y)^T = B \cdot (\theta, \eta)^T$. We use the theories of unbiased tests described in Section 7.2 and of *invariant tests*, presented in LEHMANN AND ROMANO (2005). Briefly, a test is *unbiased* if its power is nowhere less than its size, and invariance here means that the test is unaffected by some group of transformations of (X, Y).

We saw in Section 7.2 that the two-sided test $\mathbf{1}[|X^*| \geq \sigma^* z_{\alpha/2}]$ based on X^* is uniformly most powerful and unbiased for testing H_{02} versus H_{A2}. Now Lemma A - 3 can be adapted to justify ignoring Y, applying the Generalized Neyman-Pearsom lemma (LEHMAN AND ROMANO, 2005) to obtain a most powerful test of $(\theta', \eta') = (0, \eta'_A)$ versus $(\theta', \eta') = (\theta', \eta'_A)$ within the class of tests unbiased for θ'.

In terms of the transformed variables (X^*, Y), the hypotheses are

$$H'_{02} : \theta' = 0, \eta' \in \mathcal{R} \quad \text{versus} \quad H'_{A2} : \theta' \neq 0, \eta' \in \mathcal{R}. \tag{A.4}$$

We now reconsider testing H_{02} versus H_{A2} in Equation (A.3), requiring invariance of the test with respect to sign and scale changes of X^*, and show that the same two-sided test is *uniformly most powerful invariant* The testing problems (A.3) and (A.4) are seen to be invariant under arbitrary linear transformation of Y as well, but we only need to impose the stated invariance to find a uniformly most powerful test. (Actually, invariance under sign changes would be sufficient, but allowing $X^* \to cX^*$ for arbitrary $c \neq 0$ is more natural in the sequel) Thus, we limit attention to tests based on $Z = (|X^*|/\sigma^*, Y)$, a *maximal invariant*, with distribution of the first coordinate depending only on $|\theta'|/\sigma^* = |\theta|\sigma^*$.

The test in Proposition A.2 may be shown to be uniformly most powerful among unbiased tests of size α tests based only on $|X^*|/\sigma^*$—that is, uniformly most powerful among invariant tests based only on X^*. tests based only on X^*. (Note that $(X^*/\sigma^*)^2$ is distributed as chi-squared with 1 degree of freedom and noncentrality $(\theta'/\sigma^*)^2$, and apply the Neyman-Pearson Lemma.) Lemma A.1 may be modified to justify ignoring Y. We are thus led to

Proposition A - 4 *The test* $\mathbf{1}[|X^*| \geq \sigma^* z_{\alpha/2}]$ *is uniformly most powerful among invariant tests of size* α *for testing* H_{02} *versus* H_{A2} *in Equation (A.3).*

As in Section A.I.1, the dimension of Y and η is unimportant in the above arguments. Moreover, the invariance argument extends to X and θ of higher dimension, confining attention to tests invariant under nonsingular (homogeneous) linear transformation of $X^* \equiv X - B_{\theta\eta} B_{\eta\eta}^{-1} Y$, $X^* \to C^T X^*$ for nonsingular C. The maximal invariant function of (X, Y) is now (W, Y) with $W \equiv X^{*^T} (B^*)^{-1} X^*$, and with distribution of W depending only on $\theta^T B^* \theta$. That Y can be ignored is justified as before. The resulting uniformly most powerful invariant test is $\mathbf{1}[W \geq w_\alpha]$ where w_α is the upper α-point in the chi-squared distribution with d_θ degrees of freedom.

Large-sample analogs of these developments appear in Section A.II.2 below.

Appendix II: Asymptotic Efficiency with Fewer Assumptions

We show that the asymptotically efficient one-sided tests of Sections 7.1 and 7.2 are asymptotically uniformly most powerful, *without* any restriction to asymptotically normal test statistics. The idea is due to CHOI (1989); a modified version appears in CHOI, HALL & SCHICK (1996). We also sketch proofs that the asymptotically efficient two-sided tests of Sections 7.1 and 7.2 are asymptotically uniformly most powerful (*unbiased*) and asymptotically uniformly most powerful *invariant*), and that the asymptotically efficient tests for multidimensional hypotheses of Section 7.3 are asymptotically uniformly most powerful invariant, again without restriction to asymptotically normal test statistics; full proofs appear in the cited reference (1996).

The development here parallels that given in Appendix I for analogous normal-theory test problems (in contrast to proofs in the cited reference). Section A.II.1 is devoted to the one-sided testing problem and Section A.II.2 to the two-sided and multidimensional testing problems.

A.II.1 The Large-Sample One-Sided Testing Problem

Consider the one-sided testing problem described in Section 7.1. However, we now relax the *regularity* requirement of confining attention to tests based on an asymptotically normal test statistic, only retaining the local asymptotic normality assumption.

Define a test by a test function ψ_n from $\mathbf{x}_n$ to $[0, 1]$, and write $\gamma_0 = (\theta_0, \eta)$. Formally, we say that a test ψ_n is *asymptotically uniformly most powerful* among tests of *asymptotic level* α of the local hypotheses (3) if, with $\gamma_n = \gamma_0 + (h_\theta, h_\eta)/\sqrt{n}$ and whatever the value of the nuisance parameter η,

$$\limsup E_{\gamma_n}(\psi_n) \leq \alpha \quad \text{for each} \quad h_\theta \leq 0 \ \text{ and each } \ h_\eta$$

ensuring that the test has asymptotic level α and, for any other test ψ'_n of asymptotic level α,

$$\limsup E_{\gamma_n}(\psi_n) \geq \limsup E_{\gamma_n}(\psi'_n) \quad \text{for each} \quad h_\theta \leq 0 \ \text{ and each } \ h_\eta$$

ensuring that ψ is asymptotically uniformly most powerful among such tests. We shall now show that the *power bound* in Proposition 7.3 holds generally, for all tests of asymptotic level α, assuming only Local Asymptotic Normality

First note that, without loss of generality, we can assume orthogonality of the scores for θ and η; that is, $B_{\theta\eta}(\theta_0, \eta) = 0$ for every η. For, if not, let $h'_\eta = h_\eta + B_{\eta\eta}^{-1} B_{\eta\theta} h_\theta$, and it then follows that

$$L_n(\gamma, h_\theta, h_\eta) = S_n^* h_\theta + S_{\eta n}^T h'_\eta - \tfrac{1}{2} h_\theta^2 {\sigma^*}^2 - \tfrac{1}{2} h_\eta'^T B_{\eta\eta} h'_\eta + o_p(1)$$

which has the same form as that for orthogonal parameters; moreover, the local hypotheses (7.10) are the same, whether in terms of h_η or h'_η. We shall see below that tests are determined solely by $L_n(\gamma_0, h_\theta, h_\eta)$. So we assume orthogonality and use $S_{\theta n}$ and $\sigma^2 = B_{\theta\theta}$ instead of S_n^* and $(\sigma^*)^2 = B^*$.

We first focus on local hypotheses with h_η set equal to 0. Consider simple sub-hypotheses: $h_\theta = 0$ versus $h_\theta = h_{\theta A}$ (> 0), with η specified. According to the Neyman-Pearson Lemma, a most powerful test rejects for large values of the log likelihood-ratio, namely (under Local Asymptotic Normality) $S_{\theta n} h_{\theta A} - \frac{1}{2} h_{\theta A}^2 \sigma^2 + o_p(1)$ with $S_{\theta n}$ evaluated at γ_0—or equivalently for large values of $S_{\theta n}/\sigma + o_p(1)$. The test will have asymptotic size α if and only if the critical value is $z_\alpha + o_p(1)$, that is, the test is asymptotically equivalent to $\mathbf{1}[S_{\theta n}/\sigma \geq z_\alpha]$. Since this test does not depend on the choice of $h_{\theta A}$, with local power $\Phi(\sigma h_\theta - z_\alpha)$ at h_θ, it is found to be asymptotically a uniformly most powerful (α) test for (3) with $h_\eta = 0$ and η specified throughout.

Now tests not depending on η but which have asymptotic significance level α whatever the value of η are a smaller class of tests (but still with $h_\eta = 0$), and hence the power bound remains valid for them.

We now argue, very much as in the proof of the previous lemma, that the power bound of Proposition 7.3 remains valid when an unknown h_η is allowed in both hypotheses in (3). Any test equivalent to that based on the standardized score $\mathbf{1}[S_{\theta n}/\sigma \geq z_{\alpha/2}]$ with $S_{\theta n}$ evaluated at (θ_0, η) is in the class $\mathcal{T}$ of all tests of asymptotic significance level α for the null hypothesis in (3). Now let $\mathcal{T}(\eta, h_{\eta A})$ be the class of all tests of asymptotic level α for the simple null hypothesis $h_\theta = 0$, $h_\eta = h_{\eta A}$, and η specified. Let ψ_n be a member of this class which is most powerful against $h_{\theta A}$, $h_{\eta A}$, and η. According to the Neyman-Pearson Lemma, ψ_n rejects for large values of $L_n(\gamma_0, h_{\theta A}, h_{\eta A}) - L_n(\gamma_0, 0, h_{\eta A})$, equivalently for large $S_{\theta n}(\gamma_0) + o_p(1)$. Hence, the power bound continues to hold.

A.II.2 The Large-Sample Two-Sided and Multidimensional Testing Problems

We proceed as for the large-sample one-sided testing problem, but with a two-sided alternative hypothesis. We need to show that the corresponding power bound continues to be valid and that $\mathbf{1}[|S_n^*|/\sigma^* \geq z_{\alpha/2}]$ has the properties of an asymptotically uniformly most powerful unbiased test of level α and an Auniformly most powerfulI(α) test, using only the assumption of local asymptotic normality

This case differs from the one-sided one in the necessity to introduce limiting versions of the scores. Otherwise, we cannot reduce a stationarity constraint (see Section A.I.2 in Appendix I) by differentiating under the integral sign as was required in the bivariate normal case (nor even verify differentiability of local power). A *Skorokhod construction* and a *weak compactness* argument is needed to justify confining attention to tests based on limiting versions of the scores; see CHOI, HALL & SCHICK (1996). The small-sample methodology may then be applied to reach the desired conclusion.

For the invariance argument, again limiting versions of the scores are needed in order to justify applying an invariance argument to the scores. And we approach invariance somewhat differently, namely in terms of transformations of the parameters rather than transformations of the data. In the two-sided real θ case, we conclude that a test based on the magnitude of the standardized effective score, and any equivalent test, is asymptotically uniformly most powerful invariant of level α, the invariance being with respect to any smooth transformation of the parameter of interest (differentiable near θ_0 with non-vanishing derivative, implying a sign-and/or-scale change of the effective score for θ). In the multidimensional case, the same conclusion holds for a test based on the effective score quadratic form, and the power bound of Proposition 7.4 continues to hold. Invariance is again with respect to smooth transformation of the parameter of interest—that is, requiring that the resulting test of the corresponding null hypothesis not depend on the parametrization chosen for the parameter of interest; see CHOI, HALL & SCHICK (1996).

It is intuitively clear (and readily proved) that asymptotically efficient tests are not affected by any re-parameterization of the nuisance parameter. Indeed, they may be shown to be invariant to transformation of η into $\beta = \beta(\theta, \eta)$.

Exercises

1. (a) Derive the Neyman-Pearson one-sided tests of level 0.025 of the hypothesis $\theta = 1.0$ against the alternatives (i) $\theta > 1$ and (ii) $\theta < 1$, given a single observation X from an exponential density $f(x, \theta) = \theta \exp(-\theta x)$. Note that each test is uniformly most powerful for the applicable hypothesis.

   (b) Combining the two tests, i.e. rejecting the hypothesis $\theta = 1$ if either individual test rejects the null hypothesis, gives a two-sided test of level 0.05. Plot the power function of this combined test in the range $0.4 \leq \theta \leq 3.0$ and show that it achieves its minimum at $\theta \approx 1.36$ and remains below 0.05 for $1 \leq \theta \leq 1.99$, so the test is very far from unbiased.

   (c) Derive (numerically) the uniformly most powerful unbiased test for this problem and the distribution of the total level between the two parts of the alternative hypothesis.

2. For the case of an unbiased test of $\theta = 0$ versus $\theta \neq \theta_0$ for observations from a standard normal distribution (see Section 2.2) show that it is possible to choose the Lagrange multipliers λ_1 and λ_2 to achieve any specified level of the test.

3. (a) Derive an asymptotically efficient test of the hypothesis $\lambda \leq \lambda_0$ versus $> \lambda_0$ when sampling from a Poisson(λ) distribution.

   (b) Compare the asymptotically efficient test in (a) with the *uniformly most powerful test.* Be specific about each when the significance level is 5%, $\lambda_0 = 1.6$ and $n = 25$ (not really a very large sample).

   (c) Compare the approximate p-value for the asymptotically efficient test with the exact p-value of the uniformly most powerful test when $\bar{x}^* = 2.00$.

   (d) Approximate the power of the asymptotically efficient test at $\lambda = 2.1$, and compare it with the power of the uniformly most powerful test.

4. Derive an asymptotically efficient two-sided test of the hypothesis $\theta = 0$ when sampling from a

   (a) *Cauchy distribution* centered at θ and with unit scale.

   (b) *double exponential (Laplace) distribution* with density $\frac{1}{2}\exp(-|x-\theta|)$ (see *Exercise 9* in Chapter 4).

5. As discussed in Section 7.1, in general hypothesis testing, a test is said to be *unbiased* if the probability of rejecting the null hypothesis, at any parameter value where the null hypothesis is true, never exceeds the probability of rejecting at any parameter value where the null hypothesis is false. Correspondingly, we say a test of $\theta = \theta_0$ versus $\neq \theta_0$ is *locally unbiased* if the local power, $pow(h)$, has its minimum at $h = 0$.

   Show that every regular two-sided test about a real parameter θ is *locally unbiased*, whether or not there are nuisance parameters. Hence, just as in large-sample estimation, regularity plays a role similar to unbiasedness in 'small-sample' test theory.

6. The *sign test*, for testing whether a location parameter $\theta \leq \theta_0$ vs. $\theta > \theta_0$, has test statistic $T_n = (1/\sqrt{n}) \sum \text{sign}(X_j - \theta_0)$. Determine the asymptotic relative efficiency of the *sign test* with $\theta_0 = 0$ when sampling from the following distributions:

   (a) Normal $N(\theta, 1)$;

   (b) Cauchy distribution as in (a) of *Exercise 4*;

   (c) Double exponential as in (b) of *Exercise 4*.

7. *Continuation of Exercise 13 in Chapter 5: Cauchy scale problem with known location.*

   (a) Find (numerically) Rao, Wald and likelihood-ratio tests of the null hypothesis that the scale parameter satisfies $\tau \leq 1.0$ versus $\tau > 1.0$.

(b) Determine large-sample p-values for each test.

[Test statistics and p-values may not agree very closely since the sample size is not large.]

8. In a model with real parameter θ and nuisance parameter η, the score $S_{n\theta}$ has variance $B_{\theta\theta} = \sigma^2$, say, not $B^* = (\sigma^*)^2$, so why does the Rao test statistic have an estimate of σ^* in its denominator rather than an estimate of σ?

9. *Continuation of Example 7.3 of Section 7.2.* Consider a test based on the following estimate of τ: Let iqr_x and iqr_y be the sample interquartile ranges of the x's and of the y's, respectively; average them and then divide by 2. Reject the null hypothesis that $\tau \leq \tau_0$ when this estimate exceeds τ_0 by more than z_α estimated standard errors. Verify that this test is regular, and determine its asymptotic relative efficiency.

10. *Continuation of Example 7.2 of Section 7.3 and of Exercise 21 in Chapter 5): testing exponentiality in a censored 2-parameter Weibull model.*

(a) Determine an estimate of η under H_0, based on the proportion of censored observations.

(b) Determine the effective score test statistic, using the estimate of η in (a), and using average sample effective information to estimate the effective information. What is the resulting large-sample p-value?

(c) Determine the p-value from the score test, using an efficient estimate of η rather than (a).

(d) Approximate the power of (either of) these tests, when the size is 5%, at $\theta = 1.5$.

11. Here are some alternative definitions of the *effective score.* Confine attention to the case of random sampling, and assume that both θ and η are real (for simplicity).

(a) Let $s^* = s_\theta - bs_\eta$. Choose b to minimize the variance of s^*. (Here, s_θ and s_η are scores per observation.)

(b) (See Appendix II to Chapter 6 for relevant discussion). The *profile likelihood* (or 'effective likelihood') p_n^* is defined as the likelihood with nuisance parameters replaced by estimates which are asymptotically efficient when the parameters of interest are specified; hence, it is a function only of the parameters of interest, through these asymptotically efficient estimates as well as directly. Now $S_{n\theta} = n^{-1/2}\partial \log p_n/\partial\theta$. Argue (with sufficient regularity conditions) that $S_n^* + o_p(1)$ results if p_n is replaced by the profile likelihood p_n^*—that is, the effective score (for θ) is the score when likelihood is replaced by effective likelihood.

TABLE 7.1
Failure Intervals for Machine Data

				Machine 1					
31	31	33	66	99	100	105	120	128	133
138	171	282	302	302	320	366	373	555	558
				Machine 2					
10	10	32	35	37	39	51	54	59	63
65	65	77	111	119	119	120	176	178	179
180	181	197	236	245	281	297	344	385	717

12. Listed in Table 7.1 below are (ordered) observations on time intervals (days) between failures on each of two copying machines—20 observations on machine 1 and 30 on machine 2. These are assumed to be independent random samples from two Weibull distributions.

    (a) Construct an asymptotically efficient test of the hypothesis that the two shape parameters are identical, against a two-sided alternative. State your conclusions by calculating a large-sample p-value.

    (b) Describe other asymptotically efficient tests for the hypotheses in (a), without carrying out the details.

    (c) Construct an asymptotically efficient test of the hypothesis that the two scale parameters are identical, against a two-sided alternative, and assuming the two shape parameters are identical. State your conclusions by calculating a large-sample p-value.

    (d) Describe other asymptotically efficient tests for the hypotheses in (c), without carrying out the details.

    (e) Approximate the power of the test in (a) when the true difference is 1, or of the test in (c) when the true difference is 0.005. In either case, assume a significance level of 5%.

13. Show that under the null hypothesis and under local alternatives, Q_{NR}, Q_R and Q_W each differ from Q_n^0 (defined in (17)) by $o_p(1)$.

14. Show that the asymptotically efficient quadratic form tests (for multidimensional hypotheses) are *locally unbiased*. [See *Exercise 5*].

15. *Continuation of Exercise 4.* Derive the likelihood-ratio test for $\theta = 0$ versus $\neq 0$ when sampling, respectively, from a Cauchy distribution and from a double exponential distribution, each with location parameter θ.

16. *Continuation of Example 7.2 of Section 7.2.* Derive the likelihood-ratio test of exponentiality within a Weibull model.

17. *Continuation of Example 7.3 of Section 7.2.* Derive an asymptotically efficient test of the hypothesis that both location parameters are 0 ($\mu = \nu = 0$), assuming a common but unknown scale

parameter. Describe briefly what would be required to derive each of two aternative asymptotically efficient tests. Give a formula for the large-sample power.

18. *Estimates based on tests.* Suppose that θ is real and and that there is a nuisance parameter η, also real for convenience. Let $T_n(\theta_0)$ be a test statistic for testing $\theta = \theta_0$ versus $\theta > \theta_0$. Suppose that $T_n(\theta_0)$ is asymptotically standard normal, and jointly asymptotically normal with the scores, with asymptotic covariance $\omega = \omega(\theta_0, \eta)$ with the effective score for θ, and asymptotically uncorrelated with the score for η. And suppose that all of the above holds for each possible value of θ_0. Assume further the *regular test* property:
$$T_n(\theta_n) \;=\; T_n(\theta) - \omega(\theta,\eta)h_\theta + o_p(1)$$
uniformly for bounded h_θ, where $\theta_n = \theta + h_\theta/\sqrt{n}$
Now define an estimate $\tilde{\theta}_n$ as the solution (or 'near root') of $T_n(\theta) = 0$; that is
$$T_n(\tilde{\theta}_n) \;=\; o_p(1).$$
Prove that $\tilde{\theta}_n \sim AN(\theta, \sigma^2/n)$ for some σ (jointly with the scores) if and only if
$$\sqrt{n}(\tilde{\theta}_n - \theta) \;=\; \omega^{-1}T_n(\theta) + o_p(1).$$

(a) What is the asymptotic relative efficiency of such a $\tilde{\theta}_n$?

(b) Consider sampling from $f_{\theta,\eta}(x) = (1/\eta)f\{(x-\theta)/\eta\}$, for some density f which is symmetric at 0. Consider the *sign test*, defined in *Exercise 6*, but now with a nuisance parameter η. What is the corresponding estimate of θ? [The symmetry of f is needed to assure orthogonality with the score for η.]

19. *Continuation.* Now consider the random sampling case, with $(1/\sqrt{n})T_n(\theta) \sim AL(0, k_\theta(\cdot), 1)$. Show that the corresponding estimate $\tilde{\theta}$ say, defined in *Exercise 18*, satisfies $\tilde{\theta} \sim AL\{\theta, k_\theta(\cdot)/\omega, 1/\omega^2\}$, where now $\omega = \omega(\theta, \eta)$.

20. *Tests based on estimates: Wald-type tests.* Let $\tilde{\theta}_n$ be a regular estimate of θ (real) in a model with nuisance parameter η, and has large-sample variance σ^2/n for some $\sigma = \sigma(\theta, \eta)$. A test of hypotheses about θ based on the statistic $T_n = \sqrt{n}(\tilde{\theta}_n - \theta_0)/\tilde{\sigma}$, with $\tilde{\sigma}$ consistent for σ (under H_0), is called a *Wald-type test.*

(a) Describe a Wald-type test derived from an analog estimate of θ based on the sample median in a random sample (assuming the population median is one-to-one in θ).

(b) Derive a formula for the asymptotic relative efficiency of the test in (a).

(c) Exemplify (a) and (b) in a two-parameter model of your choice.

21. *Wald-type tests (continued).* Suppose that $\tilde{\theta}_n$ is a regular estimate of θ of dimension d_θ, with asymptotic variance matrix V. Let Q_n be a quadratic-form test statistic, for testing $\theta = \theta_0$ versus "not so" (with nuisance parameter η), based on $T_n = \sqrt{n}(\tilde{\theta}_n - \theta_0)$ — a *Wald-type test.*

    Determine the local power of such a test.

22. *Score-type tests; M-tests.* Consider a random sampling model with a single real parameter θ, and let $\psi(x, \theta)$ be a 'pseudo score': a function with expectation 0 for every θ. (See Section 5.6 of Chapter 5 for other assumptions and notation.) A test based on T_n proportional to the sample average of the ψ's, evaluated at a null value θ_0, may be called a *score-type-test* or an *M-test.*

    (a) Describe such a test.

    (b) Be specific for the case of sampling from $N(\theta, 1)$ when using the *Huber ψ-function* (see (23) in Chapter 5). What is the asymptotic relative efficiency of this test?

23. Consider random sampling from a Poisson distribution with mean λ. Derive and compare Wald and Wilson confidence intervals for λ. Be specific when $n = 25$, $\bar{x}^* = 2.00$ (as in *Exercise 3*), and the confidence coefficient is 90%. What is the exact (not large-sample) confidence coefficient for each of these large-sample confidence intervals?

24. *Continuation of Exercise 5.* Derive, numerically, Rao, Wald and likelihood-ratio 95% upper confidence bounds for the Cauchy scale parameter using the data provided in *Exercise 13* in Chapter 5.

25. *Continuation of Exercise 12.* Derive, numerically, an asymptotically efficient 90% confidence interval for the ratio of the two scale parameters under the assumption of a common, but unknown, shape parameter. Describe what would need to be done to obtain each of two other types of asymptotically efficient confidence intervals.

26. Consider the two-independent-Bernoulli-sequences sampling model in Appendix III of Chapter 5.

    (a) Construct a score test of the null hypothesis $\theta = 2.5$ against two-sided local alternatives, with a 5% significance level.

    (b) Construct a two-sided Wald test of this same null hypothesis. Be sure to base it on an asymptotically efficient estimate! Although, formally, this test is asymptotically equivalent to the one in (a), which would you prefer and why? What can be said about the behavior of a Wald-type test statistic (see *Exercise 20*) constructed from the maximum likelihood estimate?

(c) Construct a two-sided likelihood-ratio test of this same null hypothesis. Be sure to base it on an asymptotically efficient estimate! Although, formally, this test is asymptotically equivalent to the one in (a), which would you prefer and why? What can be said about the behavior of a likelihood-ratio type test statistic constructed from the maximum likelihood estimate?

8

An Introduction to Rank Tests and Estimates

We introduce ideas and methods for testing hypotheses using rank statistics and the associated methods of estimating characteristics of a distribution. These methods are *nonparametric*; in that, many properties do not require specific parametric models for the data. And they are insensitive to transformations of the measurement scale that leave the order (rank) of certain characteristics of the observations unchanged.

We confine attention largely to testing hypotheses about the center of a symmetric distribution and of estimating this center of symmetry—the so-called *one-sample symmetric location problem.* A primary application of such methods is to settings in which the data consist of differences between paired measurements, perhaps measures taken 'before' and 'after treatment'. Under a hypothesis of no treatment effect, differences might be expected to be distributed symmetrically around zero. If any treatment effect is additive, then the differences would remain symmetric but around a shifted center of symmetry. This problem provides a vehicle for illustrating applications of large-sample methods.

In the first three sections, we consider exact 'small-sample' properties of the *sign test* and the *Wilcoxon signed-rank test*, and of associated estimates. Much of this can be omitted, except for the definitions of the *sign test* and Wilcoxon *signed-rank* statistics, *Walsh averages*, the *Tukey representation*, and the *Hodges-Lehmann estimate.* In Sections 8.4 to 8.8, we develop large-sample properties of these tests and estimates, and sometimes relax the assumption of symmetry. In Section 8.9, we briefly describe other *linear rank* procedures for the symmetric location problem. Much of the general theory can be cast in the framework of U-statistics, that was discussed in Chapter 3, Section 3.4.

Rank tests for the two-sample problem find the widest usage, ranging from the Wilcoxon-Mann-Whitney *rank-sum test* to the *logrank test*, popular in survival analysis. Large-sample properties of many of these can be developed analogously to those developed here for the one-sample problem. (The censored-data version of the logrank test, however, is an exception, requiring different methods.) These and other popular rank statistics are very briefly surveyed in Section 8.10, with references to other sources.

DOI: 10.1201/9780429160080-8

8.1 The Symmetric Location Problem; the Sign Test

Consider a random sample $X_1, \ldots, X_n$ from a continuous distribution with median θ. We can conveniently write the distribution function as $F_\theta(x) = F(x - \theta)$ for an unspecified continuous distribution function F with median zero. In much of what follows, we also assume F to be symmetric about zero, specifically $\bar{F}(-x) = F(x)$ for all x (so that the tails match), in which case θ is the *center of symmetry* of F_θ, as well as the median and also the mean, if this exists. We also write $P_\theta(\cdot)$ for the probability measure associated with F_θ. If F is absolutely continuous, the density is $f_\theta(x) = f(x - \theta)$ for a density f which is symmetric about zero: $f(-x) = f(x)$ for all x. Absolute continuity of F would imply that, with probability one, all observations would be distinct. In practice, real data often include tied observations, due to recording error or rounding. Various notes below, and references, give practical and theoretical ways of dealing with ties.

For comparison with the rank tests to be introduced below, we first introduce a simple nonparametric test, the *sign test*, which does not require symmetry. Let $Z_j = \mathbf{1}[X_j > 0]$, the indicator of the event that $X_j > 0$. To test the hypothesis H_0 that $\theta = 0$ against the alternative $\theta > 0$, the *sign test statistic* is

$$Z^{(n)} = \sum_{j=1}^{n} \mathbf{1}[X_j > 0], \tag{8.1}$$

the number, or $\bar{Z}_n = Z^{(n)}/n$, the proportion, of positive X_j, rejecting H_0 for large values. We could equivalently sum the signs $2Z_j - 1$ of the X_j, hence the name, but the form Equation (8.1) is often more convenient. We use the alternative form in Section 8.9. The exact significance level and power of the test can be determined from the binomial distribution of $Z^{(n)}$, with $p = \frac{1}{2}$ under H_0 and $p = \text{pr}(X > 0) = \bar{F}_\theta(0)$ generally. A two-sided test is likewise available. And to test the hypothesis that $\theta = \theta_0$, first subtract θ_0 from all the X_j; equivalently, count how many X_j exceed θ_0. [Sample values equal to zero can simply be ignored, reducing n, or alternatively half the number of zeros can be added to $Z^{(n)}$ in Equation (8.1). We discount this possibility by assuming that the distribution is continuous.]

A corresponding estimate of the population median is the sample median. Its connection to the sign test will become clear in later sections (and in *Exercise 10*). In the case of symmetry, the sample median can easily be shown to be unbiased, so long as its expectation exists; in fact, it is symmetrically distributed around the population median (see *Exercises 1* and *2*).

These are the most elementary of nonparametric techniques: a test of fixed significance level is available without knowing the population distribution, hypothesizing only its median, and an estimate is available, unbiased in the case of symmetry. In fact, a family of sign tests for the median can be inverted to provide confidence bounds and intervals for the population median; see *Exercise 10*.

8.2 The Wilcoxon Signed-Rank Statistic

We now assume symmetry about θ. Then the sensitivity of a test statistic like Equation (8.1) may be improved by giving greater weight to the positive observations that are more distant from zero. Our quantification of 'distant' should preserve the nonparametric property; for example, it must be unaffected by scale changes done symmetrically away from zero. This suggests using *ranks* of the magnitudes $|X_j|$ of the X_j.

Define R_j^+ to be the rank of $|X_j|$ among $|X_1|, \ldots, |X_n|$, that is R_j^+ is the position of $|X_j|$ when the magnitudes are placed in increasing order; hence, $(R_1^+, \ldots, R_n^+)$ is a permutation of $(1, \ldots, n)$. Define the *Wilcoxon signed-rank statistic* (WILCOXON, 1945) by

$$W_n = \sum_{j=1}^{n} R_j^+ \mathbf{1}[X_j > 0], \tag{8.2}$$

the sum over the positive X_j of the associated ranks of magnitudes. Thus, for sample values -2.0, 1.2, 6.5, -0.3, 3.7, the corresponding values of R^+ are 3, 2, 5, 1, 4; and $W_n = 2 + 5 + 4 = 11$. See Table 8.1. We could (see Section 8.9) replace the indicator for positivity in Equation (8.2) by the sign of X_j, and then the statistic would, literally, be the sum of 'signed ranks' of magnitudes of the observations, analogous to the mathematical formula $\sum_j X_j = \sum_j |X_j| \text{sign}(X_j)$ but with each $|X_j|$ replaced by the corresponding R_j^+. But the form Equation (8.2) is equivalent, and mostly simpler to deal with mathematically. Note that the support of W_n consists of the integers zero to $N \equiv \frac{1}{2}n(n+1)$, the sum of the first n integers, in contrast to zero to n for the statistic in the sign test. If any values of X_j in the sample equal zero, they can be discarded and n reduced accordingly; but we ignore this possibility, as well as that of ties among the ordered magnitudes, by assuming that the distribution is continuous.

The *Wilcoxon signed-rank test* rejects the null hypothesis H_0, that $\theta = 0$, in favor of positive alternatives $\theta > 0$ when W_n in Equation (8.2) is large. The null distribution, and resulting p-values, will be considered below.

We now develop the first of two alternative representations of W_n. This representation will be especially useful for developing properties of the null distribution of W_n. We first define *antiranks*: If the rank of $|X_{D_i}|$ is i, then D_i is the antirank of i; that is D_i is the index j for which $R_j^+ = i$. Hence, $|X_{D_1}| < |X_{D_2}| < \ldots |X_{D_n}|$. Also, let $Z_j = \mathbf{1}[X_j > 0]$ and $Y_i = Z_{D_i}$, the positivity indicator for the observation whose magnitude has rank i. Then, W_n may be re-written

$$W_n = \sum_{j=1}^{n} R_j^+ Z_j = \sum_{i=1}^{n} i \cdot \mathbf{1}[X_{D_i} > 0] = \sum_{i=1}^{n} iY_i \tag{8.3}$$

simply by summing the terms in a different order. We call either of the last two expressions in Equation (8.3) the *antirank representation* of W_n. Table 8.1 illustrates these concepts using the example introduced above:

$$W_n = 2+5+4 \quad \text{(from Equation(8.2))} \quad = 2+4+5 \quad \text{(from Equation(8.3))}$$

Let $\mathbf{Z} = (Z_1, \ldots, Z_n)$, and let $\mathbf{R}^+ = (R_1^+, \ldots, R_n^+)$ be the rank vector. We find:

Lemma 8.1 *Under H_0, $\mathbf{Z}$ and $\mathbf{R}^+$ are stochastically independent.*

Proof The n pairs $(Z_1, |X_1|), \ldots, (Z_n, |X_n|)$ are independent and identically distributed because Z_j and $|X_j|$ are functions only of X_j. We show below that Z_1 and $|X_1|$ are independent, and hence all the $2n$ random variables just listed are independent. But $\mathbf{Z}$ is a function of n of these and $\mathbf{R}^+$ a function of the remaining n, implying the claimed independence.

Now $\text{pr}(Z_1 = 1, |X_1| \leq x) = \text{pr}(0 < X_1 \leq x)$ and $\text{pr}(Z_1 = 1) \cdot \text{pr}(|X_1| \leq x) = \frac{1}{2} \cdot \text{pr}(-x \leq X_1 \leq x) = \text{pr}(0 < X_1 \leq x)$, by symmetry and continuity of F. Therefore the events $\{Z_1 = 1\}$ and $\{|X_1| \leq x\}$ are independent for every x. Similarly, the events $\{Z_1 = 0\}$ and $\{|X_1| \leq x\}$ are independent for every x. Hence Z_1 and $|X_1|$ are independent random variables. □

The Z_j are clearly independent and identically distributed Bernoulli($\frac{1}{2}$) random variables under H_0. We now show that the same is true for the Y_i, where $Y_i \equiv Z_{D_i}$, the D_i's being the antiranks, although this may not be so intuitively apparent.

Proposition 8.1 *Let $Y_i = \mathbf{1}[X_{D_i} > 0]$, $i = 1, \ldots, n$, with $\mathbf{D}$ the antiranks of $\mathbf{X}$. Then, under the null hypothesis H_0 of symmetry about zero, $Y_1, \ldots, Y_n$ are independent and identically distributed Bernoulli($\frac{1}{2}$) random variables.*

Proof Let $D = (D_1, \ldots, D_n)$, and let d be a permutation of the integers 1 to n. Then

TABLE 8.1
An Example of the Signed-Rank Statistic

j	X_j	$\lvert X_j \rvert$	R_j^+	D_j	Z_j	Y_j
1	–2.0	2.0	3	4	0	0
2	1.2	1.2	2	2	1	1
3	6.5	6.5	5	1	1	0
4	–0.3	0.3	1	5	0	1
5	3.7	3.7	4	3	1	1

$$\text{pr}(Y_1 = y_1, \ldots, Y_n = y_n) = \sum_d \text{pr}(Z_{D_1} = y_1, \ldots, Z_{D_n} = y_n | D = d)\text{pr}(D = d)$$
$$= \sum_d \text{pr}(Z_{d_1} = y_1, \ldots, Z_{d_n} = y_n | D = d)\text{pr}(D = d).$$

Each conditional probability in the last sum is independent of the condition (by Lemma 8.1) and, by independence of the X_j, is the product of the corresponding probabilities; a typical term is $\text{pr}(Z_{d_j} = y_j) = \frac{1}{2}$. Hence, the right side becomes $(\frac{1}{2})^n \sum_d \text{pr}(D = d) = (\frac{1}{2})^n$.

To obtain the marginal distribution of Y_i for a specific i, we sum over the 2^{n-1} terms with $y_j = 0$ or 1 for $j \neq i$, obtaining $\text{pr}(Y_i = y_i) = \frac{1}{2}$, a Bernoulli($\frac{1}{2}$) distribution. Thus, the joint distribution of $\mathbf{Y}$ is the product of the marginals, so independence prevails. □

We now see the power of the antirank representation (8.3). In the first sum, each term is the product of two random (but independent) quantities; call them the 'coefficient' and the 'indicator'. The coefficients, the R_j^+, are dependent since they have distinct integer values. In the alternative expression, the coefficients are constants, and under H_0, the indicators are independent and identically distributed. The antirank form is therefore much easier to use for mathematical purposes, especially under the null hypothesis.

Specifically, we conclude from Proposition 8.1 that, under H_0, the Wilcoxon signed-rank statistic W_n may be represented as a weighted sum of n independent and identically distributed Bernoulli($\frac{1}{2}$) variables. Calculation of the resulting distribution does not depend on how these Bernoulli variables were defined. Thus, the distribution of W_n under H_0 is free of the distribution function F, allowing a truly nonparametric test.

No explicit formula for the density of the discrete statistic W_n is available, but numerical calculation is elementary for small n. *Table 8.2* and *Exercise 3* show the detail for $n = 5$. Recursion formulas (not given here) facilitate computation for larger n; moreover, tables have been constructed. Such tables appear in many introductory textbooks—e.g., Brown & Hollander (1977), and exact probabilities are available in R. As a consequence, exact significance levels for the associated tests may be determined: a critical value c_α is the largest integer for which $\text{pr}(W_n \geq c_\alpha) \leq \alpha$. To test against $\theta < 0$, reject when $W_n \leq N - c_\alpha$ (using the easily established symmetry of the null distribution of W_n). To test against $\theta \neq 0$, reject when $W_n \leq N - c_{\alpha/2}$ or $W_n \geq c_{\alpha/2}$.

Table 8.2 lists the 2^5 possible outcomes for the vector Y, all equally likely under the null hypothesis. The distribution of W is obtained by counting the number r giving each of its possible values, and the corresponding proportion $r/32$.

W_5	0	1	2	3	4	5	6	7	8	9	10	11	12	13	14	15
r	1	1	1	2	2	3	3	3	3	3	3	2	2	1	1	1

The resulting distribution has $E(W) = 7.5$ and $\text{var}(W) = 13.75$, see Equation (8.4) below.

TABLE 8.2
Computation of the Distribution of W_5

5	4	3	2	1	W	5	4	3	2	1	W
0	0	0	0	0	0	1	0	0	0	0	5
0	0	0	0	1	1	1	0	0	0	1	6
0	0	0	1	0	2	1	0	0	1	0	7
0	0	0	1	1	3	1	0	0	1	1	8
0	0	1	0	0	3	1	0	1	0	0	8
0	0	1	0	1	4	1	0	1	0	1	9
0	0	1	1	0	5	1	0	1	1	0	10
0	0	1	1	1	6	1	0	1	1	1	11
0	1	0	0	0	4	1	1	0	0	0	9
0	1	0	0	1	5	1	1	0	0	1	10
0	1	0	1	0	6	1	1	0	1	0	11
0	1	0	1	1	7	1	1	0	1	1	12
0	1	1	0	0	7	1	1	1	0	0	12
0	1	1	0	1	8	1	1	1	0	1	13
0	1	1	1	0	9	1	1	1	1	0	14
0	1	1	1	1	10	1	1	1	1	1	15

A p-value for the right-tailed test is $\mathrm{pr}(W_n \geq w_n^*)$, with w_n^* the observed value of W_n. A *mid-p-value*, defined as $\frac{1}{2}\mathrm{pr}(W_n = w_n^*) + \mathrm{pr}(W_n > w_n^*)$, is less conservative; see *Exercise 4*. A left-tailed p and mid-p are defined analogously; two-sided p's are obtained by doubling the smaller of the one-sided p's (justified by the symmetry under the null-hypothesis).

The mean and variance (and higher cumulants) of W_n are readily determined under H_0 from the antirank representation (8.3):

$$E(W_n) = \tfrac{1}{2}\textstyle\sum_{i=1}^n i = \tfrac{1}{4}n(n+1) = \tfrac{1}{2}N,$$

$$\mathrm{var}(W_n) = \sum_{i=1}^n i^2 \mathrm{var}(Y_i) = \tfrac{1}{4}\textstyle\sum_{i=1}^n i^2 = \tfrac{1}{24}n(n+1)(2n+1). \quad (8.4)$$

As already noted, the null distribution of W_n is symmetric around its mean, and so the coefficient of skewness coefficient is zero. The excess kurtosis of W_n turns out to be $-3.6/n + O(n^{-2})$ (see *Exercise 5*), and so the use of a normal approximation to determine significance levels may be expected to be quite good, even for moderate n. For comparison, the excess kurtosis of a standardized binomial$(n, \frac{1}{2})$ distribution is $-2/n$. As with a binomial distribution, the normal approximation is improved by a 'continuity correction', based on the fact that $\mathrm{pr}(W_n \geq w) = \mathrm{pr}(W_n \geq w - \frac{1}{2})$ for integral w, and further improved by an Edgeworth correction; see *Exercise 6*.

That W_n is asymptotically normal under H_0 may be proved by the *Lyapounov Central Limit Theorem* (see Chapter 2 and *Exercise 7*). We will derive the asymptotic normality, under H_0 and under local alternatives, by different methods in Section 8.6 (Theorems 8.2 and 8.3) below.

We now give another representation of W_n, one that is useful in the construction of estimates and confidence intervals for the center of symmetry θ, and in developing large-sample results.

Define the *Walsh averages* (Walsh, 1949) W_{ij} by

$$W_{ij} = \frac{X_i + X_j}{2} \; (i \leq j).$$

There are $N = \frac{1}{2}n(n+1)$ of these, including $N' \equiv \frac{1}{2}n(n-1)$ truly pairwise averages and n 'self-averages' X_i obtained by averaging each X_i with itself. The Wilcoxon statistic W_n can be expressed in terms of these averages, namely, as the total number of positive Walsh averages (Tukey, 1949). See Figure 8.1; note that exactly 11 of the 15 W_{ij} are positive. To this end:

Proposition 8.2 *(Tukey Representation): With* $Z_{ij} = \mathbf{1}[W_{ij} > 0]$,

$$W_n = \sum\sum_{1 \leq i \leq j \leq n} Z_{ij}. \tag{8.5}$$

Proof Let $x_{i_1} < \cdots < x_{i_p}$ be the positive x's. Then from (2), $W_n = \sum_{j=1}^{p} R^+_{i_j}$. Let I_1 be the interval $\{|x| \leq x_{i_1}\}$, I_2 the interval $\{|x| \leq x_{i_2}\}$, etc. Then

$$R_{i_1} = \#\{x_j \in I_1\}, \;\; R^+_{i_2} = \#\{x_j \in I_2\}, \quad \text{etc.}$$

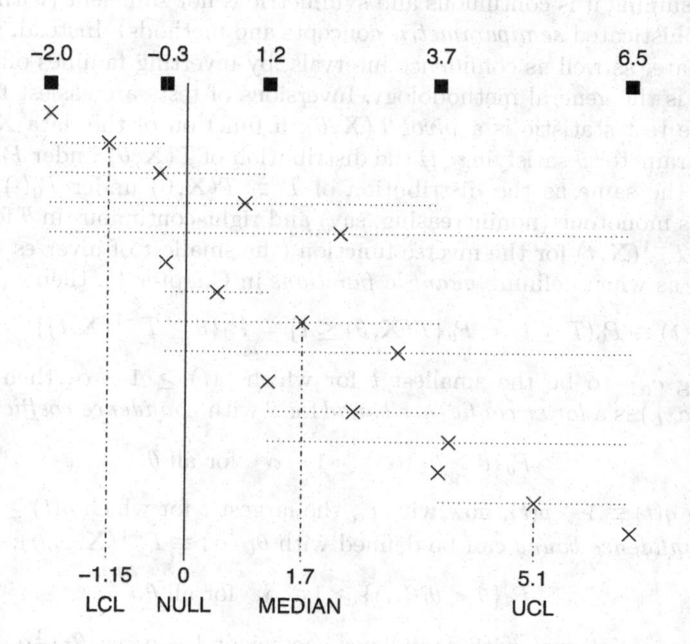

FIGURE 8.1
Walsh averages from Table 8.1.

It is also readily seen (see Figure 8.1) that

$$\begin{aligned} R^+_{i_1} &= \#\{\text{positive Walsh averages formed from the } x_j \text{ in } I_1\} \\ R^+_{i_2} &= \#\{\text{additional positive Walsh averages formed from the } x_j \text{ in } I_2\}, \end{aligned}$$

etc. Summing these expressions, we have that $\sum_{j=1}^{p} R^+_{i_j}$ equals the number of positive Walsh averages formed from the x's in I_p. But any x_j that is not in I_p contributes to no positive Walsh averages. □

Hence, the Wilcoxon signed-rank test can be carried out by counting the positive Walsh averages and comparing with a critical value from the null distribution of W_n (or its normal approximation).

8.3 Confidence Intervals for θ and the Hodges-Lehmann Estimate

To estimate the center of symmetry nonparametrically, we cannot use the estimation theory from Chapter 5, for example maximum likelihood, without assuming knowledge of the shape of the distribution from which we are sampling; assuming it is continuous and symmetric is not sufficient (without some more sophisticated *semiparametric* concepts and methods). Instead, we obtain an estimate, as well as confidence intervals, by inverting families of tests.

Here is the general methodology. Inversions of tests are easiest to express when the test statistic is a *pivot* $T(\mathbf{X}, \theta)$: a function of the data $\mathbf{X}$ and the (real) parameter θ satisfying *(i)* the distribution of $T(\mathbf{X}, \theta)$ under $P_\theta(\cdot)$ is free of θ, so the same as the distribution of $T \equiv T(\mathbf{X}, 0)$ under $P_0(\cdot)$, and *(ii)* $T(\mathbf{x}, \theta)$ is monotone (nonincreasing, say) and right-continuous in θ for each $\mathbf{x}$. Writing $T^{-1}(\mathbf{X}, t)$ for the inverse function (the smallest of inverses when T is discrete, as when defining *quantile functions* in Chapter 1), then

$$p(t) \equiv P_0(T \le t) = P_\theta\{T(\mathbf{X}, \theta) \le t\} = P_\theta\{\theta \ge T^{-1}(\mathbf{X}, t)\}. \tag{8.6}$$

Choosing $c_{\alpha L}$ to be the smallest t for which $p(t) \ge 1 - \alpha$, then $\hat{\theta}_L(\alpha) \equiv T^{-1}(\mathbf{X}, c_{\alpha L})$ is a *lower confidence bound* for θ with *confidence coefficient* $1-\alpha$:

$$P_\theta\{\theta \ge \hat{\theta}_L(\alpha)\} \ge 1 - \alpha \quad \text{for all } \theta.$$

Defining $q(t) \equiv 1 - p(t)$, now with c_α the largest t for which $q(t) \ge 1 - \alpha$, an *upper confidence bound* can be defined with $\hat{\theta}_U(\alpha) \equiv T^{-1}(\mathbf{X}, c_{\alpha U})$:

$$P_\theta\{\theta < \hat{\theta}_U(\alpha)\} \ge 1 - \alpha \quad \text{for all } \theta.$$

A *confidence interval*, with confidence coefficient $1 - \alpha$, is $\big[\hat{\theta}_L(\tfrac{1}{2}\alpha), \hat{\theta}_U(\tfrac{1}{2}\alpha)\big)$. The assumed right-continuity of a discrete T necessitates the lower limit be closed and the upper limit be open.

When T has a continuous distribution, the common value of $\hat{\theta}_L(\frac{1}{2})$ and $\hat{\theta}_U(\frac{1}{2})$ serves as an estimate, a so-called *median unbiased estimate*, being both a 50% lower and a 50% upper confidence bound. When T is discrete, a similar estimate is $\hat{\theta} = \frac{1}{2}\{\hat{\theta}_L(\frac{1}{2}) + \hat{\theta}_U(\frac{1}{2})\}$, which has these properties only approximately. An alternative, and essentially equivalent, perspective is to choose $\hat{\theta}$ as the value of θ_0 for which one-sided p-values are approximately $\frac{1}{2}$. The familiar example is the pivot $(\bar{X} - \mu)/S_X$ obtained when sampling from a normal distribution; discrete examples are less common.

To apply this general methodology here, we first note that we can test $\theta = \theta_0$, for any value of θ_0 in a location-shift model (considered here), by subtracting θ_0 from each X and then computing the resulting signed-rank statistic, say $W_n(\theta_0)$:

$$W_n(\theta_0) = \sum_{j=1}^{n} R_j^+(\theta_0)\mathbf{1}[X_j > \theta_0] \tag{8.7}$$

where $R_j^+(\theta_0)$ is the rank of $|X_j - \theta_0|$ among $|X_1 - \theta_0|, \ldots, |X_n - \theta_0|$. The same critical values obtain for each choice of θ_0; indeed, $W_n(\theta_0)$ is seen to be a pivot.

But the Tukey representation (8.5) of $W_n(\theta_0)$ leads to

$$W_n(\theta_0) = \sum\sum\nolimits_{i \le j} \mathbf{1}[W_{ij} > \theta_0]; \tag{8.8}$$

that is, $W_n(\theta_0)$ is the count of Walsh averages exceeding θ_0. To use this fact, we introduce the ordered values of the N Walsh averages $\{W_{ij}\}$, say $W_{(1)} < W_{(2)} < \cdots < W_{(N)}$. Then we find the correspondences

$$W_n(\theta_0) \le w \quad \text{if and only if} \quad \theta_0 \ge W_{(N-w)} \tag{8.9}$$

for integral w (see Figure 8.1). Moreover, since the distribution of W_n is symmetric around $\frac{1}{2}N$, on the integers 0 to N, $P_0(W_n > c) = P_0(W_n < N - c)$, required to be less than or equal to α or to $\frac{1}{2}\alpha$ for a confidence bound or interval.

In passing, we note that, apart from the inclusion of the self-averages with $i = j$ in the double summation, $W_n(\theta_0)$ is a U-statistic as discussed in Chapter 3. As will be seen shortly, the inclusion or exclusion of these n terms does not affect the asymptotics.

Applying the methodology above, and letting $r = c_{\alpha/2} - 1$, we obtain

Proposition 8.3 *(Confidence Interval for Center of Symmetry): The interval* $I \equiv [W_{(N-r+1)}, W_{(r)}]$ *is a confidence interval for the center of symmetry* θ, *with confidence coefficient* $1 - \alpha$, *where* $r = r_{\alpha/2}$ *is defined as the smallest integer for which* $P_0(W_n \ge r) \le \frac{1}{2}\alpha$.

Here is a direct verification: Since the W_{ij}'s are continuously distributed,

$$P_\theta(\theta \in I) = P_\theta(W_{(N-r+1)} \le \theta < W_{(r)}) = P_\theta(N - r < W_n(\theta) < r) \quad \text{by} \tag{8.10}$$

$$= P_0(N - r < W_n < r) = 1 - 2P_0(W_n \ge r) \ge 1 - \alpha.$$

The value of r can be determined from tables of the Wilcoxon signed-rank statistic, or by normal approximation. The latter is satisfactory for most practical purposes. Using a continuity correction, we have

$$P_0(W_n \geq r) \;=\; P_0(W_n > r - \tfrac{1}{2}) \;\approx\; \bar{\Phi}\{\tfrac{r-(N+1)/2}{\sigma_n}\}, \qquad (8.11)$$

where $\sigma_n^2 = \frac{1}{24}n(n+1)(2n+1)$ from Equation (8.4), so that $r \approx \frac{1}{2}(N+1) + \sigma_n \cdot z_{\alpha/2}$, with $N = \frac{1}{2}n(n+1)$ and z an upper percentage point in the standard normal distribution.

To obtain a point estimate of θ, first recall that the distribution of W is symmetric around $N/2$ on the integers 0 to N. Hence, if we choose the largest r for which $P_0(W_n \geq r) \leq \frac{1}{2}$, we find that $r = (N+1)/2$ when N is odd (with the probability exactly equaling $\frac{1}{2}$) and $r = N/2+1$ when N is even (with the probability strictly less than $\frac{1}{2}$ since there is positive probability at the center $\frac{1}{2}N$ when N is even). In either case, $r = \lfloor N/2 \rfloor + 1$.

The corresponding upper confidence bound is then $W_{(r)} = W_{(\lfloor N/2 \rfloor+1)}$ and the corresponding lower confidence bound is $W_{(N-r+1)} = W_{(\lfloor (N+1)/2 \rfloor)}$ (since $N - \lfloor N/2 \rfloor = \lfloor (N+1)/2 \rfloor$). When N is odd, both bounds are $W_{((N+1)/2)}$, the median of the N Walsh averages; when N is even, they are $W_{((N+1)/2)}$ and $W_{(N/2)}$ and their average is again the median (as usually defined) of the N Walsh averages.

This leads to (Hodges & Lehmann, 1963)

Definition 8.1 *The Hodges-Lehmann estimate $\hat{\theta}_{HL}$ of the center of symmetry θ of a continuous distribution is the median of the N Walsh averages.*

An alternative (approximate) derivation is as follows: Since $W_n(\theta)$ is symmetrically distributed around $\frac{1}{2}N$, one-sided p-values (or mid-p-values) will equal $\frac{1}{2}$ if θ satisfies $W_n(\theta) \approx \frac{1}{2}N$, and, by Equation (8.8), this implies that the solution θ is the median of the Walsh averages.

Exact properties of the estimate $\hat{\theta}_{HL}$ are not easily derived, except for its unbiasedness (a consequence of the following proposition). Other properties are best obtained through large-sample approximations; see Section 8.8.

Proposition 8.4 *When sampling from a continuous distribution which is symmetric about θ, the Hodges-Lehmann estimate is also symmetrically distributed around θ.*

Proof Note that $\hat{\theta}(\mathbf{x}+\theta) = \hat{\theta}(\mathbf{x}) + \theta$; hence $P_\theta(\hat{\theta} - \theta \leq u) = P_0(\hat{\theta} \leq u)$. We can therefore assume that $\theta = 0$ without loss of generality. Then $\mathbf{X} = (X_1, \ldots, X_n) =_d$ ('has the same distribution as') $(-X_1, \ldots, -X_n) = -\mathbf{X}$, and similarly then for the $\{W_{ij}\}$ and $\{-W_{ij}\}$. Therefore, $\hat{\theta}(\mathbf{X}) =_d \hat{\theta}(-\mathbf{X})$. But $\hat{\theta}(\mathbf{X}) = -\hat{\theta}(-\mathbf{X})$ (from definition of $\hat{\theta}$, for both N odd and N even) and so $\hat{\theta}(\mathbf{X}) =_d -\hat{\theta}(\mathbf{X})$. □

The assumption of symmetry about θ has enabled the conclusion that, for N even, the average of the two middle-ordered Walsh averages $\frac{1}{2}\{W_{(\frac{1}{2}N/2)} +$

$W_{(N/2+1)}\}$ is truly median-unbiased—that is, it is both a 50% lower and 50% upper confidence bound.

Corollary 8.1 *Under the same assumptions, $\hat{\theta}_{HL}$ is unbiased (if its expectation exists) and median-unbiased.*

The expectation typically *does* exist, except possibly for very small sample sizes n; see *Exercise 8*.

Confidence intervals and estimation for a population median can be derived by applying similar reasoning to the sign-test statistics. See *Exercise 10*.

8.4 Large-Sample Theory of the Wilcoxon Signed-Rank Procedures; Mean and Variance of W_n

As noted above, the representation (8.8) shows that W_n is essentially a U-statistic of the form discussed in Chapter 3, and many of the results in this and the next section can be derived from the general theory given there. However, we prefer to proceed directly.

We first derive the mean and variance of W_n quite generally, then develop an *asymptotic linear representation* of T_n, and from it derive its large-sample normality—under H_0 and under local alternatives. None of this requires the symmetric location structure—that is, we need only assume that we are sampling from a continuous distribution. When local alternatives are introduced, we also need a parametric structure satisfying Local Asymptotic Normality—in particular, *scores* and *information*. Along the way, we develop the important *projection method* of Hájek, often useful for developing *asymptotic linear representations* (as noted in Chapter 3). Throughout, the Walsh-averages form of W_n is most useful. We return to symmetry assumptions in Section 8.7.

We focus on large-sample properties of W_n as a test statistic and summarize in Section 8.8 the corresponding properties of the Hodges-Lehmann estimate that is associated with the signed-rank statistic.

The mean and variance of W_n are functions of the four quantities p_1, p_2, p_3, p_4 defined below; here, X, X_1, X_2, X_3 are independent and identically distributed random variables with continuous distribution F. No other assumptions about F, such as symmetry or its parametric form, are needed.

$$
\begin{aligned}
p_1 &\equiv \mathrm{pr}(X_1 > 0) = \bar{F}(0), \quad p_2 \equiv \mathrm{pr}(X_1 + X_2 > 0) = E\{\bar{F}(-X)\}, \\
p_3 &\equiv \mathrm{pr}(X_1 > 0,\, X_1 + X_2 > 0) = E\{\bar{F}(-X)\mathbf{1}[X > 0]\}, \\
p_4 &\equiv \mathrm{pr}(X_1 + X_2 > 0, X_1 + X_3 > 0) = E\{\bar{F}(-X)\}^2.
\end{aligned}
$$

Various alternative formulas above may be derived by conditioning arguments; e.g.,

$$\text{pr}(X_1 + X_2 > 0) = E\{\text{pr}(X_1 + X_2 > 0|X_1)\}$$
$$= E\{\text{pr}(X_2 > -X_1|X_1)\} = E\{\bar{F}(-X_1)\}.$$

Proposition 8.5 $E(W_n) = \frac{1}{2}n(n-1)p_2 + np_1$ and $\text{var}(W_n) = \frac{1}{2}n(n-1)p_2(1-p_2) + np_1(1-p_1) + 2n(n-1)(p_3 - p_1p_2) + n(n-1)(n-2)(p_4 - p_2^2)$.

Proof (partial) Write $W_n = \sum\sum_{i<j} Z_{ij} + \sum_k Z_{kk}$ where $Z_{ij} = \mathbf{1}(X_i + X_j > 0)$ and $Z_{kk} = \mathbf{1}[X_k > 0]$. Since $E(Z_{ij}) = p_2$ when $i \neq j$ and $E(Z_{kk}) = p_1$, the formula for $E(Z_n)$ follows.

For the variance formula, we will only derive the dominant (last) term; the others may be derived similarly (see Chapter 3). The variance of the sum is the sum of the variances plus the various covariance terms. There are N variance terms, making a contribution of order $O(n^2)$. There are also $O(n^2)$ covariances among the Z_{kk}'s (all zero). The covariances between the Z_{kk}'s and the Z_{ij}'s are zero whenever k is different from both i and j; there are $O(n^2)$ remaining terms of this form.

Finally, we consider in detail the covariances among the Z_{ij} terms: $\text{cov}(Z_{ij}, Z_{kl})$. Here $i < j$ and $k < l$ and $(i,j) \neq (k,l)$ (since $(i,j) = (k,l)$ yields a variance term, not a covariance). If all four indices differ, the resulting covariance is zero. Hence, we only need to consider the case of three differing indices. All possible variations yield the same covariance, namely

$$\text{cov}(Z_{12}, Z_{13}) = \text{pr}(Z_{12} = Z_{13} = 1) - \text{pr}(Z_{12} = 1) \cdot \text{pr}(Z_{13} = 1) = p_4 - p_2^2.$$

There are $n(n-1)(n-2)$ such terms (choosing 3 distinct indices from among $1, \ldots, n$). Hence, we have proved that $\text{var}(W_n) = n(n-1)(n-2)(p_4 - p_2^2) + O(n^2)$, and this will be sufficient for our purposes. □

It is apparent from the mean and variance formulas in Proposition 5 that the n self-averages $X_1, \ldots, X_n$ play a negligible role asymptotically in W_n; that is, the Z_{kk} terms in the sum (8.5). Rather than carry them along, we will now use a modified W_n, say W_n', which counts the positive Walsh averages only among those with $i < j$. And its average form

$$\bar{W}_n' \equiv \frac{1}{N'} W_n' = \frac{1}{N'} \sum\sum_{i<j} Z_{ij} \quad \text{with} \quad N' = \binom{n}{2} = \tfrac{1}{2}n(n-1), \qquad (8.12)$$

an exact U-statistic, will often be more convenient. Its large-sample behavior is identical to that of the average form of W_n, namely $\bar{W}_n \equiv \frac{1}{N} W_n$. In practice, the latter should probably be used, but it will be simpler to develop the large-sample theory for $\bar{W}_n'$.

We now have, as a consequence of Proposition 5,

Corollary 8.2

$$E(\bar{W}_n') = p_2 \quad \text{and} \quad \text{var}(\bar{W}_n') = \frac{1}{n}\tau^2 + O(\frac{1}{n^2}) \quad \text{with} \quad \tau^2 = 4(p_4 - p_2^2).$$

When X is symmetric about zero, the case of most interest, $\bar{F}(-X) = F(X)$ $p_2 = \frac{1}{2}$, $p_4 = \frac{1}{3}$ and $\tau^2 = \frac{1}{3}$.

8.5 Hájek's Projection Method

We now recapitulate *Hájek's Projection Theorem*, involving the approximation of function of n independent random variables $X_1, \cdots, X_n$ by a sum of terms, each depending on only a single X_i. The approximation is the projection of the function onto the space spanned by such sums. The quality of the approximation must be evaluated in each application. The method was described briefly in Chapter 3 in the context of the general theory of U-statistics.

To this end, we consider a function $V = V(X_1, \ldots, X_n)$ of n independent X_i's and with $E(V) = 0$ (for convenience). Let

$$p_i^*(x) \equiv E(V|X_i = x) \quad \text{and} \quad V^* \equiv \sum_{i=1}^{n} p_i^*(X_i).$$

Write (temporarily) $W \equiv \sum_i p_i(X_i)$ for arbitrary $p_i(\cdot)$.

Theorem 8.1 *(Hájek Projection): Among all* $W = \sum_i p_i(X_i)$, $\quad W = V^*$ *uniquely minimizes* $E(V - W)^2$. *Moreover,*

$$E\{p_j^*(X_j)\} = E(V^*) = 0; \;\; E(V - V^*)^2 = E(V^2) - E(V^{*2}); E(VV^*) = E(V^{*2}).$$

This defines an *orthogonal projection*, in that $V = V^* + R$ where $R \equiv V - V^*$ and, from the last statement in the Theorem, $E(V^*R) = 0$—implying that V^* and R are uncorrelated. Whether or not V^* is a satisfactory approximation to V—that is, whether or not R is small compared to V^* in some sense—needs to be determined in each application. (For comparison, recall the orthogonal decomposition of unbiased estimates given in Section 5.2 of Chapter 5.)

In many applications, including ours, the X_i are independent and identically distributed and V is invariant under permutation of its arguments—i.e., all the X_i are treated similarly. In this case, p_i^* does not vary with i, and it may be reasonable to require the same of candidate p_i's in W. We therefore drop the subscripts on p_i and p_i^* in the proof below, but the same proof is valid with subscripts restored.

Proof First note that $E\{p^*(X_j)\} = E\{E(V|X_j)\} = E(V) = 0$, and hence $E(V^*) = 0$ also.

We write $V - W$ as $(V - V^*) - (W - V^*)$, and first show that these latter two terms are orthogonal (uncorrelated):

$$E\{(W - V^*)(V - V^*)\} = \sum_{i=1}^{n} E\big[\{p(X_i) - p^*(X_i)\}(V - V^*)\big]$$

$$= \sum_{i=1}^{n} E[E\{\ldots\}|X_i\}] = \sum_{i=1}^{n} E\big[\{p(X_i) - p^*(X_i)\} \cdot E(V - V^*|X_i)\big].$$

But

$$E(V - V^*|X_i) = E\Big\{V - p^*(X_i) - \sum_{j\neq i} p^*(X_j)\Big|X_i\Big\} = 0 - \sum_{j\neq i} E\{p^*(X_j)|X_i\}$$
$$= -\sum_{j\neq i} E\{p^*(X_j)\} = 0,$$

so the orthogonality claim is verified. Hence

$$E(V-W)^2 = E\{(V-V^*)-(W-V^*)\}^2 = E(V-V^*)^2+E(W-V^*)^2. \quad (8.13)$$

Now Equation (8.13) is uniquely minimized by the choice $W = V^*$. Also, if we set $W = 0$ in Equation (8.13), we find that $E(V - V^*)^2 = E(V^2) - E(V^{*2})$. Finally, expanding the left side in this last expression yields $E(V^2) - 2E(VV^*) + E(V^{*2})$, from which it follows that $E(VV^*) = E(V^{*2})$. □

We now apply this projection to $V_n = W_n' - E(W_n') = \sum\sum_{i<j}(Z_{ij} - p_2)$. To this end, fix k and consider

$$E(Z_{ij} - p_2|X_k = x) = \begin{cases} \mathrm{pr}(x + X_j > 0) - p_2 & \text{if } i = k \\ \mathrm{pr}(X_i + x > 0) - p_2 & \text{if } j = k \\ 0 & \text{otherwise} \end{cases}$$

giving

$$E(Z_{ij} - p_2|X_k = x) = \begin{cases} \bar{F}(-x) - p_2 & \text{if } i \text{ or } j = k \\ 0 & \text{otherwise} \end{cases}$$

The function $p^*(x)$ is the sum of these over $i < j$. We must now count the number of terms in the double sum $\sum\sum_{i<j}$, that have either $i = k$ or $j = k$. Conditional expectations given X_k are zero except for these terms. If $i = k$, there are $n - k$ values for j; if $j = k$, there are $k - 1$ values for i. Therefore, there are $n - k + k - 1 = n - 1$ terms. Hence,

$$p^*(x) = E\Big\{\sum\sum\nolimits_{i<j}(Z_{ij} - p_2)\Big|X_k = x\Big\} = (n-1)\{\bar{F}(-x) - p_2\}$$

(which is k-free). Therefore, the projection of V_n is

$$V_n^* = \sum_{k=1}^{n} p^*(X_k) = \sum_{k=1}^{n}(n-1)\{\bar{F}(-X_k) - p_2\}.$$

Equivalently, the projection of $\bar{V}_n = 2V_n/\{n(n-1)\}$ is

$$\bar{V}_n^* = \frac{1}{n}\sum_{j=1}^{n} 2\{\bar{F}(-X_j) - p_2\}. \quad (8.14)$$

We next show that this projection provides the basis for an *asymptotic linear expansion* of the statistic $\bar{W}_n'$, as described in Chapter 3.

8.6 Asymptotic Linearity of $\bar{W}_n$

By the Projection Theorem,

$$E(\bar{V}_n - \bar{V}_n^*)^2 = E(\bar{V}_n^2) - E(\bar{V}_n^{*2}).$$

But $E(\bar{V}_n^2) = \text{var}(\bar{W}_n') = \tau^2/n + O(1/n^2)$ by the Corollary to Proposition 5. Note also that $E(\bar{V}_n^{*2}) = (1/n)\text{var}[2\{\bar{F}(-X) - cp_2\} = \tau^2/n$, where τ^2 and p_4 are defined in Section 8.5. From the projection (8.14), we therefore have $\bar{W}_n' = p_2 + \sum_{j=1}^n k_F(X_j) + (\bar{V}_n - \bar{V}_n^*)$, where $k_F(x) \equiv 2\{\bar{F}(-x) - p_2\}$.

Now $E(\bar{V}_n - \bar{V}_n^*)^2 = O(1/n^2)$, and hence $R_n \equiv \sqrt{n}(\bar{V}_n - \bar{V}_n^*)$ converges in mean-square to zero: $E(R_n^2) = O(1/n) = o(1)$. Therefore $R_n = o_p(1)$. Putting this together, and recalling that $p_2 = E\{\bar{F}(-X)\}$ and $p_4 = E\{\bar{F}(-X)\}^2$, yields

Theorem 8.2 *(Asymptotically Linear Expansion): $\bar{W}_n' \sim AL\{p_2, k_F(\cdot), \tau^2\}$; that is,*

$$\bar{W}_n' = p_2 + \frac{1}{n}\sum_{j=1}^{n} k_F(X_j) + o_p\Big(\frac{1}{\sqrt{n}}\Big) \tag{8.15}$$

where $k_F(X) \equiv 2\{\bar{F}(-X) - p_2\}$ with expectation zero and variance $\tau^2 = 4(p_4 - p_2^2)$.

Note The variance τ^2 is zero if and only if X is supported either on $(-\infty, 0]$ or on $[0, \infty)$, in which cases, with probability one, $\bar{F}(-X) = 1$ or $\bar{F}(-X) = 0$ respectively.

The central limit theorem now applies in Equation (8.15) to give the asymptotic distribution of $\bar{W}_n'$, when sampling from the distribution F. If F is parametric, say F_θ, and satisfies Local Asymptotic Normality, then we also have the asymptotic distribution under local alternatives (see Proposition 3 in Chapter 4):

Theorem 8.3 *(Asymptotic Distribution of W_n under Local Alternatives): Let $\tau^2(\theta) \equiv 4\{p_4(\theta) - p_2(\theta)^2\}$ and $\omega(\theta) \equiv \text{cov}\{k(X,\theta), s(X,\theta)\}$ where $k(x,\theta) \equiv 2\{\bar{F}_\theta(-X) - p_2(\theta)\}$ and $s(x,\theta)$ is the score (per observation). With $\theta_n \equiv \theta + h/\sqrt{n}$, we have that $\sqrt{n}\{\bar{W}_n' - p_2(\theta)\}$ converges in distribution under θ_n to $N\{\omega(\theta)h, \tau^2(\theta)\}$.*

Both theorems apply to the average $\bar{W}_n$, which includes the self-averages, as well as to $\bar{W}_n'$, and to other forms of these statistics after appropriate standardization. It should be re-emphasized that neither theorem requires an assumption of symmetry or of location structure.

These theorems provide a basis for constructing large-sample tests about $p_2(\theta)$, with or without an assumption of symmetry—specifically, testing $p_2 = \frac{1}{2}$, that the median of the distribution of $X_1 + X_2$ is zero. Without the assumption of symmetry, this is *not* the same as testing whether X has median zero; see *Exercise 11*. For such testing, a consistent estimate of τ^2 is needed. This can be done parametrically or nonparametrically, the latter by using empirical versions of p_2 and p_4. The first of these is just $\bar{Z}'_n$ while the latter is

$$\tilde{p}_4 = \frac{1}{n\binom{n-1}{2}} \sum_{i=1}^{n} \sum_{j<k, j\neq i, k\neq i} \mathbf{1}[X_i + X_j > 0, X_i + X_k > 0].$$

We pursue such testing only in the case of symmetry (next section), the case for which W_n was introduced. But Theorem 8.3 provides a basis for evaluating the power of a test about p_2 more generally, and evaluating the relative efficiency of such a test; see *Exercises 20* and *21*.

8.7 Asymptotic Efficiency Calculations

The large-sample version of the Wilcoxon signed-rank test rejects a null hypothesis of symmetry about zero when $\bar{W}_n - \frac{1}{2} \geq \sigma_n z_\alpha / N$ with $N = \frac{1}{2}n(n+1)$ and $\sigma_n^2 = \frac{1}{24}n(n+1)(2n+1)$. See Equation (8.11). Note that $\sigma_n/N \approx 1/\sqrt{(3n)}$. A two-sided version also rejects for small values of $\bar{W}_n - \frac{1}{2}$, but now using $\pm z_{\alpha/2}$. The local power (of either test) at a parametric alternative can be derived from Theorem 8.3. We pursue this only for symmetric alternatives, now symmetric about θ rather than about the null value zero. (For some other families of alternatives, see *Exercises 20* and *21*.)

Suppose that sampling is from a distribution with density $f_\theta(x) = f(x-\theta)$ with $f(x)$ a density symmetric about zero, with support $\mathcal{R}$ and corresponding distribution function F; hence, $f(-x) = f(x)$ and $\bar{F}(-x) = F(x)$. When $\theta = 0$, we therefore find $\bar{F}(-X) \equiv F(X)$ to be uniformly distributed on (0,1), and therefore $p_2 = E(U) = \frac{1}{2}$ and $p_4 = E(U^2) = \frac{1}{3}$, yielding $\tau^2 = 4(\frac{1}{3} - \frac{1}{4}) = \frac{1}{3}$, as noted above. Also, $\tau^2(\theta_n) = \frac{1}{3} + O(\theta_n)$ locally to $\theta_n = 0$.

The score function (per observation) at $\theta = 0$ is $s(x) = (\partial/\partial\theta)\log f(x - \theta)\big|_{\theta=0} = -f'(x)/f(x)$ with variance $B \equiv \int s^2 f dx$, the information for θ, assumed finite and positive. The kernel in the asymptotically linear expansion, at $\theta = 0$, is $k(x) = 2\{\bar{F}(-x) - p_2\} = 2F(x) - 1$. Its covariance with $s(X)$ is found by integrating by parts:

$$\begin{aligned} \omega = E\Big[\{2F(X) - 1\} \cdot \Big\{-\frac{f'(X)}{f(X)}\Big\}\Big] &= -\int \{2F(x) - 1\} f'(x) dx \\ = -\int \{2F(x) - 1\} df(x) &= 0 + \int f(x) \cdot 2f(x) dx = 2\int f(x)^2 dx. \end{aligned}$$

Theorem 8.3 now becomes (upon multiplying through by $\sqrt{3}$), taking $\theta = 0$:

Corollary 8.3 *When sampling from a distribution with density $f(x-\theta_n)$, for f symmetric at 0 and $\theta_n = h/\sqrt{n}$, $\sqrt{(3n)}(\bar{W}_n - \frac{1}{2})$ converges in distribution to $N(\sqrt{3}\omega h, 1)$ with $\omega \equiv 2\int f(x)^2 dx$.*

Hence, the local power of the one-sided test is $\Phi(\sqrt{12}\int f^2 dx \cdot h - z_\alpha)$ at $\theta_n = h/\sqrt{n}$, depending on f only through $\int f^2 dx$.

Now let us compare the performance of this rank test with an asymptotically efficient parametric test, assuming that the 'shape' f of the symmetric distribution is known. The first of these test statistics is asymptotically linear with kernel $k(x) = 2F(x) - 1$ and the other is asymptotically linear with kernel $s(x) = -f'(x)/f(x)$. From Chapter 7, the asymptotic relative efficiency is therefore

$$\rho_f^2 = \text{corr}^2\{k(X), s(X)\} = \frac{12(\int f^2)^2}{B}. \tag{8.16}$$

It may be verified that ρ_f is invariant under scale transformation: that is, if $g(x) = cf(cx)$, then $\rho_g = \rho_f$ (*Exercise 12*); hence, we can compute asymptotic relative efficiencies without concern for standard deviations or other measures of scale. Under what conditions is the Wilcoxon test efficient? From Equation (8.16), efficiency requires $s(x)$ to be proportional to $k(x) = 2F(x) - 1$ (and f symmetric about zero), which occurs for a *logistic distribution* with $F(x) = 1/(1 + e^{-x})$ (*Exercise 13*). This distribution is quite similar to the normal, but with density tails decreasing exponentially in $-|x|$ rather than in $-x^2$. Asymptotic relative efficiencies for some other common symmetric distributions are given in Table 8.3. In particular, for the normal distribution, the value for the Wilcoxon test is $3/\pi \approx 0.955$; so that, when sampling from a normal distribution, the Wilcoxon test and the t-test have the same large-sample power if the t-test is based on a sample size 4.5% smaller.

For comparison, the asymptotic relative efficiency of the *sign test* when sampling from a density symmetric about zero is

$$\text{corr}^2\{1[X > 0], s(X)\} = \frac{4f(0)^2}{B} \tag{8.17}$$

(*Exercise 14*), and the asymptotic relative efficiency of the t-test (equivalently, a test based on $\bar{x}$) is

$$\text{corr}^2\{X, s(X)\} = \frac{1}{\sigma_f^2 \cdot B} \tag{8.18}$$

where σ_f is the standard deviation of X (see Chapter 6 or *Exercise 15*). These efficiencies also appear in Table 8.3 below.

To obtain the efficiency of one test relative to another test that is not efficient, the asymptotic efficiencies of both relative to an efficient test may be divided; for example, to determine the asymptotic relative efficiency of the Wilcoxon test relative to the t test when sampling from the same specified distribution, we can divide the asymptotic efficiencies of the Wilcoxon test and the t-test relative to the asymptotically efficient test for that distribution. For example from Table 8.3, we find the resulting asymptotic relative efficiency

TABLE 8.3
Asymptotic Efficiencies of Some Common Tests

	Laplace	logistic	normal	Cauchy
$f(x)$	$\frac{1}{2}\exp(-\|x\|)$	$\frac{e^{-x}}{(1+e^{-x})^2}$	$\frac{1}{\sqrt{(2\pi)}}\exp(-\frac{1}{2}x^2)$	$\frac{1}{\pi(1+x^2)}$
$s(x)$	$\mathrm{sign}(x)$	$\frac{e^x-1}{e^x+1}$	x	$\frac{2x}{1+x^2}$
B	1	$\frac{1}{3}$	1	$\frac{1}{2}$
$\int f^2 dx$	$\frac{1}{4}$	$\frac{1}{6}$	$\frac{1}{2\sqrt{\pi}}$	$\frac{1}{2\pi}$
$f(0)$	$\frac{1}{2}$	$\frac{1}{4}$	$\frac{1}{\sqrt{(2\pi)}}$	$\frac{1}{2\pi}$
σ_f^2	2	$\frac{\pi^2}{3}$	1	∞
ARE's				
sign test	1	$\frac{3}{4}$	$\frac{2}{\pi} \approx 0.637$	$\frac{2}{\pi^2} \approx 0.203$
Wilcoxon	$\frac{3}{4}$	1	$\frac{3}{\pi} \approx 0.955$	$\frac{6}{\pi^2} \approx 0.608$
t-test	$\frac{1}{2}$	$\frac{9}{\pi^2} \approx 0.912$	1	0

of the Wilcoxon test relative to the t test when sampling from a Laplace distribution to be $\frac{3}{4}/\frac{1}{2} = 1.5$, from the logistic distribution to be $\frac{1}{9}\pi^2 = 1.097$, from the normal distribution to be 0.955, and from Cauchy distribution to be infinite. It can be proved *Exercise 17*) that this ratio is greater than or equal to 0.864 for sampling from any symmetric distribution. Hence, a Wilcoxon signed-rank test may well be preferred to a t-test for the general purpose of sampling from a distribution known only to be symmetric: the relative efficiency is at least 86% and may well be very large, much larger than 100%. (Regrettably, the opposite advice is sometimes given in applied contexts: Use a signed-rank test instead of t when there is evidence of asymmetry! With at least moderate n, the t-test can tolerate asymmetry under the null hypothesis so long as it is interpreted as a test about the population mean, whereas the Wilcoxon test without symmetry under the hull hypothesis tests neither the median nor the mean, but whether the population median of pairwise averages is zero, or equivalently $p_2 = \frac{1}{2}$.) Of course, these are only large-sample comparisons.

8.8 Large-Sample Properties of the Hodges-Lehmann Estimate

By exploiting the relationship between tests and their associated estimators, we see that the Hodges-Lehmann estimate of the center of symmetry has the asymptotically linear representation

$$\hat{\theta}_{HL} = \theta + \frac{1}{n}\sum_{j=1}^{n}\frac{F(X_j - \theta) - \frac{1}{2}}{\gamma} + o_p\left(\frac{1}{\sqrt{n}}\right) \quad \text{with } \gamma \equiv \int f^2(x)dx \quad (8.19)$$

when sampling from $f_\theta(x) = f(x - \theta)$ with support $\mathcal{R}$ and f symmetric about zero. See remarks in Section 8.9; and note the resemblance to the Bahadur representation of the sample median. Recalling that $F(X - \theta)$ is uniformly distributed over $(0, 1)$ we see that $\hat{\theta}_{HL} \sim AN\{\theta, 1/(12n\gamma^2)\}$, and its asymptotic relative efficiency is the same as that of the Wilcoxon signed-rank test, for example 95.5%, when sampling from a normal distribution. Moreover, it is robust (since its *influence function*—the kernel in the asymptotically linear expansion (8.19)—is bounded.

Evaluation of its large-sample standard error, namely $1/\{\sqrt{(12n)}\gamma\}$, requires consistent estimation of $\gamma \equiv \int f^2$. We describe such an estimate $\tilde{\gamma}_n$ below. Large-sample confidence intervals can also be constructed, by inverting a family of tests or by using the large-sample normality of $\hat{\theta}_{HL}$: an interval for the center of symmetry θ is $\hat{\theta}_{HL} \mp z_{\alpha/2}/\{\sqrt{(12n)}\tilde{\gamma}_n\}$, with large-sample confidence $1 - \alpha$. An alternative confidence interval, with endpoints based directly on Walsh averages, was given in Proposition 3.

To estimate γ, two possibilities are (i) to write $\gamma = \int f(z)dF(z)$ and substitute an estimate $\tilde{f}_n$ of f and an empirical version of F and (ii) to use $\int \tilde{f}_n(z)^2 dz$. We pursue (i).

Under the null hypothesis that $\theta = 0$, we can use any consistent density estimate for $\tilde{f}_n$ (see below) and the sample distribution function F_n for F. However, since it is assumed that f is symmetric about zero, we might better use a symmetrized version of both $\tilde{f}_n$ and F_n, namely

$$\tilde{f}_n^s(z) \equiv \tfrac{1}{2}\{\tilde{f}_n(z) + \tilde{f}_n(-z)\} \quad (8.20)$$

and

$$F_n^s(z) \equiv \tfrac{1}{2} + \tfrac{1}{2}\frac{\#\{j \mid |x_j| \le z\}}{n} \quad \text{for } z > 0 \text{ and } \equiv 1 - F_n^s(-z) \text{ otherwise.} \quad (8.21)$$

The latter can be interpreted as the sample distribution function for the sample of $2n$ observations consisting of $\pm x_1, \pm x_2, \ldots, \pm x_n$. Both Equations (8.20) and (8.21) are readily seen to be symmetric about zero. (Their dependence on the sample values x_j is not made explicit in the notation.)

However, for the calculation of a confidence interval, an estimate is needed that does not assume that the center of symmetry is at zero. We need to estimate f and F when we are sampling from $F(\cdot - \theta)$. For this, we need only replace the sample values x_j in Equations (8.20) and (8.21) by $x_j - \hat{\theta}_{HL}$.

As a density estimate $\tilde{f}_n$, one popular choice is a *kernel estimate*. For it, let $k(z)$ be a density which is symmetric around 0 and has small variance; we do this by choosing some density $l(z)$ that is symmetric about zero and letting $k(z) = (1/\epsilon)l(z/\epsilon)$. Here ϵ, called the *band width*, controls the smoothness of the resulting density estimate. Popular choices of $l(z)$ are $\frac{1}{2}\mathbf{1}[|z| < 1]$ (*uniform*), $\phi(z)$ (*normal*), $\frac{3}{4}(1-z^2)\mathbf{1}[|z| < 1]$ (*parabolic* or *Epanechnikov*), and $\frac{15}{14}(1-z^2)^2\mathbf{1}[|z| < 1]$ (*biweight*). The resulting *kernel estimate* of f is then

$$\tilde{f}_n(x) \equiv \frac{1}{n}\sum_{j=1}^{n} k(x - x_j).$$

Using a symmetrized density estimate (8.20) and symmetrized empirical distribution (8.21), both centered at $\hat{\theta}_{HL}$, the resulting estimate of $\gamma \equiv \int f^2$ is found to be

$$\begin{aligned}\tilde{\gamma}_n &= \frac{1}{2n^2}\left[n\,k(0) + \sum_{j=1}^{n} k\{2(x_j - \hat{\theta}_{HL})\}\right] \\ &+ \frac{1}{n^2}\sum\sum\nolimits_{i<j}\{k(x_j - x_i) + k(x_j + x_i - 2\hat{\theta}_{HL})\}. \qquad (8.22)\end{aligned}$$

This provides a consistent estimate of γ quite generally provided $\epsilon = \epsilon_n \to 0$ but $\epsilon_n n/\log n \to \infty$. Still, it must be remembered that the derivations in this section assume that the density of interest is symmetric about θ. (For more on density estimation, see, for example, SILVERMAN (1986).)

8.9 Other Linear Rank Statistics for the Symmetric Location Problem

In this section, we summarize general results on linear rank statistics for inference about a center of symmetry; they parallel those for the Wilcoxon signed-rank test and the associated Hodges-Lehmann estimate, and the sign test and median. To simplify the exposition we make some changes in notation, importantly we will we replace the indicator $Z_j = \mathbf{1}[X_j > 0]$ with $\tilde{Z}_j = \text{sign}(X_j) = 2Z_j - 1$ so that $X = |X|\tilde{Z}$ and $Y_i = Z_{D_i}$ by $\tilde{Y}_i = \tilde{Z}_{D_i}$ Note that $E(\tilde{Y}) = 0$ and $\text{var}(\tilde{Y}) = 1$. We will consider a general *linear rank statistic*

$$\bar{T}_n = \frac{1}{n}\sum_{j=1}^{n} a(R_j^+)\tilde{Z}_j = \frac{1}{n}\sum_{i=1}^{n} a(i)\tilde{Y}_i. \qquad (8.23)$$

Here $a(1) \le a(2) \le \cdots \le a(n)$ are specified positive numbers and the D_i are the antiranks of the $|X_i|$ (see Section 8.2) We use $\bar{T}_n$ rather than $\bar{W}_n$

because of the change from Y to $\tilde{Y}$ and our division by n^2 rather than $N = n(n-1)/2$ to form the average as will be seen below. Setting $a(i)$ to be constant gives the sign statistic; setting $a(i) = i$, gives the Wilcoxon statistic. Other choices are possible. A more general method is to introduce a non-negative, non-decreasing function ψ on $(0,1)$ with $\int \psi(u)^2 du = \tau^2 < \infty$. (say).

Then, define *scores* $a_i \equiv a(i) \equiv \psi\{i/(n+1)\}$ and set

$$\bar{T}_n = \frac{1}{n}\sum_{i=1}^{n} \psi(\frac{i}{n+1})\tilde{Y}_i$$

This can also be written

$$\bar{T}_n = \frac{1}{n}\sum_{j=1}^{n} \psi\Big(\frac{R_j}{n+1}\Big)\tilde{Z}_j.$$

Under the null hypothesis of symmetry, the $\tilde{Z}_j$ are independent of the vector $\{R_j^+\}$ and with $E(\tilde{Z}_j) = 0$. Hence

$$E(\bar{T}_n) = \frac{1}{n}E\{\sum_{j=1}^{n} \psi(\frac{R_j^+}{n+1})\}E(\tilde{Z}_j) = 0.$$

We also have

$$\mathrm{var}(\bar{T}_n) = E[\mathrm{var}(\bar{T}_n|\{R_j^+\}] + \mathrm{var}[E\{\bar{T}_n|\{R_j^+\}]$$

The first term equals $n^{-2}\sum_{i=1}^{n}\psi\{i/(n+1)\}^2$ and the second term is zero.

Now $R_j^+/(n+1)$ is essentially the sample distribution function $F_n^+(x)$ of the magnitudes $|X_j|$ evaluated at $x = X_j$. Under the hypothesis of symmetry the corresponding population distribution (for $x > 0$) is $F^+(x) = 2F(x) - 1$. Also, $n^{-1}\sum_{i=1}^{n}\psi\{i/(n+1)\}^2 \to \int \psi(u)^2 du = \tau^2$ by the definition of Riemann integration. This leads to an asymptotically linear representation of $\bar{T}_n$. For $F_n^+ \to F^+$ weakly and this convergence is uniform, since F^+ is continuous (see Chapter 2). Therefore, under the hypothesis of symmetry

$$\bar{T}_n = \frac{1}{n}\sum_{j=1}^{n} \psi\{2F(|X_j|) - 1\}\tilde{Z}_j + o_p(1/n)$$

It is easily seen that the kernel $k(x) = \psi\{2F(|x|) - 1\}\tilde{Z}$ has mean zero and variance

$$\int_{-\infty}^{\infty} \psi\{2F(|x|) - 1\}^2 f(x)dx = 2\int_{0}^{\infty} \psi\{2F(|x|) - 1\}^2 f(x)dx = \tau^2,$$

the second equality following by the substitution $u = 2F(x) - 1$, with $du = 2f(x)dx$, which takes the range $0 < x < \infty$ to $0 < u < 1$.

Were f known, up to its point of symmetry θ, then an optimal parametric test of $\theta = 0$ would be based on the score statistic for θ, which for a single observation x is $s(\theta) = -f'(x)/f(x)$, the minus sign arising because the prime here denotes differentiation in x rather than in θ. We can now exhibit a rank test, that is a function $\psi_f(\cdot)$, which achieves full asymptotic efficiency under a distribution with specified symmetric density f but still has reasonable properties under different (though still symmetric) densities. Writing F_+^{-1} for the inverse function of F^+, consider the function

$$\psi_f(u) = s \circ F_+^{-1},$$

so that

$$\psi_f(u) = \frac{f'\{F_+^{-1}(u)\}}{f'\{F_+^{-1}(u)\}}.$$

Then, substituting into $k(x)$ gives, simply, $k(x) = -f'(x)/f(x) = s(x)$ so that the rank test based on ψ_f is asymptotically fully efficient when f is the true density.

For the Laplace distribution, with $f(x) = \frac{1}{2}\exp(-|x|)$ and, for $x > 0$, $F^+(x) = 1 - \exp(-x)$, we find that $\psi_f(u) = \text{sign}(u)$ leading to the simple binomial sign test. For the logistic, with $F(x) = \{1+\exp(-x)\}^{-1}$, after noting that $s(x) = F^+(x)$, see *Exercise 13*, we obtain $\psi_f(u) = u$ giving the signed rank test.

In comparing this last result with the derivation in Section 8.6, we note two differences between the derivations of $\bar{W}_n$ and $\bar{T}_n$, namely $\bar{W}_n$ is calculated using the positivity indicator Y and $\bar{W}_n$ is calculated by dividing W_n by the total number N of pairs whereas T_n using the sign function $\tilde{Y}$ and the division is by the original sample size n. Thus, asymptotically, under the null hypothesis, $\bar{W}_n \sim \frac{1}{2} + \bar{T}_n$, justifying our use of τ^2 for the asymptotic variance of each statistic.

For a general symmetric f, the substitution $u = F^+(x) = 2F(x)-1$, giving $du = 2f(x)dx$, which takes the range $0 < u < 1$ to $0 < x < \infty$, shows that the information about θ may be written as

$$\beta_f^2 = E\{s(X)^2\} = \int_{-\infty}^{\infty} s(x)^2 f(x)dx = 2\int_0^{\infty} s(x)^2 f(x)dx \int_0^1 \psi_f(u)^2 du.$$

Also, writing $\omega = \text{cov}\{k(X), s(X)\}$ we have

$$\omega = \int_0^1 \psi(u)\psi_f(u)du = 2f(0)\psi(0+) + 2\int_{\infty}^{\infty} \psi'\{F^+(x)\}f(x)^2 dx. \quad (8.24)$$

For example, when the signed rank test (with $\psi(u) = u$) is applied to data from a general symmetric f we find that $\omega = 2\int f(x)^2 dx$, as noted in Section 7.

In general, the correlation coefficient, ρ_f say, between $k(X)$ and $s(X)$ is

$$\rho_f = \frac{\omega}{\tau\beta_f} = \frac{\int \psi\psi_f}{\sqrt{(\int \psi^2 \int \psi_f^2)}}.$$

As in Section 8.7, see equation (8.16) w then find that the *linear rank test* which rejects the hypothesis that $\theta = 0$ for large values of $\bar{T}_n$ has asymptotic relative efficiency ρ_f^2, compared with an efficient parametric test when f is assumed.

An asymptotically efficient linear rank test when sampling from a normal distribution is the *signed normal scores test* defined by

$$a(i) = \Phi_+^{-1}\left(\frac{i}{n+1}\right) \quad \text{where} \quad \Phi^+(x) = 2\Phi(x) - 1,$$

It has the nonparametric features of a rank test, without losing any asymptotic efficiency if the data really are normally distributed (*Exercise 23*).

Estimates analogous to the Hodges-Lehmann estimates can be defined by inverting the general linear rank statistic (8.23) in the previous section. A large-sample definition is to take $\hat{\theta}$ as the solution to $\bar{T}_n(\theta) = 0$ (or with $\bar{T}_n(\theta)$ defined from Equation (8.23) as in Equation (8.7). (See the remark after the definition in Section 3.) Proceeding as in *Exercises 18* and *19* in Chapter 7 (or, alternatively, using a Gâteaux derivative argument as in HETTMANSPERGER (1984)), we find that the corresponding estimate $\hat{\theta}_{HL}$ has $\hat{\theta}_{HL} \sim AL\{\theta, (1/\omega)k(\cdot - \theta), (\tau/\omega)^2\}$. Its asymptotic relative efficiency is again ρ_f^2 when the true density is f. These results are consistent with those in Section 8.8, and with corresponding results for the sample median when f is the Laplace density. For the signed normal scores case, the associated estimate is asymptotically efficient when sampling from a normal distribution, compared with $\bar{X}_n$. However, it shares the same poor robustness features of $\bar{X}_n$ since the kernel in the asymptotic linear representation of $\hat{\theta}$ is proportional to $k(x - \theta) = x - \theta$, which is unbounded.

Some bounded *score-generating-functions*, with associated robustness properties for the corresponding tests and estimates, are described in HETTMANSPERGER (1984). See HÁJEK & ŠIDÁK (1967), LEHMANN (1975) and HETTMANSPERGER (1984), and references therein, for more information about rank-based procedures for the center-of-symmetry problem.

A limitation of all of these procedures, except for the sign test and sample median, is the requirement of symmetry. They are tests about, and estimates of, a center of symmetry, rather than a population median; regrettably, this is not always made clear in the associated literature.

A signed-rank test aimed at certain types of asymmetric alternatives is developed in HALL AND WELLNER (2013); also see *Exercises 20* and *21*.

8.10 Rank-Based Procedures for Some Other Inference Problems

So far, we have confined attention to inference about the center of a symmetric distribution, based on a random sample therefrom. Here we briefly list rank-based inference procedures for other problems common in applications.

Rank-based procedures for the *two-sample shift problem* are quite analogous to those studied above. Suppose we have available two independent random samples $(X_1, \ldots, X_m)$ and $(Y_1, \ldots, Y_n)$ with respective distribution functions F and G. The hypothesis of interest is whether G and F are identical. A popular alternative hypothesis is that $G(x) = F(x - \theta)$ (possibly after a log transformation if the original values are positive), with $\theta > 0$, say. Let $R_1, \ldots, R_n$ be the ranks of the n values of Y in a combined ranking of all $m + n$ values of X and y. A *linear rank statistic* is

$$U_{mn} = \sum_{j=1}^{n} a(R_j)$$

for specified numbers $0 \leq a(1) \leq \cdots \leq a(m+n)$.

The *Wilcoxon two-sample statistic* has $a(i) = i$. Thus, it is the sum of the ranks associated with the Y-sample from a combined ranking. It may equivalently be represented as the difference between mean ranks, subtracting the average of the ranks associated with the X's from the average of the ranks associated with the Y's. It has an alternative representation, very much as for the one-sample statistic, as

$$U_{mn} = \sum_{i=1}^{m} \sum_{j=1}^{n} \mathbf{1}[Y_j > X_i].$$

In this *Tukey representation* form, it is called the *Mann–Whitney statistic*, and is likewise intuitive: considering all (X_i, Y_j) pairs, how often does Y_j exceed X_i? (Mann & Whitney 1947). The corresponding *Hodges-Lehmann estimate of shift* θ is the median of the mn pairwise differences $Y_j - X_i$, which play the role played by Walsh averages in the one-sample problem, and again provides easy interpretation, though this is limited to a shift model. The Wilcoxon-Mann-Whitney statistic has properties quite analogous to those of the Wilcoxon signed-rank statistic. Note that no symmetry assumption is required here.

The Projection Theorem enables the development of an *asymptotic bilinear representation* (linear in an average of functions of the X_i's and an average of functions of the Y_j's, as both m and n tend to infinity), with resulting asymptotic normality—without the necessity of a location-shift structure. For the location-shift problem, the asymptotic relative efficiencies of the Wilcoxon-Mann-Whitney tests, and of the corresponding Hodges-Lehmann estimate of shift, are identical to those for the one-sample signed-rank statistic under symmetry.

Other two-sample linear rank statistics are *Mood's median statistic*, with $a(i) = \mathbf{1}[i > \frac{1}{2}(m+n+1)]$—an analog of the one-sample sign statistic—and the *normal scores statistic*, with $a(i) = \Phi^{-1}\{i/(m+n+1)\}$. The latter leads to asymptotically efficient tests and estimates when both samples come from normal distributions with a common variance, but it is not robust; for robust variations, see Hettmansperger (1984).

So far, we have only considered the case of G being a shift of F. Actually, since these tests are invariant under increasing transformation of the measurement scale, we only require that, for some such transformation, applied to both distributions, G is a shift of F.

Some other possibilities have been considered, namely that of G being a monotone function of F, thereby defining a so-called *Lehmann alternative*. In such cases, other two-sample rank tests may be more appropriate. Specifically, if $G = F^\rho$, or if $\bar{G} = \bar{F}^\rho$, a two-sample *logrank test* is efficient; see SAVAGE (1956) and HALL & WELLNER (2013). A variation on this problem, allowing for censored data in the two samples, leads to the so-called *logrank statistic* popular in survival analysis, although neither 'ranks' nor 'logs' seem apparent in its conventional definition; see books on survival analysis. But the test may also be quite appropriate for uncensored data and outside a survival analysis context.

The *Kruskal-Wallis statistic* is an extension of the Wilcoxon two-sample statistic, appropriate for testing whether several independent samples come from the same population—a so-called *one-way analysis-of-variance* problem. The *Friedman rank statistic* is appropriate for a *two-way analysis-of-variance* problem. More recently, *regression rank statistics* have been introduced and studied,

There are also rank-based procedures for evaluating *randomness versus trend* in single samples, *difference in scale* between two samples, and *independence* in bivariate observations, among others. For measuring dependence, various correlation coefficients have been developed, each a product moment correlation coefficient between sets of (U_i, V_i)'s with $U_i = a(R_i)$ and $V_i = a(S_i)$ and R_i and S_i the ranks of X_i and Y_i among the X's and Y's, respectively. Taking $a(i) = i$ yields *Spearman's rho* and $a(i) = \Phi^{-1}\{i/(n+1)\}$ the *normal-scores correlation.*

Another direction of nonparametric theory is the development of tests which are *most powerful* among rank-based procedures—or among procedures invariant under certain types of data-transformation. Instead of shift models, the focus here is on various types of *Lehmann alternative models*; see FRASER (1957), LEHMANN (1999), and LEHMANN & ROMANO (2005) and see HETTMANSPERGER (1984) for *locally most powerful* versions.

For other theoretical developments, see HÁJEK & ŠIDÁK (1967), LEHMANN (1975) and HETTMANSPERGER (1984). Lehmann's book is also a good guide to practical considerations, including the treatment of ties. Many of these procedures including rank regression are now implemented in R, and other software packages.

8.11 Final Notes on Methodology

The development of *Pitman efficiencies* (asymptotic relative efficiencies for rank-based procedures) presented here is quite different from that in the books of LEHMANN (1994) and HETTMANSPERGER (1984). We already developed methodology based on local asymptotic normality—including *asymptotic linear representations*, use of *LeCam's third lemma*, and asymptotic relative efficiency calculations derived therefrom—and hence we applied this general methodology rather than follow the specialized methods in these other texts. Much of this general methodology—as described here and in HALL & MATHIASON (1990)—was adapted from the book of HÁJEK & ŠIDÁK (1967), presented there in the context of rank-based procedures.

The Projection Theorem, discussed here in the context of the signed-rank statistic, is useful in the general theory of U-statistics. For more discussion, see SERFLING (1980).

Exercises

1. Prove that the sample median is symmetrically distributed around the population median when sampling from a symmetric distribution.

2. Prove that the sample median has a finite expectation whenever $xF(x)\bar{F}(x) = O(1)$ as $x \to \pm\infty$. In particular, show that the sample median from a Cauchy distribution has a finite expectation whenever the sample size $n \geq 3$. How about the cases $n = 1$ and $n = 2$?

3. Verify the entries in Table 2, the resulting null distribution of T_5, and its mean and variance.

4. *Mid-p-values.* Consider an integer-valued test statistic T, with large-values considered to be evidence against a null hypothesis. Let p_i represent the probability that $T = i$ under the null hypothesis. The *mid-p-value*, when the observed value of T is t^*, is defined as

$$P_{t^*} = \tfrac{1}{2}\,p_{t^*} + \textstyle\sum_{i>t^*} p_i$$

(LANCASTER, 1949). This quantity is more nearly uniformly distributed than the conventional p-value, which has the factor of $\frac{1}{2}$ omitted, is less conservative, and more symmetric in its treatment of what is 'more extreme' and 'less extreme'. In particular, the standard p-value is *stochastically larger* than a uniform random variable.

Since p-values are often interpreted as if they were uniformly distributed under a null hypothesis, a mid-p-value might better serve this function.

Show that P_T has mean 12 and variance $\frac{1}{12}(1-s)$ where $s = \sum p_i^3$. Compare this with the mean and variance of $U(0,1)$. [*Hint:* Let $P = P_T + p_T \cdot (U - \frac{1}{2})$ where $U \perp T$ is $U(0,1)$, and show that P is $U(0,1)$ by first finding the conditional distribution of P given $T = t$. See BARNARD (1990).]

5. Show that the *excess kurtosis coefficient* of the Wilcoxon signed-rank statistic is $-3.6/n$ (to order $O(\frac{1}{n})$) when sampling from a distribution symmetric about zero. [*Hint:* Use the antirank representation, additivity of cumulants, and the fact that $\sum_{i=1}^n i^r \approx n^{r+1}/(r+1)$.]

6. Approximation to a tail probability using the Central Limit Theorem can often be improved through an *Edgeworth series expansion* (see BICKEL (1974)). If Z_n obeys the Central Limit Theorem and is standardized to have zero mean and unit variance and is symmetrically distributed, then

$$\text{pr}\{Z_n > z\} \approx \bar{\Phi}(z) + \frac{1}{4!} z(z^2 - 3)\varphi(z)\lambda_{4n} \tag{8.25}$$

where λ_{rn} is the rth cumulant of Z_n. (Without symmetry, there would also be terms involving λ_{3n} and its square.) Apply this to the null distribution of the Wilcoxon signed-rank statistic, also using a continuity correction. With $z = 1.645, 1.960$ and 2.576 in turn, compare the exact values with the one- and two-term approximations in (24), when $n = 10$ and when $n = 15$. At the same z-values, compare the one- and two-term approximations at $n = 25$ and $n = 100$. [*Hint:* See *Exercise 5*].

7. Use the Lyapounov Theorem (Chapter 2) to prove asymptotic normality of the Wilcoxon signed-rank statistic when sampling from a distribution symmetric around zero. [*Hint:* Apply the Theorem to summands $X_i \equiv iY_i$ in (3).]

8. (a) Show that $\hat{\theta}_{HL}$ has an existing expectation whenever the condition in *Exercise 2* is satisfied and sampling is from a symmetric distribution with $n \geq 4$. [*Hint:* Show that $X_{(2)} \leq \hat{\theta}_{HL} \leq X_{(n-1)}$ when $n \geq 4$.]

   (b) Show that $\hat{\theta}_{HL}$ does *not* have an existing expectation when sampling from a Cauchy distribution with $n \leq 3$.

9. A major setting in which symmetry about zero is an appropriate null hypothesis is in the analysis of paired data—e.g., before-and-after treatment measurements on n individuals, or a comparative study of two treatments on the same units. In the context of the former,

if the treatment has no effect, it is reasonable to assume that the difference between before and after measurements is symmetrically distributed about zero. A treatment effect would tend to yield more positive (say) measurements, and—at least if the effect is small—not disturb symmetry too much. Analyzing only the n differences (rather than using the $2n$ raw data values) is a way of eliminating some of the effects of unit heterogeneity, as in a paired data t-test.

Listed below are data from an experiment comparing tensile strength of tape-closed and sutured wounds, with both techniques evaluated on the same 8 laboratory rats:

tape	984	397	574	447	479	676	761	577
suture	587	460	787	351	277	234	516	513

The null hypothesis is 'no difference', while the alternative is that the tape-closed wounds are stronger. Even though the sample size is quite small, we will illustrate some large-sample procedures with these data. (a) Carry out a sign test. Determine the exact, and the normal-approximation, p-values. Also determine an exact mid-p-value. [See *Exercise 3.*]

(b) Carry out a Wilcoxon signed-rank test. Determine the exact (find a table of the Wilcoxon signed-rank distribution), and the normal-approximation, p-values. Also determine an exact mid-p-value.

(c) Find the Hodges-Lehmann estimate of the center of symmetry, and compare with the sample median. Does this appear to be appropriate for these data?

(d) Find a small-sample 75% confidence interval for the center of symmetry, based on the ordered Walsh averages. Compare it to the interval based on a large-sample formula for the choice of ordered Walsh averages and to the large-sample method using an estimate of the standard error of the Hodges-Lehmann estimate. (See Section 8.8.)

(e) Determine an approximate p-value using the Edgeworth correction in *Exercise 5.*

(f) Repeat (a)–(c), using some convenient computer software, such as R.

10. (a) Show that $[X_{(n-r+1)}, X_{(r)}]$ is a confidence interval for the population median θ, with confidence coefficient $1 - \alpha$, when sampling from a population with df $F(\cdot - \theta)$, F unspecified (continuous) with median 0; here, $r = r_{\alpha/2}$ is the largest integer for which $P_0(S_n \geq c) \leq \frac{1}{2}\alpha$ for the sign-test statistic S_n. [*Hint:* Use the pivot $T_n(\theta) = \sum \mathbf{1}[X_j > \theta]$. Note that no assumption of symmetry is required.)

(b) Show that, for n odd, the sample median is the corresponding median-unbiased estimate of the population median. (This is true only as an approximation when n is even; why?)

(c) Consider the data in *Exercise 9*, and find a 70% confidence interval for the median of differences, tape-closed minus sutured.

11. Suppose $X_1, \ldots, X_n$ is a random sample from the distribution F. Show by example that the distribution of $X_1 + X_2$ can have median zero even though the median of F is not zero. Hence, the Wilcoxon signed rank test cannot be considered as a test of the hypothesis: $med(F) = 0$ without imposing an assumption of symmetry on F. [*Hint:* Take $X + c$ to be standard exponential, so that X is standard exponential on the interval $(-c, \infty)$—with $c \approx 0.83917$.]

12. Show that ρ_f in (15) is invariant under scale transformation.

13. Show that the score $s(x)$ for a logistic distribution is $2F(x) - 1$.

14. Verify Equation (8.17).

15. Verify Equation (8.18).

16. Choose three asymptotic relative efficiencies in the bottom section of the table in Section 8.7, from three different columns and from three different rows, and verify the values given.

17. From results in Section 8.7, the asymptotic relative efficiency of the Wilcoxon signed-rank test relative to a t-test when sampling from a symmetric location family with density $f(x - \theta)$ is $ARE = 12\sigma_f^2 \cdot [\int f^2]^2$ where σ_f is the standard deviation of f.

(a) Show that the asymptotic relative efficiency above is valid even if the information B in f is infinite (assuming an absolutely continuous distribution, symmetric about zero, with $\int f^2$ and σ_f finite). [*Hint:* Try mixing f in proportions $1 - \epsilon$ and ϵ with a density g having finite information as well as the other conditions and letting $\epsilon \to 0$.]

(b) Consider the parabolic density $f(x, \theta = f_0(x - \theta)$ where $f_0(x) = \frac{3}{4}(1 - x^2)$ on $(-1,1)$ (a shifted beta density), and show that the asymptotic relative efficiency of the Wilcoxon test relative to the t-test for this family is $108/125 \approx 0.864$. Also, show that B is infinite for this distribution. [*Note:* Since the support $(\theta - 1, \theta + 1)$ of the distribution $f(x, \theta)$ is not free of θ, it does not satisfy the Basic Assumptions.]

18. (Continuation). (Hodges & Lehmann, 1963). Prove that this density is *least favorable* in that this is the smallest possible asymptotic relative efficiency for the Wilcoxon test relative to the t-test when sampling from a symmetric location family. [*Hint:* The mathematical problem is to find the density $f(x)$ that minimizes $\int f(x)^2 dx$ subject to constraints

$$f(x) \geq 0, \quad \int f(x)dx = 1, \int xf(x)dx = 0. \quad \int x^2 f(x)dx = 1/5,$$

Introduce Lagrange multipliers $\lambda_1 > 0$ and $\lambda_2 < 0$ corresponding to the second and fourth constraints and consider maximizing $\int f^2(x)dx - \lambda_1 \int f(x)dx + \lambda_2 \int x^2 f(x)dx$. After combining the three integrals, the minimization can be achieved pointwise in x, considering the cases $\lambda_1 - \lambda_2 x^2 \leq 0, > 0$ separately and using the nonnegativity of $f(x)$. It may be verified that the solution satisfies the third constraint and leads to values of λ_1 and λ_2 with the asserted signs].

19. Verify Equation (8.22).

20. Much of the theory in Section 8.7 was about testing the null hypothesis of symmetry about zero against symmetry about $\theta > 0$. However, the signed-rank test may be used against asymmetric alternatives as well. Here we consider a particular *Lehmann-alternative* parametric family.

Suppose sampling is from a distribution with survival function

$$\bar{F}_\theta(x) = \bar{F}(x)^{\exp(-\theta)}, \quad \theta \in \mathcal{R},$$

where $F(x) = F_0(x)$ is a specified absolutely continuous distribution function which is symmetric about zero, for example the standard normal distribution funtion $F = \Phi$ with $\bar{F} = 1 - F$. We consider testing whether $\theta = 0$, implying symmetry about zero, versus $\theta > 0$ where symmetry no longer holds.

(a) Show that the family $\{F_\theta\}$ is stochastically increasing in θ.

(b) Describe Wald and score tests of $\theta = 0$ versus $\theta > 0$ in this one-parameter parametric problem.

(c) Show that the asymptotic relative efficiency of the Wilcoxon signed-rank test, when sampling from this distribution, is 3/4, whatever the distribution function F, including $F(x) = \Phi(x/\sigma)$ for any σ. [*Hint:* Again, the asymptotic relative efficiency is the squared correlation between kernel and score, but the score is now different.] Hence, the signed-rank test is less efficient against such Lehmann alternatives to mean-zero normality than it is to location shifts from mean-zero normality.

Note: When such a model is true for some unknown F, a *semi-parametric model*, it may be shown that a signed linear rank test using scores $a_n(r) \equiv \frac{1}{2}\log\{(n+1+r)/(n+1-r)\}$ a so-called *signed log-rank test*; see HALL & WELLNER (2013) is asymptotically efficient.

21. As in *Exercise 19*, consider sampling from an exponential family with density

$$f_\theta(x) = \exp\{\theta x - \psi(\theta)\} \cdot f(x)$$

where the density f is a symmetric around zero and $\psi(\theta)$ is its *cumulant generating function* assumed finite for θ in some neighborhood of zero.

(a) Describe an asymptotically efficient test, for testing $\theta = 0$ versus $\theta > 0$. In particular, is the t-test asymptotically efficient?

(b) Give a formula for the asymptotic relative efficiency of the Wilcoxon signed-rank test.

(c) Show that this asymptotic relative efficiency is less than or equal to unity, with equality if and only if f is the density of a uniform distribution.

(d) Suppose that f_ϵ is the convolution of a symmetric distribution on the 2-point set $\{-1, +1\}$ with a symmetric absolutely continuous distribution with finite cumulant generating function and variance $\epsilon \downarrow 0$. Show that the asymptotic relative efficiency of the Wilcoxon signed-rank test relative to the t-test has the limit $3/4$. [*Note*: This is worse than reported in *Exercise 17* for symmetric shift alternatives.]

22. Show that the signed normal scores test has asymptotic relative efficiency unity when sampling from a normal distribution.

23. Carry out a large-sample signed normal scores test, using the data in *Exercise 9*.

24. Find the parameter estimate associated with the signed normal scores statistic, using the data in *Exercise 9*. Compare it to the mean of the differences, numerically and in interpretation.

25. Show that linear rank statistics for the two-sample problem are invariant under common monotone increasing transformations of the measurement scales. Under what transformations of the x- and y-scales are the associated parameter estimates invariant?

9

Introduction to Multinomial Chi-Squared Tests

We introduce *multinomial chi-squared tests*, emphasizing applications to testing *goodness-of-fit*. Along the way, we briefly mention variations and other applications (and references thereto). These concepts were initially developed by Karl Pearson over a century ago (PEARSON, 1900).

Much of the material is derived from the writings of DAVID S. MOORE, especially MOORE (1977, 1978). Other papers are referenced therein. Chapter 10 of the textbook by LARSEN & MARX (2018) gives an excellent descriptive introduction. Some related materials appear in Chapter 6 of BICKEL & DOKSUM (2015). The chapter by Moore in D'AGOSTINO & STEPHENS (1986) gives a thorough review.

An essential tool is the *generalized inverse* of a matrix, an extension of the definition of an inverse to a matrix that is singular. We start with a brief introduction to the *singular multinormal distribution*, and then develop theory related to the *multinomial distribution*, especially large-sample results. These are applied to obtain a *chi-squared test* of a simple hypothesis based on multinomial data. Most practical applications, however, have composite null hypotheses; for *goodness-of-fit* testing, we present Moore's approach to dealing with these, an approach generalizing ideas of NIKULIN (1973) and RAO & ROBSON (1974). We conclude with a review of some applied considerations, some extensions, and some variations.

9.1 Generalized Inverses, the Singular Multinormal Distribution, and Quadratic Forms

We first recall the definition of a *generalized inverse*, *g*-inverse for short, of a matrix. We confine attention to *positive semi-definite* symmetric matrices, since in what follows matrices for which such inverses are required will always be variance matrices.

Suppose that Σ is such a matrix ($m \times m$), of rank $r \leq m$. Then there exists a symmetric matrix Σ^- for which

$$\Sigma\Sigma^-\Sigma = \Sigma. \tag{9.1}$$

DOI: 10.1201/9780429160080-9

Any such Σ^- is a *generalized inverse* of Σ. If $r = m$, Σ is non-singular and $\Sigma^- = \Sigma^{-1}$, but otherwise Σ^- is not unique, and its rank need not equal r. See RAO (1973) or books on matrix algebra (e.g., SEARLE, 1982) for more information about generalized inverses, including their construction. (Symmetry of Σ^- is an unnecessary but harmless requirement; see *Exercise 3.*)

For a simple example ($m = 2$), consider the singular matrix

$$\Sigma = \begin{pmatrix} 1 & -1 \\ -1 & 1 \end{pmatrix}$$

and consider

$$\begin{pmatrix} a & b \\ b & c \end{pmatrix}$$

as a potential generalized inverse Σ^-. We find that the four identities implied by Equation (9.1) are all satisfied if and only if $a - 2b + c = 1$; in particular,

$$\begin{pmatrix} 1 & 0 \\ 0 & 0 \end{pmatrix}$$

is a generalized inverse. See *Exercise 1.* As this example illustrates, there is a rich supply of generalized inverses for any singular matrix, but often one can be found that may simplify further analyses. Following RAO (1973), we present a simple construction for a general $m \times n$ matrix, assumed written in partitioned form as

$$A = \begin{pmatrix} B & C \\ D & E \end{pmatrix}$$

where B is a non-singular matrix of rank r equal to the rank of A. Any matrix with rank r can be written in this form after permutation of its rows and/or columns. Since the full matrix A has the same rank as B, each of the last $m - r$ rows of A must be a linear combination of the first r rows. So there is a matrix G such that $(D, E) = G(B, C)$. Then $D = GB$ and $E = GC$. Since B is invertible, $G = DB^{-1}$ and so $E = DB^{-1}C$. It is now straightforward to verify (*Exercise 2*) that $AA^-A = A$ when

$$A^- = \begin{pmatrix} B^{-1} & \mathbf{0} \\ \mathbf{0} & \mathbf{0} \end{pmatrix}.$$

A random m-vector $\mathbf{Y}$ is *multinormally distributed* $N_m(\boldsymbol{\mu}, \Sigma)$ *of rank* r, for some m-vector $\boldsymbol{\mu}$ and some symmetric positive semi-definite matrix Σ of rank $r \leq m$, if there exist independent standard normal variables $Z_1, \ldots, Z_r$ and an $r \times m$ matrix C of rank r for which $C^TC = \Sigma$ and $\mathbf{Y} = \boldsymbol{\mu} + C^T\mathbf{Z}$. Since $\mathbf{Z}$ is $N_r(\mathbf{0}, I_r)$ with I_r the $r \times r$ identity matrix, the mean of $\mathbf{Y}$ is seen to be $\boldsymbol{\mu}$ and the variance matrix is $C^TC = \Sigma$. The distribution is said to be *singular* if $r < m$, that is if Σ is singular. This generalizes the definition of multinormality in Chapter 1 where Σ was required to be of full rank. In the singular case, there exist linear functions of the Y_i's that have zero variance, actually $m - r$ linearly independent functions.

For example, with $m = 2$, suppose that $Y_1 = \mu_1 + Z$ and $Y_2 = \mu_2 - Z$ with Z standard normal. Then (Y_1, Y_2) is bivariate normal with mean (μ_1, μ_2), unit variances and correlation coefficient -1—that is, with Σ as in the simple example above. Here $C = (1, -1)$. Note that $Y_2 = \mu_1 + \mu_2 - Y_1$, so the distribution in the (y_1, y_2)-plane is concentrated on the line $y_2 = \mu_2 - \mu_1 - y_1$ where it has a univariate normal distribution.

Another possibly familiar example is that of $\mathbf{Y}$ being a vector of *residuals* in a fitted linear model with normally distributed errors—or even just the residuals from the sample mean in a random normal sample, see *Exercise 8*. Such residuals sum to zero, and hence their joint distribution—and their variance matrix—is singular.

An alternative useful characterization of multinormality, whether singular or non-singular (see Chapter 1), is:

$$\mathbf{Y} \text{ is } N_m(\boldsymbol{\mu}, \Sigma) \quad \text{if and only if} \quad \mathbf{a}^T\mathbf{Y} \sim N(\mathbf{a}^T\boldsymbol{\mu}, \mathbf{a}^T\Sigma\mathbf{a}) \quad \text{for every} \quad \mathbf{a} \in \mathcal{R}^m.$$

There is no special difficulty working with singular normal distributions, so long as we have no need to write down a density: for that, it is necessary to transform linearly down to r dimensions (as in the simple example above). Equivalently, from the defining representation,

$$\text{pr}\{\mathbf{Y} \in A\} = \int \cdots \int_{\{\mathbf{z}|\boldsymbol{\mu}+C^T\mathbf{z}\in A\}} \prod_{j=1}^{r} \{\varphi(z_j)dz_j\}.$$

We will often consider *quadratic forms*

$$Q(\mathbf{Y}, B) \;=\; \mathbf{Y}^T B \mathbf{Y} \quad \text{for } \mathbf{Y} \in \mathcal{R}^m \text{ and } B \text{ a symmetric } m \times m \text{ matrix.}$$

When $\mathbf{Y}$ is non-singular $N_m(\mathbf{0}, \Sigma)$, then the corresponding C is also non-singular, and

$$Q(\mathbf{Y}, \Sigma^{-1}) = \mathbf{Y}^T\Sigma^{-1}\mathbf{Y} = \mathbf{Z}^T C(C^T C)^{-1} C^T \mathbf{Z} = \mathbf{Z}^T\mathbf{Z},$$

the sum of squares of m independent and identically distributed standard normals. Hence, this quadratic form has a chi-squared distribution with m degrees of freedom (Chapter 1). A similar result holds in the singular case, and, fortunately, the choice of Σ^- does not matter:

Proposition 9.1 *If $\mathbf{Y} \sim N_m(\mathbf{0}, \Sigma)$, where Σ has rank r, then $Q(\mathbf{Y}, \Sigma^-)$ is invariant to the choice of Σ^- and is distributed as chi-squared with r degrees of freedom.*

Checking the simple example introduced earlier, now with $\mu_1 = \mu_2 = 0$, we find that $Q = Z^2$, whatever the choice of generalized inverse.

Proof Write $\mathbf{Y} = C^T\mathbf{Z}$ with C an $r \times m$ matrix, so that $Q(\mathbf{Y}, \Sigma^-) = Q(\mathbf{Z}, C\Sigma^- C^T)$. We show below that $C\Sigma^- C^T = I_r$. Then $Q = \mathbf{Z}^T\mathbf{Z}$, and the conclusions follow.

Consider the $r \times r$ matrix $B = C\Sigma^- C^T - I_r$. Then $C^T BC = \Sigma\Sigma^-\Sigma - \Sigma = 0$ $(m \times m)$ since Σ^- is a generalized inverse of Σ. The matrix C has r linearly independent columns (since its rank is r), which we will take to be the first r columns, and write $C = (C_r, E)$ in block form with C_r being nonsingular. Then the $r \times r$ block in the upper left corner of $C^T BC$ is $C_r^T BC_r$, and this vanishes if and only if B is a matrix of zeroes. □

Proposition 9.1 may be extended to the noncentral case: When the mean of $E(\mathbf{Y}) = \boldsymbol{\mu}$ and $\boldsymbol{\mu} = \Sigma\boldsymbol{\nu}$ for some $\boldsymbol{\nu}$, then Q is distributed as noncentral chi-squared with noncentrality parameter $Q(\boldsymbol{\mu}, \Sigma^-) = \boldsymbol{\mu}^T\Sigma^-\boldsymbol{\mu}$, whatever the choice of Σ^-. See MOORE (1978); for related results, see *Exercise 9.* But we will make little use of the noncentral case in this introductory presentation.

As an application, suppose $\tilde{\boldsymbol{\theta}}_n$ is a regular estimate of an m-dimensional parameter $\boldsymbol{\theta}$, so $\mathbf{Y} \equiv \sqrt{n}(\tilde{\boldsymbol{\theta}}_n - \boldsymbol{\theta})$ is $AN(\mathbf{0}, \Sigma)$, say. If Σ is singular, then $Q(\mathbf{Y}, \Sigma^-)$ is asymptotically distributed as chi-squared, as is $Q(\mathbf{Y}, \tilde{\Sigma}_n^-)$ if $\tilde{\Sigma}_n^-$ is consistent for Σ^-. The latter quadratic form may serve as a test statistic. However, as noted before Theorem 9.2 in Section 9.6, a generalized inverse of a consistent estimate of Σ might not be consistent for Σ^-; the individual elements of Σ^- need to be consistently estimated.

9.2 The Singular Multinomial Distribution and a Special Quadratic Form

Consider a random vector $\mathbf{U}^T = (U_1, \ldots, U_m)$ where each $U_i = 0$ or 1 and $U_i = 1$ for exactly one i; hence, $\sum_{i=1}^m U_i = 1$. Let $p_i \equiv \mathrm{pr}(U_i = 1)$; then $\sum_{i=1}^m p_i = 1$. The vector $\mathbf{U}$ typically represents the classification of an object into one of m exclusive and exhaustive categories.

We say that $\mathbf{U}$ has a *singular multinomial distribution* with parameters $(1; \mathbf{p})$. It is *singular* because the coordinates sum to unity. A *nonsingular multinomial distribution* is obtained by omitting any one coordinate, say U_m; we write $\mathbf{U}'$ and $\mathbf{p}'$ for the m' $(= m-1)$-dimensional vectors with the last coordinate omitted. With $m = 2$, U' (real) is a *Bernoulli* random variable, an indicator for a 'success', whereas $\mathbf{U}$ (two-dimensional) includes the indicators both for 'success' and for 'failure'.

Now consider n independent and identically distributed vectors $\mathbf{U}$, and sum them to form $\mathbf{V}$. The coordinates of $\mathbf{V}$ are now counts in various categories. The vector $\mathbf{V}$ has a *singular multinomial distribution* with parameters $(n; \mathbf{p})$, with coordinates of $\mathbf{V}$ now summing to n. (Some settings require that additional constraints be imposed, but we focus here on the case of a single constraint). Then $\mathbf{V}'$, which omits the last coordinate, has a *nonsingular multinomial distribution.* With $m = 2$, $\mathbf{V}$ contains counts of both success and of failure, adding to n, whereas V' is the usual binomial, $B(n; p_1)$, random variable.

Since both $\mathbf{V}$ and $\mathbf{v}'$ are discrete, there is no difficulty in writing down their densities. With ν_j denoting nonnegative integers, we have for both $\mathbf{V}$ and $\mathbf{V}'$:

$$g_{\mathbf{p}}(\mathbf{v}) = \prod_{j=1}^{m} p_j^{v_j} \quad \text{with} \quad \sum_{j=1}^{m} v_j = n$$

and

$$g'_{\mathbf{p}'}(\mathbf{v}') = \left(\prod_{j=1}^{m-1} p_j^{v_j} \right) \cdot \left(1 - \sum_{j=1}^{m-1} p_j \right)^{n - \sum_{j=1}^{m-1} v_j} \qquad \text{with} \quad \sum_{j=1}^{m-1} v_j \le n.$$

In the singular case, the m coordinates of the parameter $\mathbf{p}$ are constrained to sum to unity, whereas in the nonsingular case, the $m - 1$ coordinates of $\mathbf{p}'$ must sum to at most unity.

The means, variances and covariances of the V_i's are most easily obtained by first considering the U_i's:

$$E(U_i^2) = E(U_i) = p_i; \quad E(U_i U_j) = E(0) = 0 \quad \text{for } i \ne j.$$

This readily yields $\text{var}(U_i) = p_i(1 - p_i)$ and $\text{cov}(U_i, U_j) = -p_i p_j$ for $i \ne j$. Hence, in vector and matrix notation,

$$E(\mathbf{V}) = n\mathbf{p}, \quad \text{var}(\mathbf{V}) = n\left\{D(\mathbf{p}) - \mathbf{p}\mathbf{p}^T\right\} = n\Sigma,$$

say, where $D(\mathbf{p})$ is a diagonal matrix with diagonal elements $\mathbf{p}$; Σ is of rank $m' = m - 1$ (since, writing $\mathbf{1}$ for a column vector of 1's, $\Sigma\mathbf{1} = \mathbf{0}$, that is, the sum of every row of Σ is zero). The mean and variance formulas are same for $\mathbf{V}'$ as for $\mathbf{V}$.

Now convert the coordinates of $\mathbf{V}$ and $\mathbf{V}'$ to proportions, and append subscripts n: $\bar{\mathbf{V}}_n = (1/n)\mathbf{V}_n$, and $\bar{\mathbf{V}}'_n = (1/n)\mathbf{V}'_n$. Applying the Cramér-Wold device, the central limit theorem yields (with or without primes)

$$\mathbf{Y}_n = \sqrt{n}(\bar{\mathbf{V}}_n - \mathbf{p}) \ \to_d \ N_m(\mathbf{0}, \Sigma).$$

As a corollary, we have (for any choice of Σ^-)

$$Q_n(\mathbf{Y}_n, \Sigma^-) = \mathbf{Y}_n^T \Sigma^- \mathbf{Y}_n \ \to_d \ \chi^2_{m-1}, \tag{9.2}$$

the central chi-squared distribution on $m - 1$ degrees of freedom. Note that $\mathbf{Y}_n^T \mathbf{1} = 0$, that is, the sum of the deviations between observed proportions and corresponding probabilities is zero, which causes the singularity.

We now determine a generalized inverse Σ^- of Σ.

Lemma 9.1 *Suppose that* $\Sigma = D(\mathbf{p}) - \mathbf{p}\mathbf{p}^T$, $\mathbf{p}^T\mathbf{1} = 1$, *and* $p_i > 0$ *for every* i. *Then a g-inverse for* Σ *is*

$$\Sigma^- = D(\mathbf{p})^{-1}, \tag{9.3}$$

a diagonal matrix with ith diagonal element $1/p_i$.

Proof Note that

$$\mathbf{p}^T D^{-1} = \mathbf{1}^T \quad \text{so} \quad \mathbf{p}\mathbf{p}^T D^{-1} = \mathbf{p}\mathbf{1}^T \quad \text{and} \quad \mathbf{p}\mathbf{p}^T = \mathbf{p}\mathbf{1}^T D.$$

Therefore

$$\begin{aligned} \Sigma\Sigma^-\Sigma &= (D - \mathbf{p}\mathbf{p}^T)D^{-1}(D - \mathbf{p}\mathbf{p}^T) \\ &= (I - \mathbf{p}\mathbf{1}^T)(D - \mathbf{p}\mathbf{p}^T) \\ &= D - \mathbf{p}\mathbf{p}^T - \mathbf{p}\mathbf{p}^T + \mathbf{p}\mathbf{1}^T\mathbf{p}\mathbf{p}^T \end{aligned}$$

which equals Σ since $\mathbf{1}^T\mathbf{p} = 1$. □

Using this diagonal inverse (9.3), the quadratic form $Q_n(\mathbf{Y}_n, \Sigma^-) = \mathbf{Y}_n^T D(\mathbf{p})^{-1}\mathbf{Y}_n$ is simply a sum of squares. Explicitly, writing $Y_{in} = \sqrt{n}(\bar{V}_{in} - p_i) = (V_{in} - np_i)/\sqrt{n}$, we conclude from Equation (9.2):

Theorem 9.1 *With the choice* $\Sigma^- = D(\mathbf{p})^{-1}$,

$$Q_n(\mathbf{Y}_n, \Sigma^-) = \sum_{i=1}^{m} \frac{(V_{in} - np_i)^2}{np_i} \to_d \chi^2_{m-1}.$$

Theorem 9.1 also holds with primes appended, except now Σ^- is replaced by $(\Sigma')^{-1}$ where Σ', the variance of $\mathbf{Y}'_n$, equals $D' - \mathbf{p}'(\mathbf{p}')^T$. This holds since it may be shown that

$$(\mathbf{Y}'_n)^T(\Sigma')^{-1}\mathbf{Y}'_n = \mathbf{Y}_n^T\Sigma^-\mathbf{Y}_n$$

(see *Exercise 11*). However, we find the simpler singular form, without primes, to be more convenient.

Note on Methodology: MOORE works with the vector with coordinates $\sqrt{n}(\bar{V}_{in} - p_i)/q_i$ where $q_i = \sqrt{p_i}$, rather than with our Y_{in}. Its variance matrix is then $I - \mathbf{q}\mathbf{q}^T$, and a generalized inverse is simply I. We find prefer to avoid the resulting 'sprinkling' of $\sqrt{p_i}$'s, here and in what follows. Incidentally, this generalized inverse, that in Lemma 9.1 above, and those in lemmas yet to come, can each be derived from projection arguments (see MOORE, 1977); we confine attention to algebraic verifications.

9.3 Multinomial Chi-Squared Test: Simple Null Hypothesis

Consider a (singular) multinomial m-vector $\mathbf{V}_n$ with parameters n and $\mathbf{p}$, and $\bar{\mathbf{V}}_n = (1/n)\mathbf{V}_n$. Let $\mathbf{p}^0$ be specified, with all coordinates positive (and summing to unity). Then a test of the simple null hypothesis

$$H_0 : \mathbf{p} = \mathbf{p}^0$$

is

$$\mathbf{1}[(Q_n^0 \geq w_\alpha] \text{ where } Q_n^0 = Q_n\left\{\mathbf{Y}_n^0, D(\mathbf{p}^0)^{-1}\right\} \text{ and } \mathbf{Y}_n^0 = \sqrt{n}(\bar{\mathbf{V}}_n - \mathbf{p}^0) \tag{9.4}$$

and w_α is defined by $\text{pr}(\chi^2_{m-1} \geq w_\alpha) = \alpha$. By Theorem 9.1, the test has asymptotic size α. A corresponding (large-sample) p-value is $\text{pr}(\chi^2_{m-1} \geq w)$, evaluated at w equal to the observed value of Q_n^0. Rejection for large values of Q_n^0 is motivated by the fact that, if $\mathbf{p}_n = \mathbf{p}^0 + \mathbf{h}/\sqrt{n}$ (with $\mathbf{h}$ for which $\mathbf{h}^T\mathbf{1} = 0$), then LeCam's third lemma can be used (details omitted here; use the nonsingular multinomial distribution) to prove that the large-sample distribution of Q_n^0, under the local alternative $\mathbf{p}_n$, is noncentral chi-squared with noncentrality parameter $\delta^2 = \mathbf{h}^T D^{-1}\mathbf{h}$. And the noncentral chi-squared distribution is stochastically increasing in δ^2, and hence stochastically larger than the central chi-squared distribution (see Chapter 1). Hence, rejection is more likely under this local alternative than under H_0. We will not develop this local alternative theory here, however.

Note that

$$Q_n^0 = \sum_{i=1}^m \frac{Y_{in}^{0^2}}{p_i^0} = \sum_{i=1}^m \frac{(V_{in} - np_i^0)^2}{np_i^0},$$

often expressed as the sum of "observed minus expected counts squared, divided by expected counts". Note too that the sum is over all m categories. Each term in the sum is q_i^0 times the square of an asymptotically standard normal random variable. The dependence among these random variables results in the loss of one degree of freedom, and the weights (q_i's) are just what is needed to achieve the chi-squared distribution in the limit. These remarks are, in effect, an informal interpretation of Lemma 9.1.

As an example, consider testing whether a distribution on the integers $\{1, \ldots, m\}$ is discrete uniform, for example is a die fair? or is a lottery fair? Each $p_i^0 = 1/m$ and $Q_n^0 = (m/n)\sum_{j=1}^m\{V_{jn} - (n/m)\}^2$. A large-sample (large n) p-value is the tail area beyond Q_n^0 under the chi-squared density with $m-1$ degrees of freedom.

9.4 Goodness-of-Fit Testing: Simple Null Hypothesis

Now suppose that $X_1, \ldots, X_n, \ldots$ are independent and identically distributed with density $f(x, \theta)$. This density may be discrete, absolutely continuous or of mixed form, and the dimensions of X and of the parameter θ are unimportant. Partition the support $\mathcal{X}$ of X into m sets $A_1, \ldots, A_m$ (called 'bins' or 'cells'), and let

$$p_i(\theta) \equiv P_\theta(X \in A_i) = \int_{A_i} dF(x, \theta).$$

Let V_{jn} be the count of X_i's ($i \leq n$) in cell A_j, so that $\mathbf{V}_n$ is singular multinomial with parameters n and $\mathbf{p}(\theta)$. It is convenient to refer to $\mathbf{V}_n$ as 'binned data'.

Now consider the simple null hypothesis that $\theta = \theta_0$, specified; this implies $\mathbf{p}(\theta) = \mathbf{p}(\theta_0) = \mathbf{p}^0$, say (and we assume that the partition is chosen so that each $p_i(\theta_0) > 0$). The statistic Q_n^0 in Equation (9.4) provides a basis for a test of this simple hypothesis, a so-called *goodness-of-fit* test of whether the completely specified model $f(x, \theta_0)$ is correct. It will have power for detecting any alternative model that leads to a different $\mathbf{p}$. Thus, we may test whether the data come from a specific normal distribution or from a specific Poisson distribution, etc. This is often called the *Pearson chi-squared test*, having been introduced by Karl Pearson in 1900.

As an example, consider testing whether a sample of 60 measurements comes from a standard normal distribution. We might divide the measurement scale into eight intervals with division points ± 1.5, ± 1.0, ± 0.5 and 0, with probabilities determined from standard normal tables. Or we might use seven equiprobable intervals, with division points ± 1.07, ± 0.65, ± 0.18; these would lead to expected counts of 60/7 in each interval. (For some practical guidelines on the choice of m and choice of the intervals, see the comments at the end of this chapter.)

In practice, however, it is more often of interest to test a *composite* goodness-of-fit hypothesis. We take this up next.

9.5 Goodness-of-Fit Testing: Composite Null Hypothesis

We now consider the *goodness-of-fit* problem of testing whether a random sample comes from a particular parametric family model, for example any normal distribution, or any Poisson distribution. These are composite null hypotheses since the values of the parameters are not specified and must be estimated from the data on-hand. The alternative hypothesis H_A continues to be vague, simply that the null hypothesis is false. We write, for some (closed) subset Θ of $\mathcal{R}^d$,

$$H_0: \quad \text{For some } \theta \in \Theta, \text{ the density is } f(\cdot, \theta).$$

The classical solution to this problem (FISHER, 1922, 1924) is to replace θ_0 in $\mathbf{p}(\theta_0)$ when computing the test statistic by a so-called *minimum chi-squared estimate* of θ. To this end, write $Q_n(\theta)$ for the quadratic form $Q_n(\mathbf{Y}_n, D^{-1})$ with $\mathbf{p} = \mathbf{p}(\theta)$ in both $\mathbf{Y}_n$ and in D. The *minimum chi-squared estimate* $\tilde{\theta}_n$ is defined by

$$Q_n(\tilde{\theta}_n) = \min_{\theta \in \Theta} Q_n(\theta).$$

Alternatively, replace $\mathbf{p}(\theta)$ in D by the sample proportions $\bar{\mathbf{V}}_n$ before minimizing Q_n, leading to the so-called *modified minimum chi-squared statistic*, which leads to a mathematically simpler minimization problem. Either of these minimized Q_n's may be shown to converge in distribution to chi-squared with $m - d - 1$ degrees of freedom. See the book by CRAMÉR (1946), for example. The resulting goodness-of-fit test is the *Fisher-Pearson chi-squared test*, based on comparison of observed and expected counts, with the expected counts estimated by minimum chi-squared methods.

There are two shortcomings here: (i) There is a loss of d degrees of freedom, one for each unknown parameter, and this tends to lessen power. (ii) The resulting estimates $\tilde{\theta}_n$ are typically inefficient, and often difficult to calculate. Why not use $Q_n(\hat{\theta}_n)$ based on an efficient estimate $\hat{\theta}_n$ derived from the raw data $X_1, \ldots, X_n$? The difficulty here is that the resulting Q_n converges in distribution to something other than chi-squared, something quite inconvenient (the limiting distribution typically depends on θ); however, it has been proved to be stochastically sandwiched between chi-squared distributions with $m - d - 1$ and $m - 1$ degrees of freedom, allowing upper and lower bounds on associated significance levels. Thus, the loss in degrees of freedom is partially restored, but in an inconvenient or imprecise manner.

We now turn to a more recent approach to this composite-null-hypothesis problem, as developed by MOORE (1977), generalizing earlier work by NIKULIN (1973) and RAO & ROBSON (1974). It turns out that by adding a non-negative correction term to $Q_n(\hat{\theta}_n)$, the chi-squared with $m - 1$ degrees of freedom limiting distribution can be preserved.

First, one caveat: If the actual data *are* multinomial, but with cell probabilities depending on a parameter θ, then the multinomial model is constrained by more than just the single constraint that probabilities sum to unity. In effect, it is a model with an $(m - d - 1)$-dimensional parameter, the number m of cell probabilities p_i is effectively reduced by the $d + 1$ constraints. As discussed in Chapter 6, standard large-sample procedures for testing and estimation may be adapted to this situation—in particular, likelihood-ratio tests and maximum likelihood estimation. Such tests are chi-squared tests with $m - d - 1$ degrees of freedom, and are asymptotically equivalent to the minimum chi-squared tests above, just as maximum likelihood estimates are asymptotically equivalent to the minimum chi-squared estimates. Models of this type have been studied extensively by J. Neyman and his associates, often under the name BAN ('best asymptotically normal') estimation. Examples arise in genetics, in particular; there are counts of various types of offspring and genetic models lead to cell probabilities as functions of certain unknown parameters. BICKEL & DOKSUM (2015) provided an interesting introduction to some of this. For further reading, see NEYMAN (1949). But we move on to settings in which the raw data are *not* multinomial.

9.6 The Rao-Robson-Moore Test

We now replace the role of $\mathbf{Y}_n$ by

$$\hat{\mathbf{Y}}_n = \mathbf{Y}_n(\hat{\theta}_n) = \sqrt{n}\{\bar{\mathbf{V}}_n - \mathbf{p}(\hat{\theta}_n)\} \tag{9.5}$$

with $\hat{\theta}_n$ an asymptotically efficient estimate of θ based on the raw data. However, the resulting variance matrix is no longer $D - \mathbf{p}\mathbf{p}^T$, and therefore D^{-1} is not the correct matrix to use in the quadratic form Q_n.

We assume that the $p_i(\theta)$'s have continuous partial derivatives; let $\dot{P} = \dot{P}(\theta)$ be the $m \times d$ matrix of these, with (i,j)-element $\partial p_i / \partial \theta_j$. Two important consequences are: first, that

$$\hat{\mathbf{Y}}_n = \mathbf{Y}_n - \dot{P}B^{-1}\mathbf{S}_n + o_p(1) \tag{9.6}$$

where $\mathbf{S}_n$ is the *score* $(1/\sqrt{n})\sum_{j=1}^n \mathbf{s}(X_j, \theta)$ and B the *information*, namely the variance matrix of $\mathbf{s}(X, \theta)$, the vector of scores *per observation*; and secondly, that

$$\operatorname{cov}(\mathbf{Y}_n, \mathbf{S}_n) = \dot{P}. \tag{9.7}$$

The first consequence (9.6) follows from a Taylor expansion and the assumed asymptotic efficiency of $\hat{\theta}_n$:

$$\hat{\mathbf{Y}}_n = \sqrt{n}\{\hat{\mathbf{V}}_n - \mathbf{p}(\theta)\} - \sqrt{n}\{\mathbf{p}(\hat{\theta}_n) - \mathbf{p}(\theta)\} = \mathbf{Y}_n - \dot{P}(\theta)\sqrt{n}(\hat{\theta}_n - \theta) + o_p(1)$$

(by continuity of $\dot{P}$), and this equals (9.6) by the Characterization Theorem for asymptotically efficient estimates (Chapter 5).

For the second consequence (9.7), the desired covariance is just the matrix with elements $\gamma_{ij} = \operatorname{cov}\{\mathbf{1}[X \in A_i], s_j(X,\theta)\}$ where $A_1, \ldots, A_m$ are the cells of the partition of the sample space of X and $s_j(x,\theta)$ is the score (per observation) for the jth coordinate of θ. This is so since the coordinates of $\mathbf{Y}_n$ and $\mathbf{S}_n$ are $\sqrt{n}$ times averages of independent identically distributed random variables $\mathbf{1}[X \in A_i] - p_i$ and $s_j(X,\theta)$, respectively. Now

$$\begin{aligned}\gamma_{ij} &= \int \mathbf{1}[x \in A_i]\left\{\frac{(\partial/\partial\theta_j) f(x,\theta)}{f(x,\theta)}\right\} f(x,\theta)dx \\ &= \frac{\partial}{\partial\theta_j}\int \mathbf{1}[x \in A_i] f(x,\theta)dx = \frac{\partial}{\partial\theta_j} p_i(\theta)\end{aligned}$$

(or the corresponding sums), proving (9.7).

It now follows from Equations (9.6) and (9.7) that the large-sample variance of $\hat{\mathbf{Y}}_n$ is

$$\Sigma^* = D(\mathbf{p}) - \mathbf{p}\mathbf{p}^T - \dot{P}B^{-1}\dot{P}^T, \tag{9.8}$$

the covariance term on the right side of Equation (9.6) being twice the negative of the second variance term. Under standard regularity conditions that assure Local Asymptotic Normality, continuous partial derivatives of $\mathbf{p}(\theta)$, and positive $p_i(\theta)$ that can be differentiated under the integral sign, we therefore have

Proposition 9.2 *Under H_0, $\hat{\mathbf{Y}}_n$ in Equation (9.5) converges in distribution to $N_m(\mathbf{0}, \Sigma^*)$. If Σ^* in Equation (9.8) has rank $m-1$, then*

$$Q\{\hat{\mathbf{Y}}_n, (\Sigma^*)^-\} = \hat{\mathbf{Y}}_n^T(\Sigma^*)^-\hat{\mathbf{Y}}_n \to_d \chi^2_{m-1}. \tag{9.9}$$

We now determine a generalized inverse of Σ^*. We impose the condition

$$B > \dot{P}^T D(\mathbf{p})^{-1}\dot{P}, \tag{9.10}$$

in the sense that the difference is positive definite. This condition can be interpreted as meaning that there is strictly more 'information' about all components of θ in the raw data than in the 'binned' multinomial data (details omitted). Condition Equation (9.10) also implies that Σ^* in Equation (9.8) has rank $m-1$ (see *Exercise 10*) so that Equation (9.9) holds.

Lemma 9.2 *If Equation (9.10) holds, so that $G = B - \dot{P}^T D^{-1}\dot{P}$ is nonsingular, then*

$$(\Sigma^*)^- = D^{-1} + D^{-1}\dot{P}G^{-1}\dot{P}^T D^{-1} \tag{9.11}$$

is a generalized inverse of Σ^* in Equation (9.8).

Proof: We give an algebraic proof, verifying that the asserted matrix is a generalized inverse. First, note that, with $\mathbf{1}$ denoting the $d \times 1$ column vector $(1, \ldots, 1)^T$, $\mathbf{1}^T\dot{P} = 0$ and $\mathbf{p}^T D^{-1} = \mathbf{1}^T$, so that $\mathbf{p}^T D^{-1}\dot{P} = 0$. Then

$$\Sigma D^{-1} = (D - \mathbf{p}\mathbf{p}^T)D^{-1} = I - \mathbf{p}\mathbf{1}^T$$

and

$$\Sigma(\Sigma^*)^- = I - \mathbf{p}\mathbf{1}^T + \dot{P}G^{-1}\dot{P}^T D^{-1}.$$

Now consider

$$(\Sigma^* - \Sigma)(\Sigma^*)^- = -\dot{P}B^{-1}\dot{P}^T(D^{-1} + D^{-1}\dot{P}G^{-1}\dot{P}^T D^{-1}).$$

Adding these expressions gives

$$\begin{aligned}\Sigma^*(\Sigma^*)^- - (I - \mathbf{p}\mathbf{1}^T) &= \dot{P}G^{-1}\dot{P}^T D^{-1} - \dot{P}B^{-1}\dot{P}^T(D^{-1} + D^{-1}\dot{P}G^{-1}\dot{P}^T D^{-1}) \\ &= (\dot{P} - \dot{P}B^{-1}\dot{P}^T D^{-1}\dot{P})G^{-1}\dot{P}^T D^{-1} - \dot{P}B^{-1}\dot{P}^T D^{-1}\end{aligned}$$

$$= \dot{P}B^{-1}(B - \dot{P}^T D^{-1}\dot{P})G^{-1}\dot{P}^T D^{-1} - \dot{P}B^{-1}\dot{P}^T D^{-1}$$
$$= \dot{P}B^{-1}GG^{-1}\dot{P}^T D^{-1} - \dot{P}B^{-1}\dot{P}^T D^{-1} = 0.$$

When $G = 0$, so that $B = \dot{P}D^{-1}\dot{P}$, the problem reduces to become equivalent to that in which only the grouped multinomial data are available, not the original raw data. The limiting chi-squared distribution then has $m - d - 1$ degrees of freedom, as in the *minimum chi-squared* case. The intermediate case, where G is positive semi-definite but not positive definite is more difficult, and rarely arises in practice but can be handled by replacing G^{-1} by a generalized inverse G^- and proceeding as above. See *Exercise 22.* Incidentally, other choices of the generalized inverse for Σ^* may lead to numerically different values for the quadratic form in Equation (9.9), but all must have the same asymptotic chi-squared distribution; and it can be shown that the left sides of Equation (9.9) differ by only $o_p(1)$. It only remains to substitute consistent estimates for the elements in various matrices in Equation (9.11); for example, simply replace θ by $\hat{\theta}_n$. (However, it may be easier to use average sample information instead of $B(\hat{\theta}_n)$.) Note, however, that in contrast to the nonsingular case, it is *not* sufficient to use a generalized inverse of a consistent estimate of Σ^*; see *Exercise 13* for a (nonrandom) counterexample. Instead, we must consistently estimate a (any) generalized inverse $(\Sigma^*)^-$.

We summarize in

Theorem 9.2 *Let D_n, $\dot{P}_n$ and B_n be consistent estimates of D, $\dot{P}$ and B. Suppose that $p_i(\theta) > 0$ for each i, each has continuous partial derivatives, and Equation (9.10) holds. Define $\hat{Q}_n^*$ by*

$$\hat{Q}_n^* = Q_n\{\hat{\mathbf{Y}}_n, (\Sigma_n^*)^-\} = Q_n(\hat{\mathbf{Y}}_n, D_n^{-1}) + Q_n(\mathbf{W}_n, G_n^{-1}) = \hat{Q}_n + \hat{\Delta}_n, \quad (9.12)$$

say, where $\mathbf{W}_n = \dot{P}_n^T D_n^{-1}\hat{\mathbf{Y}}_n$ and $G_n = B_n - \dot{P}_n^T D_n^{-1}\dot{P}_n$. Then, under H_0, $\hat{Q}_n^*$ converges in distribution to chi-squared with $m - 1$ degrees of freedom.

A corresponding (large-sample) p-value is $\text{pr}(\chi^2_{m-1} \geq w)$, evaluated at w equal to the observed value of $\hat{Q}_n^*$.

Note that the first term $\hat{Q}_n$ in Equation (9.12) is the goodness-of-fit chi-squared statistic Q_n^o when H_0 is simple, but with the cell probabilities replaced by estimates of θ that are asymptotically efficient under the model for the original, ungrouped, data, namely

$$\hat{Q}_n = \sum_{j=1}^{m} \frac{(V_{jn} - n\hat{p}_j)^2}{n\hat{p}_j}.$$

The second term $\hat{\Delta}_n$ in Equation (9.12) is therefore what is needed to retain the chi-squared limit with $(m-1)$-degrees of freedom after substitution of the estimated p_j's. Both terms are non-negative, and so use of only the first term (but still with $m - 1$ degrees of freedom) provides a conservative test, over-estimating the p-value.

When $d = 1$, the second term in Equation (9.12) is

$$\hat{\Delta}_n = \frac{1}{nG_n}\left\{\sum_{j=1}^{m} \frac{V_{jn}}{\hat{p}_j}\left(\frac{d\hat{p}_j}{d\theta}\right)\right\}^2 \quad \text{where} \quad G_n = B_n - \sum_{j=1}^{m} \frac{1}{\hat{p}_j}\left(\frac{d\hat{p}_j}{d\theta}\right)^2. \tag{9.13}$$

An example of Equation (9.13) is provided by a test of whether some count data come from a Poisson distribution. Here, we may take $\{0\}, \{1\}, \ldots, \{m-2\}$ for the first $m-1$ cells and $\{x|x \geq m\}$ for the remaining cell. In order to have an adequate chi-squared approximation, the resulting $n\hat{p}_j$'s should exceed one, and most should exceed five. The details of the statistic (9.13), in this case, are left to the reader (*Exercise 16*).

The distinction between limiting distributions based on minimum chi-squared estimates from grouped data and on asymptotically efficient estimates based on full data is rarely made in introductory textbooks and is often not of practical importance. Since continuous data can never be recorded exactly, the contrast is actually between greater or lesser degrees of approximation. However, the idealized theory is a useful point of reference. Certainly, irrespective of the specific distributional considerations, a substantial difference between the parameter estimate $\tilde{\theta}$ based on grouped data and the corresponding estimate $\hat{\theta}$ based on the original data would itself suggest a poor model fit.

9.7 An Example: Testing for Normality

We now apply Theorem 9.2 to the problem of testing for normality:

$$H_0: \text{ The raw data come from } N(\mu, \sigma^2) \text{ with } \mu \text{ and } \sigma \text{ unspecified.} \tag{9.14}$$

We illustrate with some data presented in the first edition of the textbook by Larsen & Marx (1986) (Case Studies 7.2.1 and 9.4.2) on head breadths of 84 male Etruscan skulls from 8th century Italy, uncovered in various archeological digs, and utilized in anthropological studies. The example appears in modified form in later editions of the book. The 84 measurements (in mm.), see Table 1 below, are integers ranging from 131 to 154, with a mean of 143.77 and a standard deviation of 5.9705. This data is also available in R. The authors grouped them into five intervals, with cell boundaries of 134.5, 139.5, 144.5 and 149.5. The observed frequencies in these five groups were 5, 10, 33, 24 and 12, respectively. Is this consistent with a normal distribution?

Call the cell boundaries c_i (with $c_0 = -\infty$ and $c_5 = +\infty$). The probabilities of the cells are

$$p_i = \Phi(r_i) - \Phi(r_{i-1}) \quad \text{with} \quad r_i = \frac{c_i - \mu}{\sigma}. \tag{9.15}$$

The elements of $\dot{P}$ are therefore

$$\frac{\partial p_i}{\partial \mu} = -\frac{1}{\sigma}\{\varphi(r_i) - \varphi(r_{i-1})\} = -\frac{1}{\sigma}\Delta\varphi_i$$

say, and, with $\psi(c) = c\varphi(c)$,

$$\frac{\partial p_i}{\partial \sigma} = -\frac{1}{\sigma}\{\psi(r_i) - \psi(r_{i-1})\} = -\frac{1}{\sigma}\Delta\psi_i.$$

The 2×2 symmetric matrix $\dot{P}^T D^{-1} \dot{P}$ has diagonal elements

$$\frac{1}{\sigma^2}\sum_{j=1}^{m}\frac{(\Delta\varphi_j)^2}{p_j} \quad \text{and} \quad \frac{1}{\sigma^2}\sum_{j=1}^{m}\frac{(\Delta\psi_j)^2}{p_j}$$

The off-diagonal element is

$$\frac{1}{\sigma^2}\sum_{j=1}^{m}\frac{\Delta\varphi_j\Delta\psi_j}{p_j}$$

Standard computations show that the information matrix for $\theta = (\mu, \sigma)$ is

$$B = \text{diag}(\frac{1}{\sigma^2}, \frac{2}{\sigma^2}).$$

Hence $(n\sigma^2)^{-1}G = E$, say, is symmetric with diagonal elements

$$\frac{1}{n} - \sum_{j=1}^{m}\frac{(\Delta\varphi_j)^2}{np_j} \quad \text{and} \quad \frac{2}{n} - \sum_{j=1}^{m}\frac{(\Delta\psi_j)^2}{np_j} \quad \text{and} \quad \sum_{j=1}^{m}\frac{\Delta\varphi_j\Delta\psi_j}{np_j} \tag{9.16}$$

is the off-diagonal element.

Throughout, μ and σ must be estimated by the sample mean $\bar{x}_n$ and sample standard deviation s_n; in particular, we replace r_i by $\hat{r}_i = (c_i - \bar{x}_n)/s_n$. Furthermore

$$W_{1n} = \frac{\sqrt{n}}{s_n}\sum_{j=1}^{m}\Delta\hat{\varphi}_j\frac{V_{jn}}{n\hat{p}_j} \quad \text{and} \quad W_{2n} = \frac{\sqrt{n}}{s_n}\sum_{j=1}^{m}\Delta\hat{\psi}_j\frac{V_{jn}}{n\hat{p}_j},$$

where we use the fact that $\sum \Delta\hat{\varphi}_j = \sum \Delta\hat{\psi}_j = 0$.

We thus have, finally, after some simplification, that $\hat{Q}_n^*$ in Equation (9.12) has

$$\hat{Q}_n = \sum_{j=1}^{m}\frac{(V_{jn} - n\hat{p}_j)^2}{n\hat{p}_j} \quad \text{where} \quad \hat{p}_j = \Phi(\hat{r}_j) - \Phi(\hat{r}_{j-1})$$

and

$$\hat{\Delta}_n = (\mathbf{U}_{1n}, \mathbf{U}_{2n})E_n^{-1}(\mathbf{U}_{1n}, \mathbf{U}_{2n})^T$$

where

$$\mathbf{U}_{1n} = \sum_{j=1}^{m} \Delta\hat{\varphi}_j \frac{V_{jn}}{n\hat{p}_j}, \quad \mathbf{U}_{2n} = \sum_{j=1}^{m} \Delta\hat{\psi}_j \frac{V_{jn}}{n\hat{p}_j}$$

and E_n is defined in Equation (9.16) but with $\hat{\varphi}$, $\hat{\psi}$ and $\hat{p}_j$ substituted for φ, ψ and p_j. Note that only a 2×2 matrix E_n needs to be inverted.

Calculation of these various terms using the data presented above yields $\hat{Q}_n = 3.716$. The 'correction term' $\hat{\Delta}_n$ is 2.052, yielding $\hat{Q}_n^* = 5.77$. R-code for performing these calculations is provided in the Appendix. With four degrees of freedom, this yields a p-value of 0.22. The substantive conclusion is that the data are consistent with the null hypothesis of sampling from a normal distribution. In contrast, LARSEN & MARX (1986) interpreted $\hat{Q}_n$ relative to chi-squared with two degrees of freedom, yielding a p-value of 0.17. Incidentally, a more sensitive test could be achieved by using more intervals, say of width 3 or 4 mm. instead of width 5 mm. And the use of intervals with probabilities closer to equality, rather than equally spaced intervals, would increase the power; see comments at the end of this chapter.

If $\hat{Q}_n$ had been much greater, its value alone could indicate lack of fit, and computation of $\hat{\Delta}_n$ would have been unnecessary.

Other tests of normality have been devised including the *Shapiro-Wilk test* and the *Lin-Mudholkar test.* For details of these and references to other tests, consult LIN & MUDHOLKAR (1980). See also the book edited by D'AGOSTINO & STEPHENS (1986).

9.8 Further Remarks on Multinomial Chi-Squared Tests

Number and choice of cells: Empirically derived guidelines are that m and the cells should be chosen so that most expected counts are at least five and virtually all are at least one. This usually assures that the chi-squared approximation is adequate. Also, as a general principle, some efficiency is lost when there are any large expected numbers: make the cell probabilities roughly equal, when feasible.

According to the theory presented, the cells should be chosen without reference to the data, but this is rarely practical, especially when there are unknown parameters. Fortunately, there is supporting theory (not presented here) for use of *data-based* cells: cell boundaries may be data-based so long as they converge in probability (and certain continuity conditions hold); see MOORE & SPRUILL (1975). For example, in testing for normality, we may use cell boundaries of the form $\bar{X}_n + c_i S_n$ for some constants c_i. (See the Exercises for other examples.)

However, the number m of cells cannot grow rapidly with sample size n; MOORE & SPRUILL claim that m must be at most $O(\sqrt{n})$. That is, although 'the larger the sample size, the larger the number of cells used' is a tempting and natural working rule, some restraint is called for. MOORE (in Chapter 3 of D'AGOSTINO & STEPHENS, 1986) suggests, as a very rough rule-of-thumb, $m \approx 2n^{2/5}$ when the cell probabilities are roughly equal, and presumably smaller when they are not.

It is a fact that cannot be circumvented, however, that the resulting test depends on the choice of cells.

Power; outliers: A chi-squared goodness-of-fit test evaluates whether a histogram is consistent with a parametric model density. However, it does not recognize, or respond to, *outliers*. These will usually be merged into cells with non-outlying observations by the grouping process, and hence concealed. Screening for potential outliers will usually need to be done separately.

Some studies have shown improvements in power by using the Rao-Robson-Moore statistic rather than minimum chi-squared. Still, all chi-squared statistics for goodness-of-fit have only moderate power. Tests specifically designed to test compatibility with a specific distribution, say the normal, Poisson or exponential may have greater power. However the broad applicability chi-squared tests is a major advantage.

Power against any specific *local* alternative may be evaluated by extending the aforementioned theory; see MOORE & SPRUILL (1975).

Contingency tables: When testing the independence of two methods of classifying data, a common technique is to prepare a two-way table of frequencies of the sample data, a so-called $r \times c$ *contingency table* (r rows and c columns). The resulting singular multinomial distribution of the cell counts is constrained by given row and column totals, in addition to the total sum constraint; this results in a variance matrix (and limiting normal distribution) with rank equal to $rc - (r-1) - (c-1) - 1 = (r-1)(c-1)$. The resulting multinomial chi-squared statistic likewise has this number of degrees of freedom. If cell probabilities are functions of a parameter θ, and raw data rather than just cell counts are available, the theory of Section 9.6 may extend. And higher dimensional tables often appear in applications. For extensive theory, see for example BISHOP, FIENBERG & HOLLAND (1975).

The Akritas-Hjort chi-squared test: AKRITAS (1988) introduced a modification of the multinomial chi-squared goodness-of-fit test with m, rather than $m - 1$, degrees of freedom in the limiting chi-squared distribution, whether or not there are parameters to be estimated. It has the further advantage that it is easily extended to accommodate censored data. Experience with this test is limited, however. A generalization of this method appears in HJORT (1990).

Other goodness-of-fit tests: Chi-squared tests are not the only method for evaluating the fit of parametric model to data, but they do have quite general applicability. No other general method handles unknown parameters as easily. Special tests for some common models, such as normal, exponential and

TABLE 9.1
Etruscan Skulls Data

141	146	144	141	141	136	137	149	141	142
142	147	148	155	150	144	140	140	139	148
143	143	149	140	132	158	149	144	145	146
143	135	147	153	142	142	138	150	145	126
135	142	140	148	146	149	137	140	154	140
149	140	147	137	131	152	150	146	134	137
142	147	158	144	146	148	143	132	149	144
152	150	148	143	143	142	141	154	141	144
142	138	146	145						

Poisson, are discussed in D'Agostino & Stephens (1986); and Moore (1986) provides a good overview of chi-squared tests.

White (1982) developed a test comparing estimators of the Fisher Information matrix obtained from the derivatives of the scores and from the sample covariance matrix of the scores. When the model is correct, any discrepancy between these estimates is due only to random variation. With a d-dimensional parameter, there are $k = \frac{1}{2}d(d+1)$ distinct elements in the information matrix (due to symmetry), and a Wald-type quadratic form in these discrepancies provides a test statistic.

Other general tests with considerable attention in the statistical literature are those of Kolmogorov, Cramér-Von Mises, and Anderson-Darling, all based on the empirical process; see Shorack & Wellner (1986) for descriptions and references.

Appendix: Data and Program for the Etruskan Skulls Example

The full data for the 84 skulls are listed below. These are also available in R.

Here $\bar{x} = 143.77$, $S_X = 5.9705$. The following R script calculates the Moore chi-squared statistic for fit of the normal distribution, using cut points 134.5, 139.5, 144.5 and 149.5. The program is easily modifiable to allow different data and cut points.

$$< m < -5$$

$$< n < -84$$

$$< c < -c(134.5, 139.5, 144.5, 149.5)$$

$$< xb < -143.77$$

$$< s < -5.9705$$

```
< v <-c(5,10,33,24,12)
< r <-(c - xb)/s
< p <-c(0,0,0,0,0)
< pa <-c(0,0,0,0,0)
< pb <-c(0,0,0,0,0)
< p[1] <-pnorm(r[1])
< pa[1] <-dnorm(r[1])
< pb[1] <-r[1] * pa[1]
< for(i ∈ 2 : 4){
< p[i] <-pnorm(r[i]) - pnorm(r[i - 1])
< pa[i] <-dnorm(r[i]) - dnorm(r[i - 1])
< pb[i] <-r[i] * dnorm(r[i]) - r[i - 1] * dnorm(r[i - 1])}
< p[5] <-1 - pnorm(r[4])
< pa[5] <- - dnorm(r[4])
< pb[5] <- - r[4] * dnorm(r[4])
< np <-n * p
< a11 <-sum(pa * pa/p)
< a12 <-sum(pa * pb/p)
< a22 <-sum(pb * pb/p)
< e11 <-1/n - (1/n) * a11
< e22 <-2/n - (1/n) * a22
< e12 <- - (1/n) * a12
< det <-e11 * e22 - e12 * e12
< e11i <-e22/det
< e22i <-e11/det
< e12i <- - e12/det
< Q <-sum((v - np) * (v - np)/(np))
< U1 <-sum(pa * v/(np))
< U2 <-sum(pb * v/(np))
< D <-U1 * U1 * e11i + 2 * U1 * U2 * e12i + U2 * U2 * e22i
< Q
< D
< q()
```

Exercises

1. Verify that
$$\begin{pmatrix} a & b \\ c & d \end{pmatrix}$$
is a generalized inverse of
$$\begin{pmatrix} 1 & -1 \\ -1 & 1 \end{pmatrix}$$
if and only if $a - 2b + c = 1$

2. Verify that if
$$A = \begin{pmatrix} B & C \\ D & E \end{pmatrix}$$
where B is a nonsingular matrix with rank equal to the rank of A, then
$$A^{-} = \begin{pmatrix} B^{-1} & \mathbf{0} \\ \mathbf{0} & \mathbf{0} \end{pmatrix}$$
is a generalized inverse of A.

3. Verify that if Σ^{-} is a generalized inverse of a symmetric matrix Σ, then so is its transpose $(\Sigma^{-})^{T}$. Hence, argue that, without loss of generality, we may always choose Σ^{-} to be symmetric.

4. Inverse matrices are often introduced in order to solve systems of linear equations, $\mathbf{Y} = A\mathbf{x}$. When A is $m \times m$ and non-singular, $\mathbf{x} = A^{-1}\mathbf{Y}$ is the unique solution, valid for all $\mathbf{Y} \neq \mathbf{0}$.

   More generally, suppose that A is $m \times n$ with $m \leq n$, and let A^{-} be a generalized inverse of A, i.e. satisfying $AA^{-}A = A$, as in 9.1. Show that, if a solution to $\mathbf{Y} = A\mathbf{x}$ exists, then $\mathbf{x} = A^{-}\mathbf{Y}$ is a solution. [Actually, all solutions are of this form if, by a solution, is meant a solution whatever the choice of $\mathbf{Y} \neq \mathbf{0}$.]

5. Suppose that $Y_1 = \mu_1 + \sigma Z$ and $Y_2 = \mu_2 - \tau Z$ for Z standard normal. Show that
$$\begin{pmatrix} a & b \\ b & c \end{pmatrix}$$
is a generalized inverse of the variance matrix of (Y_1, Y_2) if and only if $a\sigma^2 - 2b\sigma\tau + c\tau^2 = 1$.

6. A generalized inverse Σ_R^{-} of Σ for which $\Sigma_R^{-}\Sigma\Sigma_R^{-} = \Sigma_R^{-}$ (as well as $\Sigma\Sigma_R^{-}\Sigma = \Sigma$) is said to be *reflexive*.

   (a) Show that, for any generalized inverse Σ^{-}, the matrix $\Sigma^{-}\Sigma\Sigma^{-}$ is a reflexive generalized inverse of Σ.

   (b) Show that the generalized inverse in Lemma 9.1 is not reflexive, but that $D^{-1} - J$, with J a matrix of 1's, is a reflexive generalized inverse. Moreover, this inverse yields the same quadratic form as in Theorem 9.1.

7. Let P be an *orthogonal matrix* ($P^T = P^{-1}$) that diagonalizes the symmetric positive semi-definite matrix Σ of rank r: $P^T \Sigma P = D$ with the upper left $r \times r$ block in D being a diagonal matrix D_r with positive diagonal elements and the remaining elements in D are zeroes. (The existence of such a P is a standard fact in matrix algebra.) Let E be any matrix with upper left block D_r^{-1}—all other elements being arbitrary.

   (a) Show that PEP^T is a generalized-inverse of Σ, but not necessarily symmetric, and with rank between r and m (inclusive) depending on E.

   (b) Give some sufficient conditions on E to make the generalized inverse in (a) *reflexive* (see *Exercise 6*).

8. Consider a random sample $(X_1, \ldots, X_n)$ from $N(\mu, \sigma^2)$ and let $Y_i = X_i - \bar{X}_n$. Find a matrix C for which $\mathbf{Y} = C^T \mathbf{Z}$ for an $(n-1)$-vector $\mathbf{Z}$ of independent identically distributed standard normal variables Z_i's. [*Hint*: Try constructing, column by column, $C = (U, \mathbf{u})$ with U upper triangular.] Find a generalized inverse of $\Sigma = C^T C$, and the corresponding quadratic form $Q(\mathbf{Y}, \Sigma^-)$.

9. In the notation of Section 9.1, let $\mathbf{W} = C\Sigma^- \mathbf{Y}$

   (a) Show that $\mathbf{W} \sim N_r(C\Sigma^- \boldsymbol{\mu}, I_r)$, and so $\mathbf{W}^T\mathbf{W} \sim \chi_r^2(\boldsymbol{\mu}^T \Sigma^- \Sigma \Sigma^- \boldsymbol{\mu})$.

   (b) Show that, if Σ^- is a reflexive generalized inverse (see *Exercise 7*), then $Q(\mathbf{Y}, \Sigma^-)$ is distributed as noncentral chi-squared with noncentrality parameter $Q(\boldsymbol{\mu}, \Sigma^-)$.

10. In the notation of Chapters 4–6, let $\mathbf{S}_n$ be the score vector for an m-dimensional parameter θ with $\mathbf{S}_n \sim AN(\mathbf{0}, B)$ but with B singular (not allowed in earlier chapters). Propose a score test statistic of the simple null hypothesis $\theta = \theta_0$ and give its large sample null distribution. How might this be extended to the case where $\theta = (\omega, \lambda)$ where ω is of interest λ is a nuisance parameter?

11. In the notation of Section 9.2, show that $\mathbf{Y}_n'^T (\Sigma')^{-1} \mathbf{Y}_n' = \mathbf{Y}_n^T \Sigma^- \mathbf{Y}_n$ (using Equation (9.3)). *Hint:* First show that

$$\mathbf{Y}_n^T \Sigma^- \mathbf{Y}_n = (\mathbf{Y}_n')^T \left\{ D'^{-1} + \frac{\mathbf{1}\mathbf{1}^T}{1 - \mathbf{1}^T \mathbf{p}'} \right\} \mathbf{Y}_n'.$$

12. In a genetics experiment, the counts of progeny with phenotypes aa, aA and AA are recorded to be 20, 27 and 13, respectively. According to a simple Mendellian model, the probabilities of each type are $\frac{1}{4}$, $\frac{1}{2}$ and $\frac{1}{4}$, respectively. Determine a large-sample p-value for the fit of the data to this model.

13. (a) Show by example that $\Sigma_n \to \Sigma$ (singular) does not imply that $\Sigma_n^- \to \Sigma^-$. You may use the example of a 2×2 matrix $\Sigma_n = \text{diag}(1, 1/n)$. (In fact, a 1×1 matrix $\Sigma_n = \text{diag}(1/n)$ will do!)

    (b) Suppose that $\mathbf{Y}_n \sim N_2(\mathbf{0}, \Sigma_n)$ with $\Sigma_n = \text{diag}(1, 1/n) \to \Sigma = \text{diag}(1, 0)$, and choose $\Sigma^- = \text{diag}(1, c)$ for arbitrary c. Contrast the distributions of $Q_n = \mathbf{Y}_n^T \Sigma_n^- \mathbf{Y}_n$ and $Q_n' = \mathbf{Y}_n^T \Sigma^- \mathbf{Y}_n$, and the limiting distributions of Q_n/n and Q_n'/n.

14. Show that Equation (9.10) implies that $D - \dot{P}B^{-1}\dot{P}^T = A$, say, is nonsingular and that Σ^{*-} in Equation (9.11) equals A^{-1}. [*Hint:* Multiply Equation (9.11) by A.] This alternative expression for $(\Sigma^*)^-$ requires inversion of an $m \times m$ matrix A instead of the $d \times d$ matrix C, as in Equation (9.11).

15. Using the result in *Exercise 12*, show that Equation (9.10) implies that Σ^* in (8) has rank $m - 1$.

16. (a) Write out the details of Equations (9.13) and (9.16) for the Poisson case considered in Section 9.6.

    (b) Larsen & Marx (1986) presented data on an analysis of the outbreaks of wars per year over a 432-year period (1500–1931). The study identified 299 such outbreaks worldwide; 223 years had 0 outbreaks, 142 had 1, 48 had 2, 15 had 3, 4 had 4, and none had more than 4. They raise the question of whether these data are consistent with a Poisson distribution. (A fuller model might hypothesize that outbreaks occur according to a homogeneous Poisson process, resulting in a Poisson-distributed number of outbreaks per year.) Determine a Rao-Robson-Moore p-value for this hypothesis. What do you conclude?

17. Returning to the Etruscan skull data in Table 9.1, we have regrouped the data into cells of width 3 cm. The nine successive cells are ≤ 133, $134 - 136$, $137 - 139$, $\ldots$, > 154, with observed counts of 4, 4, 7, 21, 15, 15, 10, 5 and 3, respectively. Determine the resulting p-value, testing for normality.

18. In the normal goodness-of-fit example, some simplifications in the formulas for $\hat{\Delta}_n$ were noted; some are quite general, resulting from the fact that

$$\sum_{j=1}^{m} \frac{\partial p_j}{\partial \theta_i} = 0.$$

    Why is this true?

19. (a) Describe how to carry out a goodness-of-fit test of bivariate normality, partitioning the sample space using a rectangular grid.

(b) Consider instead a parameter-based partition, with each cell having concentric elliptical boundaries, with probabilities derived from the chi-squared distribution. (See notes in Section 9.9 for guidance.)

20. In *Exercise 12* of Chapter 7, two samples of data were given, each assumed to come from a Weibull distribution.

    (a) Construct and carry out a chi-squared test of the fit of the Weibull the second sample to a Weibull distribution.

    (b) Without giving all details, describe how to test whether the two samples come from independent Weibull distributions with a common shape parameter.

    (c) Carry out the numerical details for (a) or for (b).

21. As an illustrative example, suppose that $X_1, X_2, \ldots, X_n$ are a random sample from an exponential distribution with unknown parameter λ. Let $Y_i = \mathbf{1}[X_i < c]$. for some specified constant c. Show that if λ is estimated from the Y_i, the resulting chi-squared statistic comparing observed and expected frequencies of the two values of Y is degenerate at zero, with zero degrees of freedom, and hence useless for assessing the fit of the assumed exponential distribution. Show also that if λ and the expected frequencies of the two values of Y are estimated from the sample mean $\bar{X}$ of the raw data $\{X_i\}$, then the resulting chi-squared statistic is not degenerate and find its asymptotic distribution. Comment on the relation to the Rao-Robson-Moore test.

22. Suppose that the matrix $G = B - \dot{P}D^{-1}\dot{P}^T$ is positive semi-definite of rank r, with $0 < r < d$. Show that $(\Sigma^*)^-$ with G^{-1} replaced by a generalized inverse G^- of G in Equation (9.11) is still a generalized inverse of Σ^*. What is its rank? [*Hint:* Substitute the hypothesized $(\Sigma^*)^-$ into the expression

$$(D - \mathbf{pp}^T - \dot{P}D^{-1}\dot{P})(\Sigma^*)^-(D - \mathbf{pp}^T - \dot{P}D^{-1}).$$

and expand. Note that since $\mathbf{p}^T D^{-1}\dot{P} = 0$ all terms in involving $\mathbf{pp}^T$ vanish except for $\mathbf{pp}^T D^{-1}\mathbf{pp}^T$ which reduces to $\mathbf{pp}^T$. In the remaining terms, substitute for $\dot{P}D^{-1}\dot{P} = B - G$ wherever this occurs and use the property $GG^-G = G$.]

References

Abramowitz, M., & Stegun, I. A., eds. (1964). *Handbook of Mathematical Functions.* National Bureau of Standards Applied Mathematics Series **55**, U.S. Government Printing Office, Washington, DC.

Akritas, M. G. (1988). Pearson-type goodness-of-fit tests: the univariate case. *J. Am. Stat. Assoc.* **83**:222–230.

Anscombe, F. J. (1948). The transformation of Poisson, binomial and negative-binomial data. *Biometrika* **35**:246–254.

Apostol, T. M. (1957). *Mathematical Analysis.* Addison-Wesley, Reading, MA.

Bahadur, R. R. (1966). A note on quantiles in large samples. *Ann. Math. Stat.* **37**:577–580.

Barnard, G. A. (1990). Must clinical trials be large? The interpretation of *p*-values and the combination of test results. *Stat. Med.* **9**:601–614.

Bera, A. K., & Bilias, Y. (2001). Rao's score Neyman's $C(\alpha)$ and Silvey's LM tests: an essay on historical developments and some new results. *J. Stat. Plan. Inference* **97**:9–44.

Bickel, P. (1974). Edgeworth expansions in nonparametric statistics. *Ann. Stat.* **2**:1–20.

Bickel, P., & Doksum, K. A. (2015). *Mathematical Statistics: Basic Ideas and Selected Topics, Vols. I, II,* 2nd ed. Pearson Prentice Hall, Upper Saddle River, NJ.

Bickel, P. J., Klaassen, C. A. J., Ritov, Y., & Wellner, J. A. (1993). *Efficient and Adaptive Estimation for Semiparametric Models.* The Johns Hopkins University Press, Baltimore.

Bishop, Yvonne M. M., Fienberg, Stephen E., & Holland, Paul W. (1975). *Discrete Multivariate Analysis: Theory and Practice.* The MIT Press, Cambridge, MA.

Boos, D. D. (1992). On generalized score tests. *Am. Stat.* **46**:327–333.

Box, G. P., & Cox, D. R. (1968). An analysis of transformations (with discussion). *J. R. Stat. Soc. B* **26**:211–252.

Breiman, L. (1968). *Probability.* Addison-Wesley, Reading, MA.

Brown, B. Wm. Jr., & Hollander, M. (1977). *Statistics. Wiley, New York.*

Brown, L. D., Cai, T. T., & DasGupta, A. (2002). Interval estimation for a binomial proportion. *Stat. Sci.* **16**:101–133.

Casella, G., & Berger, R. L. (2002). *Statistical Inference*, 2nd ed. Duxbury, North Scituate, MA.

CHOI, S. (1989). On asymptotically optimal tests. Ph.D. dissertation, University of Rochester, Rochester, NY.

CHOI, S., HALL, W. J., & SCHICK, A. (1996). Asymptotically uniformly most powerful tests in parametric and semiparametric models. *Ann. Stat.* **24**:841–861.

COX, D. R. (2006). *Principles of Statistical Inference.* Cambridge University Press, Cambridge, UK.

COX, D. R., & HINKLEY, D. V. (1974). *Theoretical Statistics.* Chapman & Hall, London.

CRAMÉR, H. (1946). *Mathematical Methods of Statistics.* Princeton University Press, Princeton, NJ.

D'AGOSTINO, R. B., & STEPHENS, M., eds. (1986). *Goodness-of-Fit Techniques.* M. Dekker, New York.

DIACONIS, P., HOLMES, S., & MONTGOMERY, R. (2007). Dynamical bias in the coin toss. *SIAM Rev.* **49**:211–235.

FELLER, W. (1968). *An Introduction to Probability Theory and Its Applications*, Vol. 1, 3rd ed. Wiley, New York.

FINNEY, D. J. (1941). On the distribution of a variate whose logarithm is normally distributed. *J. R. Stat. Soc. Suppl.* **7**:155–161.

FISHER, R. A. (1922). On the interpretation of χ^2 from contingency tables, and the calculation of P. *J. R. Stat. Soc.* **85**:87–94.

FISHER, R. A. (1924). The conditions under which χ^2 measures the discrepancy between observation and hypothesis. *J. R. Stat. Soc.* **87**:442–450.

FISHER, R. A. (1925). Theory of statistical estimation. *Proc. Camb. Phil. Soc.* **22**:700–725.

FRASER, D. A. S. (1957). *Nonparametric Methods in Statistics.* Wiley, New York.

FREEMAN, M. F., & TUKEY, J. W. (1950). Transformations related to the angular and square root. *Ann. Math. Stat.* **21**:607–611.

GELBAUM, B. R., & OLMSTED, J. M. H. (1964). *Counterexamples in Analysis.* Holden-Day, San Fransisco, CA.

GHOSH, J. K. (1971). A new proof of the Bahadur representation of quantiles and an application. *Ann. Math. Stat.* **42**:1957–1961.

GNEDENKO, B. V. (1962). *Theory of Probability.* Chelsea, New York, USA.

GODAMBE, V. P., ed. (1991). *Estimating Functions.* Oxford University Press, Oxford, UK.

HÁJEK, J. (1970). A characterization of limiting distributions of regular estimates. *Z. Wahrs. Verw. Geb.* **14**:323–330.

HÁJEK, JAROSLAV, & ŠIDÁK, ZBYNĚK (1967). *Theory of Rank Tests.* Academic Press, New York.

HALL, W. J., & MATHIASON, D. J. (1990). On large-sample estimation and testing in parametric models. *Rev. Int. Stat. Inst.* **58**:77–97.

HALL, W. J., & WELLNER, J. A. (2013). Efficient testing and estimation in two Lehmann alternatives to symmetry-at-zero models. In *From Probability to Statistics and Back, High-Dimensional Models and Processes, A*

Festschrift in Honor of Jon A. Wellner, Vol. 9, eds. M. Banerjee, F. Bunea, J. Huang, V. Koltchinskii, & M. H. Maathuis, pp. 197–212. Institute of Mathematical Statistics Collections, Cleveland.

HAMPEL, F. R. (1974). The influence curve and its role in robust estimation. *J. Am. Stat. Assoc.* **69**:383–393.

HARDY, G. H. (1952) *A Course of Pure Mathematics*, 10th ed. Cambridge Univsersity Press, Cambridge, UK.

HASTINGS, C. (1955). *Approximations for Digital Computers.* Princeton University Press, Princeton, NJ.

HETTMANSPERGER, T. P. (1984). *Statistical Inference Based on Ranks.* Wiley, New York.

HJORT, N. L. (1990). Goodness of fit tests in models for life history data based on cumulative hazard rates. *Ann. Stat.* **18**:1221–1258.

HODGES, J. L. JR., & LEHMANN, E. L. (1963). Estimates of location based on rank tests. *Ann. Math. Stat.* **33**:482–497.

HOEFFDING, W. (1951). "Optimum" nonparametric tests. In *Proceedings of the Second Berkeley Symposium on Mathematical Statistics and Probability*, ed. J. Neyman, pp. 83–92. University of California Press, Berkeley, CA.

IBRAGIMOV, I. A., & HAS'MINSKII, R. Z. (1981). *Statistical Estimation: Asymptotic Theory.* Springer-Verlag, New York.

JOHNSON, N. L., KEMP, A. W., & KOTZ, S. (2008). *Univariate Discrete Distributions*, 3rd ed. Wiley, New York.

JOHNSON, N. L., KOTZ, S., & BALAKRISHNAN, N. (1994). *Continuous Univariate Distributions, Vol. 1*, 2nd ed. Wiley, New York.

JOHNSON, N. L., KOTZ, S., & BALAKRISHNAN, N. (1997). *Continuous Univariate Distributions, Vol. 2*, 2nd ed. Wiley, New York.

JONKER, M., & VAN DER VAART, A. (2014). On the correction of the asymptotic distribution of the likelihood ratio statistic if nuisance parameters are estimated based on an external source. *Int. J. Biostat.* **10**:123–142.

KENDALL, M. G., STUART, A., & ORD, J. K. (1977). *Kendall's Advanced Theory of Statistics, Vol. 1*, 5th ed. Griffin, London.

KIEFER, J. (1967). On Bahadur's representation of sample quantiles. *Ann. Math. Stat.* **38**:1323–1342.

KIEFER, J. (1970). Old and new methods for studying order statistics and sample quantiles. In *Proceedings Conference on Nonparametric Techniques in Statistical Inference*, ed. M. L. Puri, pp. 299–319. Cambridge University Press, Bloomington.

KOLASSA, J. E. (1997). *Series Approximation Methods in Statistics*, 2nd ed. Lecture Notes in Statistics **88**, Springer, New York.

KRAFT, C. H., & LECAM, L. M. (1956). A remark on the roots of the maximum likelihood equation. *Ann. Math. Stat.* **27**:1174–1177.

LANCASTER, H. O. (1949). Statistical control of counting experiments. *Biometrika* **39**:419–422.

LARSEN, R. J., & MARX, M. L. (1986). *An Introduction to Mathematical Statistics and Its Applications*, 2nd ed. Prentice-Hall, Englewood Cliffs, NJ.

LARSEN, R. J., & MARX, M. L. (2018). *An Introduction to Mathematical Statistics and Its Applications*, 6th ed. Pearson, Hoboken.

LECAM, L. (1960). Locally asymptotically normal families of distributions. *Univ. Calif. Publ. Stat.* **3**:37–98.

LEHMANN, E. L. (1975). *Nonparametrics: Statistical Methods Based on Ranks.* Holden-Day, San Francisco, CA.

LEHMANN, E. L. (1980). Efficient likelihood estimates. *Am. Stat.* **34**:233–235.

LEHMANN, E. L. (1999). *Elements of Large Sample Theory*. Springer, New York.

LEHMANN, E. L., & CASELLA, G. (2006). *Theory of Point Estimation*, 2nd ed. Springer, New York.

LEHMANN, E. L., & ROMANO, J. P. (2005). *Testing Statistical Hypotheses*, 3rd ed. Springer, New York.

LIN, C.-C., & MUDHOLKAR, G. S. (1980). A simple test for normality against asymmetric alternatives. *Biometrika* **67**:455–461.

MANN, H. B., & WHITNEY, D. R. (1947). On a test of whether one of two random variables is stochastically larger than the other. *Ann. Math. Stat.* **18**:50–60.

MARZETTA, T. L. (1993). A simple derivation of the constrained multiple parameter Cramer-Rao bound. *IEEE Trans. Signal Process.* **41**:2247–2249.

MATHIASON, D. J. (1982). Large sample test procedures in the presence of nuisance parameters. Ph.D. dissertation, University of Rochester, Rochester, NY.

MOORE, D. S. (1977). Generalized inverses, Wald's method and the construction of chi-squared tests of fit. *J. Am. Stat. Assoc.* **72**:131–137.

MOORE, D. S. (1978). Chi-square tests. In *Studies in Statistics*, ed. R. V. Hogg. Mathematical Association of America.

MOORE, D. S. (1986). Tests of chi-squared type. In *Goodness-of-Fit Techniques*, eds. R. B. D'Agostino and M. Stephens. M. Dekker, New York.

MOORE, D. S., & SPRUILL, M. C. (1975). Unified large-sample theory of general chi-squared statistics for tests of fit. *Ann. Stat.* **3**:599–616.

MORAN, P. A. P. (1970). On asymptotically optimal tests of composite hypotheses. *Biometrika* **57**:47–55.

NEYMAN, J. (1949). Contribution to the theory of the χ^2 test. In *Proceedings of the First Berkeley Symposium on Mathematical Statistics and Probability*, ed. J. Neyman, pp. 239–273. University of California Press, Berkeley, CA.

NEYMAN, J. (1959). Optimal asymptotic tests of composite statistical hypotheses. In *Probability and Statistics* (Harald Cramér Volume), ed. U. Grenander, pp. 212–234. Wiley, New York.

NEYMAN, J., & PEARSON, E. S. (1928). On the use and interpretation of certain test criteria for purposes of statistical inference. *Biometrika* **20A**:175–240, 263–294.

NIKULIN, M. S. (1973). Chi-square test for continuous distributions with shift and scale parameters. *Theory Probab. Appl.* **18**:559–568.

PACE, L. (2011). Probabilistically proving that $\zeta(2) = \pi^2/6$. *Am. Math. Monthly* **118**:641–643.

PEARSON, K. (1900). On the criterion that a given system of deviations from the probable in the case of a correlated system of variables is such that it can be reasonably supposed to have arisen from random sampling. *Phil. Mag. Ser. 5* **50**:157–172.

PRATT, J. W. (1960). On interchanging limits and integrals. *Ann. Math. Stat.* **31**:74–77.

RAO, C. R. (1948). Large sample tests of statistical hypotheses concerning several parameters with applications to problems of estimation. *Proc. Camb. Phil. Soc.* **44**:50–57.

RAO, C. R. (1973). *Linear Statistical Inference and Its Applications*, 2nd ed. Wiley, New York.

RAO, C. R. (2001). Two score and 10 years of score tests. *J. Stat. Plann. Infer.* **97**:3–7.

RAO, K. C., & ROBSON, D. S. (1974). A chi-square statistic for goodness-of-fit tests within the exponential family. *Comm. Stat.* **3**:1139–1153.

REEDS, J. A. (1985). Asymptotic number of roots of Cauchy location likelihood equations. *Ann. Stat.* **13**:775–784.

RESNICK, S. I. (1999). *A Probability Path.* Birkhauser, Boston, MA.

ROUSSAS, G. G. (1972). *Contiguity of Probability Measures: Some Applications in Statistics.* Cambridge University Press, Cambridge, UK.

SAVAGE, I. R. (1956). Contributions to the theory of rank order statistics—the two-sample case. *Ann. Math. Stat.* **27**:590–615.

SEARLE, S. R. (1982). *Matrix Algebra Useful for Statistics.* Wiley, New York.

SERFLING, ROBERT J. (1980). *Approximation Theorems of Mathematical Statistics.* Wiley, New York.

SILVERMAN, B. W. (1986). *Density Estimation for Statistics and Data Analysis.* Chapman & Hall, New York.

SHORACK, G. R. (2000). *Probability for Statisticians.* Springer-Verlag, New York.

SHORACK, G. R., & WELLNER, J. A. (1986). *Empirical Processes with Applications to Statistics.* Wiley, New York.

SIEGMUND, D. (1985). *Sequential Analysis: Tests and Confidence Intervals.* Springer-Verlag, New York.

SILVEY, S. D. (1959). The Lagrange multiplier test. *Ann. Math. Stat.* **30**:389–407.

SILVEY, S. D. (1975) *Statistical Inference.* Chapman and Hall, London.

SMALL, C. G., WANG, J., & YANG, Z. (2000). Eliminating multiple root problems in estimation. *Stat. Sci.* **15**:313–332.

SOLARI, M. E. (1969). The "maximum likelihood" solution of the problem of estimating a linear functional relationship. *J. R. Stat. Soc. B* **31**:372–375.

STEIN, C. (1956). Efficient nonparametric testing and estimation. *Proceedings of the Third Berkeley Symposium on Mathematical Statistics and Probability* **1**, University of California Press, Berkeley, CA.

STRASSER, HELMUT (1985). *Mathematical Theory of Statistics: Statistical Experiments and Asymptotic Theory.* Gruyter, Berlin.

STUART, A. (1955). A paradox in statistical inference. *Biometrika* **42**:527–529.

TUKEY, J. (1949). The simplest signed rank tests. Princeton University Statistical Research Group, Memo Report No. 17.

WALD, ABRAHAM (1947). *Sequential Analysis.* Wiley, New York.

WALD, A. (1943). Tests of statistical hypotheses concerning several parameters when the number of observations is large. *Trans. Am. Math. Soc.* **54**:426–482.

WALSH, J. E. (1949). Some significance tests for the median which are valid under very general conditions. *Ann. Math. Stat.* **20**:64–81.

WEDDERBURN, R. W. M. (1974). Quasi-likelihood functions, generalized linear models and the Gauss-Newton method. *Biometrika* **61**:439–447.

WEFELMEYER, W. (1987). Testing hypotheses on independent, not identically distributed models. In *Mathematical Statistics and Probability Theory*, Vol. A, eds. M. L. Puri, et al., pp. 267–282. Reidel, Berlin.

WHITTAKER, E. T., & WATSON, G. N. (1927). *A Course in Modern Analysis.* Cambridge University Press, Cambridge, UK.

WHITE, H. (1982). Maximum likelihood estimation of misspecified models. *Econometrica* **50**:1–25.

WILCOXON, F. (1945). Individual comparisons by ranking methods. *Biometrics* **1**:80–83.

WILKS, S. S. (1938). The large-sample distribution of the likelihood ratio for testing composite hypotheses. *Ann. Math. Stat.* **9**:60–62.

WILSON, E. B. (1927). Probable inference, the law of succession, and statistical inference. *J. Am. Stat. Assoc.* **22**: 209–212.

WILSON, E. B., & HILFERTY, M. M. (1931). The distribution of chi-square. *Proc. Nat. Acad. Sci.* **17**:684–688.

Author Index

Subject Index

Note: **Bold** page numbers refer to tables and *italic* page numbers refer to figures.